Poseidon's Progress

ALSO BY IVER P. COOPER
AND FROM MCFARLAND

Arming the Warship: Naval Weapons Technology and Gunnery from the Spanish Armada to the Cold War (2024)

Airships: Their Science, History and Future (2024)

Poseidon's Progress

The Quest to Improve Life at Sea

Iver P. Cooper

McFarland & Company, Inc., Publishers
Jefferson, North Carolina

Portions of this work were previously published in the *Grantville Gazette* and are used by permission. Use of released U.S. Navy imagery does not constitute product or organizational endorsement of any kind by the U.S. Navy.

Library of Congress Cataloging-in-Publication Data

Names: Cooper, Iver P. author.
Title: Poseidon's progress : the quest to improve life at sea / Iver P. Cooper.
Description: Jefferson, North Carolina : McFarland & Company, Inc., Publishers, 2024 | Includes bibliographical references and index.
Identifiers: LCCN 2024017342 | ISBN 9781476694467 (paperback : acid free paper) ∞
ISBN 9781476652023 (ebook)
Subjects: LCSH: Seafaring life—History. | Shipbuilding. | BISAC: HISTORY / Military / Naval | HISTORY / Maritime History & Piracy
Classification: LCC G540 .C6778 2024 | DDC 910.4/5—dc23/eng/20240511
LC record available at https://lccn.loc.gov/2024017342

British Library cataloguing data are available

ISBN (print) 978-1-4766-9446-7
ISBN (ebook) 978-1-4766-5202-3

© 2024 Iver P. Cooper. All rights reserved

No part of this book may be reproduced or transmitted in any form or by any means, electronic or mechanical, including photocopying or recording, or by any information storage and retrieval system, without permission in writing from the publisher.

Front cover image: *Fire at Sea*, Artist Clarkson Stanfield (British, Sunderland 1793–1867 London), 1820–1846, watercolor and gouache (bodycolor) on beige paper, 42.5 × 31.5 cm (gift of Robert Tuggle, in memory of Charles Ryskamp, 2014, The Metropolitan Museum of Art)

Printed in the United States of America

McFarland & Company, Inc., Publishers
Box 611, Jefferson, North Carolina 28640
www.mcfarlandpub.com

For Lee, Louise
and Jason

Table of Contents

Preface 1

Chapter 1. Taking a Breath 3

Chapter 2. Quenching Thirst 25

Chapter 3. Filling the Stomach 48

Chapter 4. Alert Above, Asleep Below 69

Chapter 5. More Creature Comforts 94

Chapter 6. Keeping Dry (and Afloat) 114

Chapter 7. Lest You Drown 141

Chapter 8. Seeing and Being Seen in the Dark 173

Chapter 9. Lest It Spread: Fires and Infections 201

Conclusion 227

References 229

Index 253

Preface

A ship is not just a mode of transport—it is a life support system. Ideally, it provides fresh air at a comfortable temperature and humidity, healthy food and drink, adequate lighting at night and below deck (but without setting the ship on fire or interfering with navigation) and sleeping and sanitation facilities. It must be kept dry and afloat. Provision must be made for disasters, whether a man overboard or the ship itself sinking or run aground. Fires and contagious diseases must be fought if they arise.

The quest to improve life at sea has progressed, albeit haltingly, on two parallel fronts. The first is analyzing the problems and finding and implementing scientific solutions. The second is by changes in laws and regulations so as to require healthier practices.

This book examines life at sea, and the quest to improve it, from both a scientific and a historical perspective. Each chapter topic could easily support a book of its own, so the coverage here is necessarily selective. Chronologically, the emphasis is on the 16th through mid–20th centuries, although there is some coverage of earlier and later developments. Geographically, the emphasis is on British and American practice, but there are some references to Dutch, French, Scandinavian, Russian, Chinese and Japanese experience, too. Finally, this book looks more closely at warships than at merchant ships, and at merchant ships more than fishing and pleasure craft.

Chapter 1

Taking a Breath

When we take a cruise on a yacht, we look forward to escaping the polluted air of our cities and breathing in the fresh, salt-tanged air of the open ocean. And if we need to go below deck, we turn on the air-conditioning. But in earlier times, the situation below deck, especially on warships or in the steerage compartment of passenger ships, could be dire.

In 1739, "the sailors on board the fleet at Spithead were so dangerously ill for want of fresh air that they were put ashore to recover their health" (Edwards 1881, 13).

An even more extreme example from 1848: The *Londonderry* encountered a storm and "had to confine two hundred steerage passengers in a poorly ventilated compartment which afforded less than 7 cubic feet of air space per person. More than 70 out of the 200 were found dead when the compartment was opened on the following morning" (Pryor 1918, 48).

Even in the late 20th century, shipboard ventilation problems occasionally became life-threatening. During the 1970s, Coast Guard records revealed "11 incidents involving 32 deaths by asphyxia. All deaths occurred in the unventilated holds of ships in warm waters and during warm months of the year" (Glass 1980).

We will first look at what scientists say about what makes air "bad," but without going into physiological details. Next, we'll consider the history of natural ventilation on shipboard, as well as some physical constraints on its effectiveness. Then we'll turn to the history of artificial shipboard ventilation, and also consider a few issues concerning ventilation systems and cargo ventilation. Finally, we'll circle back to the science to examine, in some realistic shipboard contexts, (1) the time until asphyxiation in the absence of ventilation and (2) the development of standards for adequate ventilation.

Bad Air

With every breath, we consume oxygen and produce carbon dioxide. Even a single person, trapped in a hermetically sealed room, will ultimately suffer a physiological crisis as a result. The greater the number of people in the room, and the smaller the room, the sooner the crisis occurs. (At the end of this chapter we'll crunch a few numbers for an actual shipboard situation.)

Oxygen. Atmospheric air is 21 percent oxygen. What actually matters, physiologically speaking, is the partial pressure of oxygen, which decreases with increasing altitude. (This is why the effects of oxygen deprivation have been studied on behalf of climbers and pilots.) At the top of Mount Everest, overall atmospheric pressure is just one-third of its sea level value, and so the partial pressure of oxygen is just 7 percent (Lance 2021, 44). That is why it is a struggle to breathe there.

It is considered advisable to maintain an oxygen partial pressure of at least 0.16 atm (16 percent atmospheric pressure). Tests on pilots showed that no one passed out at 0.10 atm oxygen, but that at 0.063, the "time of useful consciousness" is just two minutes (Lance 2020, 44ff). The *United States Navy Diving Manual* (USNDM 1979) is more pessimistic. It says the first symptoms appear at 0.14 atm and that some people will lose consciousness at 0.10 (3–14).

Carbon dioxide. The level of carbon dioxide in the atmosphere was under 300 parts per million (ppm) in 1800–1910, and is perhaps 450 now (Burton; Gatewood, 147). It can be "as high at 600–900 ppm in metropolitan areas" (OSHA 2018).

Carbon dioxide is an asphyxiant. So how much must carbon dioxide levels rise in order to affect human performance? Not much. The U.S. Occupational Safety and Health Administration (OSHA) says that at 10,000 ppm (1 percent), there are "typically no effects, possible drowsiness." But a study showed that cognitive scores dropped 21 percent as a result of an increase from 600 to 1,000 ppm (Romm 2015, Grossman 2016). OSHA considers a 4 percent concentration to be "immediately dangerous to life or health." At a 5 percent concentration (50,000 ppm), hypercapnia and respiratory acidosis develop. OSHA indicates that typical symptoms are dizziness, confusion, headache and shortness of breath. At 8 percent, OSHA expects "dimmed sight, sweating, tremor, unconsciousness and possible death." At 10 percent, convulsions, coma and death are likely (Permentier 2017).

Other Noxious Substances. There can be other problematic constituents in the air. Incomplete combustion of coal or other fuel results in production of carbon monoxide, which is highly poisonous. Coal in bunkers can emit methane. Noxious gases can be released by the decomposition of the ship's timbers and of organic materials in the bilge water. And the air can contains smoke from candles and spores from molds (Ellis 1948; Bennett 2005, 32). Lastly, and perhaps most importantly, the air can carry droplets or droplet nuclei containing pathogenic organisms (Atkinson 2009, Annex C).

Natural Ventilation

Ship ventilation was not an issue until ships were built with closed decks or other enclosed structures and the crew spent significant amounts of time inside.

From papyri illustrations and tomb models, we know that closed-deck ships date back to ancient Egypt (Langstrom 1983, 24–7).

Ventilation was probably a ship design issue for ancient Greek, Phoenician, Rhodian and Roman oared warships. We know that on ancient Greek warships, the soldiers were on the top deck and the oarsmen below. Ancient Greek oarsmen were called *thalamioi* because they sat in the hold (*thalamos*) (Jordan 2000).

There was thus the problem of ensuring that the oarsmen had enough air. Langstrom says that one solution was to give the ship an "open bow" (49). On the Lenormant relief (circa 410 BC), depicting a trireme, we can see that the top bank oarsmen (*thranitoi*) were on a galleried lower deck, so fresh air shouldn't have been an issue for them (Fields 2007). Necessarily, the ships also had oarholes, and depending on their size, some fresh air could have been admitted through them for the benefit of the lower banks of oarsmen. In Field's illustration (D), the oarholes are large, but those of the lowest bank are sealed with a leather gasket. Since all three banks of rowers on a trireme were in a single hold—i.e., not separate decks, just sitting on benches at different heights—most of the fresh air for the lower oarsmen probably circulated down from the side openings of the top bank, and possibly also from a deck hatch.

Hatches offer one means of ventilating the area below deck. While the original purpose of the closed deck might have been to provide reserve buoyancy, it is likely that merchants decided early on that the space below could be used for cargo, and that meant that hatchways were needed to access it. Dating hatches is tricky, as pictures of ships usually show them in profile. But archaeologists found a clay model of an ancient Greek sailing vessel which shows a small hatch at the bow and a larger hatch at the stern (Johnston 1985, 26). However, holds were probably not used for sleeping quarters until much later. Simple superstructures for rain shelter date back to ancient Egypt (Langstrom, 12, 19) and even when they evolved into more robust and complete constructions, they were probably better ventilated than holds. A relief of a Roman merchantman (third century CE) shows two windows on one side of the aft cabin (Langstrom, 48).

On a 16th-century Spanish galleon, there was a between-deck area about 1.75 meters high. "If the ship sailed armed for war, this space would lodge the troops and the crew, but if the ship were also used as a merchantman, it would also hold cargo" (Perez-Mallaina 1998, 134). Indeed, in the latter case, it is likely that the crew (and passengers) avoided the between-deck area entirely, sleeping either on the main deck, or—if well-heeled or well-connected—in makeshift chambers in the sterncastle (136–8).

For warships, gunports offer an additional type of ventilation opening. Cannon were first placed on European warships during the late Middle Ages. Later, it was recognized to be advantageous from a ship stability standpoint to place guns, especially heavy guns, below the main deck. This led to the introduction of the gunport, which was possibly inspired by a cargo handling feature found on a few merchant ships. For example, there is a 15th-century illustration of a sailing ship with a hatch on its side, which Langstrom speculates was used for "loading timber or horses" (78).

Henry VIII's flagship, the *Henry Grace à Dieu* (1514), featured below-deck gun ports with a hinged door, although there is "little pictorial evidence supporting the incorporation of lids all along a lower deck before 1546" (Steele 2005, 144). The *Mary Rose* (1511) likewise was given lidded gunports, but some at least weren't cut until 1541 (De Castro 2022, 336).

One may presume that the gunners stationed on the enclosed gun decks soon discovered that they could get fresh air if the gun ports were left open. Thomas Lurting wrote in his memoir that on the *Mary* in 1662, "I went under the half-deck, and laid me down between two guns, on the boards, and slept very well" (1710, 27).

The history of the porthole—in the limited sense of an opening made solely for the purpose of admitting light and fresh air—is considerably murkier. The earliest clear-cut reference I have found was by Stephen Hales, who said in 1745 that "in hospital-ships … they cover the port-holes with linen." It is likely that they appeared first in the forecastle and sterncastle, rather than the main hull.

Gatewood (1909, 145) wrote, "a ship is much like a bottle of air. From its lower spaces the movement by diffusion is through relatively small … hatches in the deck above." The larger the ship, the greater the difficulty of ventilating it naturally. Why? The volume of air to be kept fresh is proportional to the hull volume, and the air intake to the hull's surface area. Surface area increases more slowly than volume.

The larger the ship, the deeper the crew's quarters are buried, and the staler the air they are breathing (Ellis 1948).

Natural ventilation is provided by wind (air movement from high to low pressure) and differences in air density between indoor and outdoor air (the "stack" effect).

The simplest method of ventilation is to leave open hatches, portholes and gun ports. However, this is feasible only if the waterline is well below the openings and the sea is quiet. Even then, there is the risk of a rogue wave. A quiet sea usually means that there is at most a light wind, in which case the rate of air inflow is likely to be low even if, fortuitously, the opening faces into that wind. As the seas become rougher, first the ports are closed and then also the hatches.

The next question is how one persuades the air to enter the hull through those openings, especially the hatches. The wind, after all, is blowing horizontally, and the hatches face upward. Moreover, the air below the hatch is likely to be warmer than the air above. Warm air rises, so hatches work better as exhaust openings (with fresh air taken in from the sides) than as intakes. That said, there were expedients to force hatches to admit fresh air.

Sails may cause some deflection of air. Macdonald (1881, 60) observes that "a great body of air is often sent down through the waist of a ship from the hollow of the mainsail or main try-sail, and it is common enough to spread small sheets of canvas, even between-decks, to deflect a current of air through a hatch or other convenient opening, to ventilate the space below."

In 1727, the Danish admiralty ordered that "each ship of 50 guns and above, and frigates which have two decks, must be given a 'wind hose'" (Brammer 2007, 12). That was what the British navy called, rather cryptically, a "windsail," and it improved ventilation somewhat. Initially, this was just an ordinary sail rolled into a tube, whose inlet was pointed in the direction of the apparent wind (Edwards 1881, 6). Purpose-built versions had canvas spread on wooden hoops to form a tube, sometimes with a flared inlet (Tomlinson 1886, 235). Note that in either case, the tube ran through an ordinary hatch, and hence could not be used in stormy weather.

Moreover, if there was no apparent wind (the vector difference between the true wind and the ship velocity), then it was no better in providing fresh air than an open hatch of the same size. Despite their limitations, wind ("air") sails were rigged on the *Fram* in 1893 (Nansen 1897, 305, 323), and even into the 20th century (Aonghais 2014, 190).

Steele (1794) says that windsails were made out of #1 or #2 canvas (the heaviest grades). The cloths, each nine yards long, were sewn together with a one-inch overlap to make the complete ventilator, and ash hoops were sewn to the inside at six-foot intervals. "The length of a windsail is taken nine feet above the deck to three or four feet below the lower hatchway."

The ordinary windsail was sometimes called a "shark's mouth," and the inlet orientation had to be adjusted whenever the wind shifted. Macdonald describes a fixed variation in which the inlet faced upward and four vertical cloth flanges, secured by guys, were suspended above it to deflect the wind downward, whatever its direction. Its original form had the

Canvas wind sail standing vertically abaft the bridge on the British heavy cruiser *Cornwall* (1926–42). Photo taken in 1930–39. (Naval History and Heritage Command, Washington, D.C. Photo catalog NH 52115. Photo cropped by author.)

British cruiser *Eclipse* (1894–1921), with all cowl vents turned to face starboard. Date unknown. Courtesy of the Naval Historical Foundation, Washington, D.C. From album of British warships (Naval History and Heritage Command. Photo catalog NH 75944).

disadvantage that this effectively reduced the aperture area by 75 percent. Raymond (1866) proposed cutting the flanges so the lower or inner parts were slack, and thus the space that the wind passed down was enlarged.

A variation on this theme is a cowl vent. In its simplest form, this is made of a rigid material (e.g., copper) and rises vertically from the deck or superstructure, turns horizontal, and flares out. It can be rotated to face and take in the wind (or to leeward to act as an exhaust vent), either manually or by some sort of weathervane mechanism.

To avoid the need for a pivoting mechanism, the cowl vent could be vertical but surmounted with a side vent structure having one or more tiers of conical canvas or metal deflectors to redirect horizontal airflow downward or upward airflow outward. The cones sloped downward for a downtake and upward for an uptake, so the vent operation could not be reversed (Macdonald 1881, 67).

A wind scoop is similar to a cowl vent, but instead of being a permanent structure, it is fitted over a hatchway as needed, and is usually made of cloth or, nowadays, plastic.

The functional equivalent of a cowl vent was depicted in Agricola's *De re Metallica* (1556). Air entered through a hole in the side of a pivotable barrel and then descended through a pipe at the center of the bottom of the barrel. On top of the barrel was a weathervane attached so that the wind would cause the barrel to rotate and thus expose the entrance hole to the full force of the wind.

Ideally, a shipboard ventilation system not only conducts fresh air into the innards of the ship, it does so in a way that keeps water out of the vents. For a long time, the solution was an imperfect one: increasing the height of the ventilator inlets. But a high enough sea could still get in.

In the Dorade Box (1930s), air entered a cowl vent. This curved down into a box with drain holes at the bottom through which any inadvertently acquired water could flow out. The air rose inside the box and descended again through an offset tube whose inlet was inside the box but near its top, and whose outlet was below deck (Adkins 2012, 92).

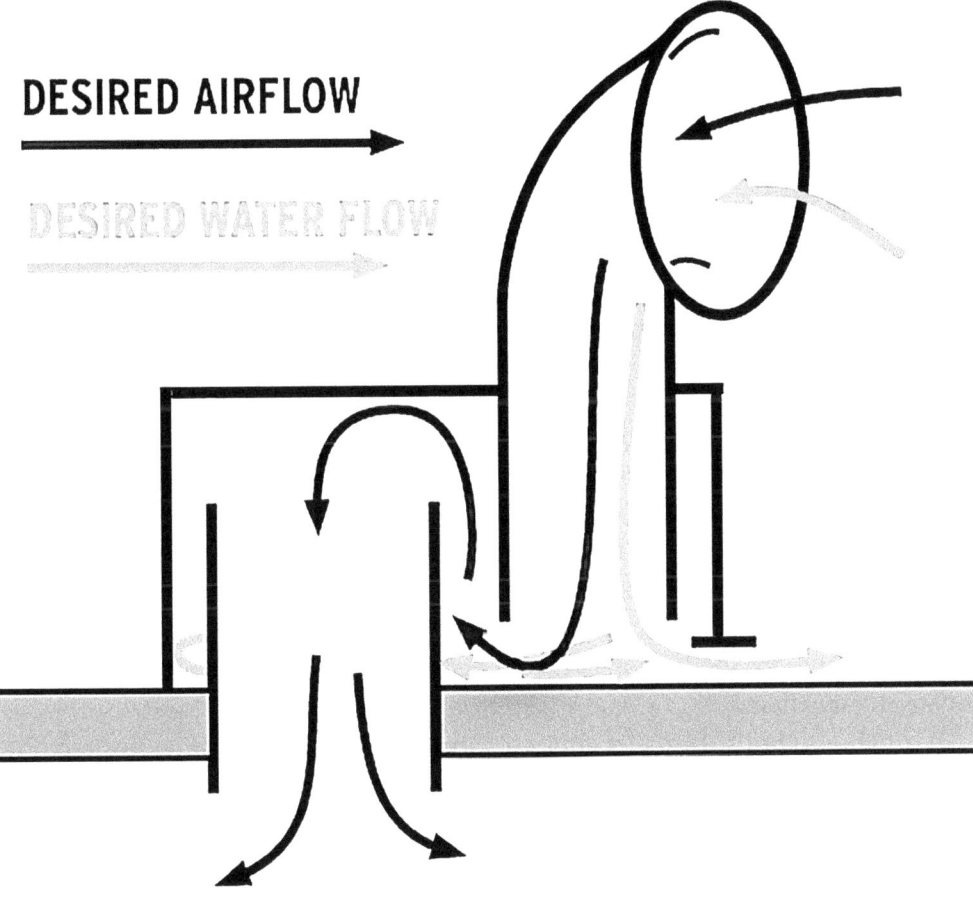

Dorade Box schematic created by Wikimedia user Aaronx, used under CC0 license, converted to black-and-white by author. "The [dark grey] arrows show the desired airflow (into the cabin) and the [light grey] arrows show the desired water flow (out of the integral scupper onto the deck). The [medium] grey area indicates the deck of the boat."

"The original Liberty ship did not have any mechanical ventilation" (Markert 1944, 216). It did have cowl intakes for the engine and boiler rooms, and "mushroom" and "torpedo" exhausts for the crew quarters (210). The hold ventilation of the Victory-class cargo ships was also natural.

The "mushroom" ventilator looks exactly like a mushroom, and the outgoing air flows up the stem, sideways to the cap, down and out. The torpedo ventilator has a double cone shape, with the meeting place of the base of the cones encircled by an intake or exhaust ring. A pipe connected the ring to the interior of the cones, and another ran from the cones to the interior of the ship.

During World War II, the openings associated with natural ventilation were problematic at night as they allowed the inside lights to shine out, attracting enemy attack (Markert 1944, 202). One solution was to place a "blackout ventilator" over the opening. For example, the Colt ventilator "allowed the free flow of air through their angled slats, which also trapped light" (Science Museum Group). They were used in "factories, ships and fighting vehicles" (Ramani 2015, 246). On the battleship *Royal Oak* at Scapa Flow in 1939, "because no underwater attack was expected, side scuttles were open for ventilation, with light-excluding ventilators to maintain blackout conditions" (Friedman 2015, 352). However, on some ships, portholes were simply blacked out at night by means of "dead light steel plates" (Vantine 2011, 13).

The British warship studied by Cope had mechanical as well as natural ventilation. However, Cope (1910) comments that the "bluejackets' normal" is "with scuttles shut, hatches open, no fans running."

Effectiveness of Natural Ventilation

We can think of the hull of the ship as being somewhat like a long, flat-roofed, rectangular building. Studies (Jing 2012) show that when the wind is perpendicular to the long side of such a building, it creates a positive pressure on the windward side and a negative pressure on the roof and leeward side. Also, if the roof is parapeted, the parapets reduce the suction at the roof edges, and I expect that bulwarks would have a similar effect.

If ports on both sides of a ship are open, cross-ventilation is generated. The depth of penetration of the air flow into the room can be up to five times its ceiling height (Bhatia 2014). If we figure that the ceiling height on an enclosed gun deck is at least five feet, than that means the fresh area can travel 25 feet from the gunports. The maximum beam on the HMS *Victory* (1765), a first-rate ship-of-the-line, was slightly less than 52 feet. So that's skirting the limit for cross-ventilation, but of course the hull narrows toward the ends.

The recommended total opening area (for buildings) on each side is 1 percent of the floor area (Bhatia). (However, that assumes that the space is not overcrowded;

the more people, the more CO_2 production.) On the middle gun deck of the *Victory*, there were 14 gunports on each side, sized for 24-pounders. On the lower gun deck, there were 15, sized for 42-pounders (but actually carrying 32-pounders at Trafalgar) (Eastland 2011, 40–42).

Du Monceau's 1764 treatise on naval architecture (which influenced Murray in England) gave the following dimensions for gunports (4):

Table 1-1: Gunport Dimensions

Shot wt (lbs)	Height (ft)	Width (ft)
48	2.75	3–3.17
36	2.67	3–3.08
24	2.42	2.67–2.83

I don't have figures for the surface area of the *Victory*'s gun deck. We know the maximum beam but not the average beam (which will depend on the deck). The length of the gun deck is 186 feet, so the area of the rectangle formed by the maximum beam (52) and that length is 9,672 square feet. The actual gun deck area must be less. On the middle gun deck, the gunport openings (with Du Monceau's dimensions) add up to 181 square feet, which is 1.9 percent of the rectangle area. For the lower gun deck, if we conservatively use the 36-pounder ports, we have 240 square feet worth of openings, which is 2.5 percent of the same rectangle.

If ports on one side must be closed, because heeling brings them dangerously close to the waterline, then we are left with single-sided ventilation. This is much less effective than cross-ventilation (depth of penetration up to 2.5 times ceiling height), and the recommended total opening area is then 5–10 percent of the floor area, with the same caveat (Bhatia).

Some merchant vessels also had enclosed gun decks and could carry passengers there. The *Mayflower*, an early 17th-century English merchantman, had a maximum beam of about 24 feet. On its gun deck, with four gunports on each side, the ceiling height was 5.5 feet. The 102 pilgrims spent most of the journey there. They could not enter the powder room aft, so they had an area about 58 feet long (assuming a length between perpendiculars of 80 feet). Total volume was thus no more than 7,656 cubic feet, and of course the air volume would be less, as some of the space was occupied by the Pilgrims and their belongings. The "encompassing" rectangle has an area of 1,392 square feet, and the actual deck area would have been less.

I do not know its gunport dimensions, but a mid–17th-century Dutch source said that gunport height is six times the ball diameter and width is five times (Hoving 2012, 104). A four-pound cannonball has a diameter of about three inches, so figure two square feet per gunport. The eight gunports alone simply could not admit enough air to adequately ventilate the *Mayflower*'s gun deck under the best of circumstances. But the circumstances weren't the best, because the passengers would have put up wood or cloth dividers to give each family some privacy, and that would interfere with the air flow. The passengers must have been dependent on open hatches, and perhaps sojourns to the open deck, for fresh air.

Air density decreases with increasing temperature and humidity. Since it is usually hotter indoors than outdoors, cold air will enter through open ports on the flanks of the ship and warm interior air will escape through the hatches topside. This is a "stack" effect. (If it is hotter outdoors than indoors, the flow will be in reverse.)

The rate of stack-induced air flow is proportional to the size of the openings (inlets or outlets, whichever is smaller) and the square root of the product of (1) the vertical distance separating inlets and outlets and (2) the ratio of the inside-outside temperature difference to the average of the two temperatures (Bhatia 2014; Brandan 2018).

On a calm day, a 15.5-inch porthole, five feet above the deck, with a hatchway near it that was four feet higher, and a 20°F temperature difference between internal and external air, reportedly delivered 13,000 cubic feet per hour (Cope 1910, 443).

Unfortunately, for structural reasons, ports and hatches are often found in close proximity, which limits the volume effectively ventilated (Cope 1910; Gatewood 1909, 145). In the building trade, this is called "short-circuiting."

Mechanical Artificial Ventilation

Both the fan and the bellows were invented in ancient times. They were first used in terrestrial dwellings, mines, and factories. But ultimately they were used on ships.

Fans. John Desaguliers (1683–1744) was a clergyman with an interest in natural philosophy. In 1714, he was appointed as demonstrator (in effect, Isaac Newton's assistant) at the Royal Society.

Desaguliers developed a fan system that was first applied in mines (1727) and in the Houses of Parliament (1745) (Worshipful Company of Fan Makers). His "fanning" or "blowing" wheel, "as first invented, was 7 feet in diameter, and 1 foot wide, and had twelve radii or partitions approaching within 9 inches of the axis, leaving a circular opening 18 inches in diameter." This was encased so that its only communication with the air was through a blowing pipe on the upper part and a suction pipe connected to the axial opening. When a crank was turned by a single worker, the air was centrifuged to the periphery and pushed out the blowing pipe (Tomlinson 1886, 236ff). This is in essence a centrifugal fan, and that type of fan was used previously to ventilate mines, as discussed by Agricola (1556).

Desaguliers attempted, unsuccessfully, to persuade the British Admiralty to adopt it for the navy. The reasons for his failure are instructive. The Surveyor of the Royal Navy, Sir Jacob Ackworth, chose to arrange the demonstrations at a time when a strong wind was blowing and demanded that Desaguliers show that his engine would "throw out as much air as our wind-sails you see do." Desaguliers

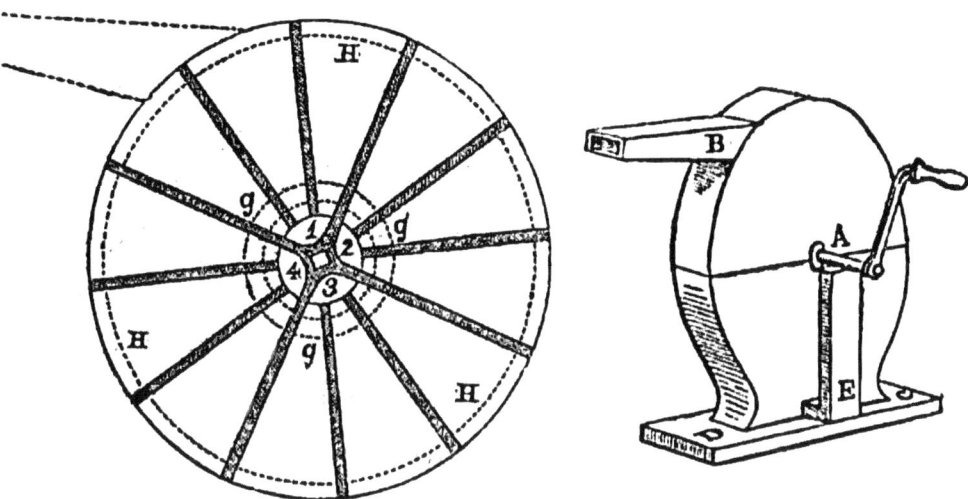

Figures 101 and 103 from Charles Tomlinson, *A Rudimentary Treatise on Warming and Ventilation* (1886). The turn of the handle around axis A rotates the vanes. In supply mode, fresh air is sucked in through a pipe connected to the central opening (1–4) in the wheel (opposite side from the handle), and pushed by the vanes into the blowing pipe B. The fan may also be operated in reverse (exhaust mode).

vainly "pointed out that the blowing wheel was to be used when the wind-sails were useless" (Edwards 1881, 7).

The familiar 20th-century room fan is an axial fan; the air enters and leaves on a course parallel to the axis. Axial fans don't produce a lot of pressure, but if resistance is low, they can move a lot of air. Desaguliers's fan is centrifugal: the air enters at the center and moves radially outward. Axial fans are suited to moving air inside a large room, whereas centrifugal fans are better for moving air in ductwork, where resistance is higher. Axial and centrifugal fans were both in use on cargo ships in World War II (Markert 1944, 206).

Bellows. The Admiralty was ultimately more sympathetic to the proposal of the clergyman-scientist Stephen Hales (1677–1761). He first described his ventilator to the Royal Society in 1741 (Harris 1916, 116). The first sea trial of the Hales ventilator was in 1743. The captain rather ungraciously concluded that since no one on his ship became ill, "he had no opportunity of judging the usefulness of it" (Krulder 2021).

In 1756, Admiral Edward Boscawen conducted an experiment in which one of the ships of his fleet, the *Royal George*, sailed with Hales's ventilators. It was subsequently reported that the crew of the *Royal George* returned to England in great health, whereas "every other Ship in his Fleet … had from forty to a hundred and twenty sick at a Time" (Krulder). This prompted the Admiralty to order installation of the Hales ventilator on all of its warships.

What Hales presented was a balance-beam double bellows. There was a lever with a center pivot, worked by two men, and connected to two adjacent bellows. In each bellows, a vertical rod was connected to one of the lever arms, and the rod in turn to a hinged wooden board called a "midriff." Each vertical movement of the rod brought in air through one valve and expelled it through another (Bernan 1845, 51ff). A similar ventilator was simultaneously invented by the merchant Martin Triewald (1691–1747) and adopted by the Swedish and French navies.

The original Hales ventilator was placed on the orlop deck. For a first-class ship, the ventilator lever was 12 feet long (Ritchie 1843). Sizes ranged from 10 feet long, 4.5 feet wide and 2 feet deep, with a 12-square-inch trunk, for the ventilator for a 100-gun ship, down to 8 by 4 by 1.5, with a 9-square-inch trunk, for one for a 20-gun ship (Rees 1819, 26).

Of course, men were needed to work the Hales ventilator. Captain Thomas of the frigate *Success* wrote in 1749, "our rule for ventilating was half an hour every four hours.... All agreed the ventilators were of great service. The men did not need to be urged to work them" (Harris 1916, 448).

According to Tomlinson (1886, 248), the air-moving power of the Hales bellows was in fact far inferior to that of the fan.

Motive power. The great disadvantage of these early mechanical ventilation systems was that they were human-powered. Once steamships were introduced, the steam engines could be used to not only propel the ship, but also drive a mechanical ventilator. And even a pure sailing ship could have a donkey steam engine used for ventilation.

However, steam also created some problems. The usual fuel was coal, and that meant that the ship had coal bunkers. Flammable gases would be emitted from the coal, especially when there was a drop in atmospheric pressure or an increase in temperature. This could result, ultimately, in the formation of an explosive hydrocarbon-air mixture. Hence, provision had to be made for ventilation of coal bunkers (Atwood 1904, 109ff).

An engine, rather than driving a ventilator directly, may be used to run a generator or dynamo that supplies electricity (in the form of alternating or direct current, respectively) to a motorized ventilator. This makes it easy to power a ventilator that is in a part of the ship remote from the engine. Rather than needing a large, heavy drive train, you have electric wires conducting the generator to the motors (*Encyclopaedia Brittanica* 1911, "Ventilation").

A dynamo ("magneto") was installed in HMS *Minotaur*, albeit to power a searchlight projector (Grove 1900, 531). With time, the use of electricity expanded to supporting numerous shipboard systems, and the battleship USN *Kearsarge* (1900) carried "55 motors and has a generating plant of 350-kilowatt capacity" (570).

In 1900, Grove said, "the electric driving of ventilating fans is now common in most navies" (598). Grove notes that electric fans "can run at least twice the speed of reciprocating engines for the same power without noise or vibration" (582), and that electric motors have higher mechanical efficiency (574). However, it appears that the engine and boiler rooms on at least some warships were then still ventilated by steam-driven fans (582).

Richards, commenting on Grove's presentation, said that "when it was decided to use electrical fans for ship ventilation, it was found necessary completely to alter the usual system of ventilation, and instead of a few large fans 5 to 6 feet in diameter, to use a number of much smaller fans" (611).

Each of the living spaces on Victory-class ships (first produced in January 1944) were "fitted with one or two mechanical porthole blackout ventilators," with air movement (despite the blackout louvers) "in excess of 300 cfm" (Markert 1944, 218). These were also installed on the newer Liberty-class ships and retrofitted onto some that were already in service.

Heat-Induced Artificial Ventilation

Another inventor, the brewer and coffeehouse proprietor Samuel Sutton (1695–1714), proposed the use of waste heat from the ship's oven. His invention was prompted by reports of the Spithead incident mentioned at the beginning of this chapter (Zuckerman 1976–77, 230). Sir Jacob Ackworth gave him the runaround for a week and then refused to arrange for a test (232). Enter the physician Richard Mead (1673–1754), who believed that scurvy was attributable to the breathing of bad air (223). Mead secured a trial for Sutton on the hulk at Deptfort in 1741.

In Sutton's original version, the normal source of air for the oven was blocked and instead pipe was laid from the rooms to be ventilated to the ash pit below the fire. With the oven in operation, air would be drawn up through the oven's chimney, and that would draw air from the connected rooms, which in turn would draw it from elsewhere (Edwards 1881, 13–4; Stuart 1845, 49ff).

Subsequently, his "air-pipes" were installed on the warship *Norwich*. In 1743, the Admiralty informed Sutton that his air-pipes were "dangerous," as the captain of the *Norwich* had to stop up two of them because "the fire came down between decks."

To assuage the concern that sparks could travel down through the pipes, Sutton proposed a modified version in which the tubes passed through the furnace and into the chimney. The air in the pipes would be warmed and this would cause it to rise, creating a partial vacuum as before (Ritchie 1843).

Mines were similarly ventilated by "constructing a ventilating furnace at the bottom of an air shaft," but their efficiency was low and they could not be used in mines producing fire-damp (methane) (EB 1911, "Ventilation"). The second objection applies to ships only if they are carrying coal as cargo or fuel.

An obvious advantage of the Sutton system over those of Desaguliers or Hales is that no sailors must be dedicated to the task of operating the ventilator. So long as the oven is in use, there is air movement in the pipe. Indeed there would be some movement for a time after its fire is put out, because of the retained heat. Nonetheless, while a few ships used the Sutton system, it was eclipsed by Hales.

In 1843, Ritchie proposed that in a steam-powered vessel, heat-induced ventilation may be achieved by passing the tubes through or near the boiler rather than the oven (Ritchie, 178). However, it appears that steam was subsequently used to drive fans rather than to induce air movement by warming air in close proximity.

Ventilation System

We've looked at the types of artificial ventilation apparatus, but to be effective ship-wide, they need to be tied together into a ventilation system.

Modern ventilation systems are classified as exhaust, supply, or balanced. An exhaust (vacuum) system actively removes air, slightly reducing the air pressure, which increases the rate at which makeup air diffuses in through cracks or passive vents. A supply (plenum) system is the opposite: outside air is forced into the room, slightly increasing air pressure, and outflow is passive. A balanced system has both active supply and active exhaust. The Sutton system was of the exhaust type, while the Desaguliers and Hales systems could be used for either supply or exhaust. Individual vents may be operated as either supply vents or exhaust vents, depending on the fan operation.

In a house, air enters through cracks even if there are no open windows and doors, but a ship is much more airtight. Hence, one must either keep inlet and outlet vents open (if water infiltration can be avoided), or just recirculate the air through the hull. In the latter case, there is no overall reduction in carbon dioxide levels as a result of the air movement, although there may be a modest evaporative cooling effect.

The ventilation system (openings, ducts and air-moving machines) must be sized to the ship and its crew in order to be effective. Also, one can expect that the rate of air movement declines as one moves away from the inlets, so it is desirable with natural ventilation to have many short ducts rather than a few long ones, and to make sure that all occupied parts of the ship are serviced by the distribution network.

After 1785, some ships were equipped with wooden (later brass) tubes that connected the "between decks" area (where the men slept) with the air over the gunwale of the forecastle (Tomlinson 1886, 250).

If air movement is room to room, without dedicated ducts, then the rooms further "downstream" will be intermittently ventilated, depending on when intervening doors are opened. The "air streak" was a space above the orlop deck and the bottom of the hold that was left unplanked to facilitate the circulation of air over the timbers. The purpose of this was to inhibit the rotting of the timbers, not to help bring fresh air to the crew, but Wyman (1846, 338ff) suggested that it could be "boxed" so that an exhaust fan could be used to create an air current within it.

The ventilation system must contend with the frictional resistance to the air current, which is proportional to the surface area of the ventilation passages and the square of the air current velocity. The required power for air circulation is thus proportional to the cube of the air current velocity (EB 1911, "Mining"). The volume of air moved by a ventilator is going to be the product of the air speed and the cross-sectional area of the ventilator.

Markert comments that shipboard "ducts are constructed of considerably heavier gauges than are commonly found on land. Vibration, flexing of the vessel, and corrosion justify this practice" (1944, 212).

Cargo Ventilation

Some cargoes can generate internal heat as a result of chemical reactions (fermentation, reaction with water, air, etc.), and it may be necessary to cool them by ventilation before they spontaneously combust (McDonald 1957, 4). Cargoes prone to spontaneous combustion include hay, coal, linseed oil, cotton, fishmeal, and pistachio nuts (Stevens 1871, 130ff).

Cargoes may also generate hazardous gases, such as methane from coal, which must be removed by ventilation. (For this reason, it is probably a good idea that the cargo hold ventilation system be separate from that serving other areas of the ship.)

Depletion of oxygen by cargo capable of absorbing or reacting with oxygen may also be an issue. "An experienced customs officer was found lifeless at the bottom of the unattended cargo hold on a ship loaded with woodchips. The oxygen content in the cargo atmosphere was below 2%" (Sundal 2017). Another fatal incident occurred on a ship transporting zinc concentrate. Zinc can react with oxygen. The chief mate entered the hold and closed the door behind him. The day after his death by asphyxiation, the oxygen content in the air at the stairwell was found to be 2.6 percent (MARS Reports).

Many cargoes are damaged if they become too moist. Fruits get moldy; iron rusts. That unwanted moisture can come from other cargo. Even seemingly dry commodities may contain considerable natural moisture, such as Polish rye (17–19 percent), wheat (10–14 percent), cotton (8–11.5 percent), silk (11 percent), raw wool (9.5–20 percent), "dry" hides (14–20 percent), and wood (11–16 percent kiln-dried, 16–24 percent air-dried, and over 30 percent unseasoned) (Duly 1950). Moreover, some cargoes, such as hides and textiles, have a propensity for absorbing water from the air even when the latter is not fully saturated (McDonald 1957, 4).

The cargo is thus, potentially, a "huge reservoir of water," and the "interstitial air" inside such cargoes may be at a very high relative humidity. This interstitial air, together with the outside air at the time of loading, determines the initial storage atmosphere. If it is raining when the cargo is loaded, that will tend to increase its humidity as the raindrops evaporate.

A tightly packed cargo hold will only gradually change temperature from the loading temperature, lagging behind the change in sea and air temperature as the ship journeys to its destination.

The dew point temperature is the temperature below which the air will yield up water vapor, causing it to condense on available surfaces. The more humid the air, the higher the dew point temperature.

If cargo is taken aboard at a tropical port and sails for a temperate one, the storage hold temperature may drop below its dew point, causing the moisture in the surface layers of the cargo to condense out. This is called "cargo sweat." This "sweat" may also appear if the deck above is cooled by wind or rain, or even just nighttime drops in air temperature, and heat is thus pulled out of the hold.

In the 20th century, it was customary for ships to mechanically ventilate their cargo holds to carry cargo sweat away, and it was a legal defense against a charge of water damage to cargo that a ship had to turn off its ventilators because of heavy weather (Duly 1950).

On the other hand, if the ship is loaded in a temperate port and passes through tropical waters, it is likely that the sea air available for ventilation is warmer than the surface of the cargo. In general, sea air is moist, so if the sea air is used for ventilation, without dehumidification, it is likely that the surface temperature of the cargo is less than the dew point of the ventilating sea air. Hence, ventilating the cargo hold under these circumstances will result in the precipitation of water from the ventilation air onto the cargo. This can seep in and cause damage.

The wise cargo master will monitor the dew point temperatures of the sea air and the cargo hold, turning ventilation on and off as needed.

Unventilated Rooms: Countdown to Respiratory Distress and Asphyxiation

How's that for a cheery title? The point is to show just how important ventilation is to shipboard health by calculating how quickly you would asphyxiate if confined to an unventilated room. There are several considerations: the initial air supply below deck per person, and its oxygen and carbon dioxide levels; the rates of oxygen consumption and carbon dioxide production per person; and the oxygen concentration (minimum) and carbon dioxide concentration (maximum) at which asphyxiation (or some lesser distress) occurs.

Initial air supply. The berth deck on the USS *Constitution* was 3,393 square feet (USS *Constitution* Museum). The height from the berth deck to the gun deck above it was six feet, four inches (*All Hands*), so its "box" volume was 21,489 cubic feet. In 1793 it carried 273 enlisted and 60 marines, for a total of 333. That yields 64.53 cubic feet per person berthed there (officers would sleep in the wardroom).

For passenger-carrying ships traveling from the United States to Europe, the Steerage Act (1819) set a limit of two passengers for every five "tons burden." The latter was a crude estimate of the volume of a ship's hold; assuming that a ton burden was about 40 cubic feet (this is something of a quagmire), that would mean 100 cubic feet per passenger. However, the law had the rather large loophole that some of the ship's burden might actually be occupied by cargo.

This loophole was addressed by the Carriage of Passengers Act (1855). Not only did it reduce the maximum ratio to one passenger for every two tons burden, it set a minimum floor space of 16 square feet of "clear superficial deck" for each passenger on the main or poop decks, and 18 on the lower deck, if any. (If the height between decks was a spacious 7.5 feet or more, only 14 square feet was required.)

The ceiling height of steerage was typically six to eight feet (Solem 2007). Six times 16 is 96, and 7.5 times 14 is 105. So let us assume that if the ship adhered to the 1855 Act, it was providing about one hundred cubic feet per passenger.

In the late 19th-century United States Navy, the sleeping spaces for the crew provided about 165 cubic feet per sailor (Pryor 1918, 48). Note however that at least one-third of the crew was always on watch (49), so the effective allotment was 495 cubic feet.

In 1910, the naval surgeon Cope estimated that on a contemporary British warship, the average cubic feet per sailor on the stokers' and seamen's mess decks were 207.7 and 216.4, respectively. In contrast, it was 427.5 for sick bay and 450.0 for the officers' cabins.

Breathing volume and rate. Both the volume of air inhaled and exhaled with each breath and the number of breaths taken per minute vary with age, sex, conditioning, and, most importantly for adults, physical activity. If you are engaging in heavy exercise—say, stoking coal in the boiler room, or operating a cannon on the gun deck—you are likely to be breathing more deeply and more frequently than when you are sleeping in a hammock.

The total volume of air processed per minute is called the "respiratory minute volume." Because of all the variables that affect respiration, as well as definitional issues, different studies report somewhat different numbers.

For an average healthy adult male at rest, the tidal volume—the volume of air inhaled and exhaled during a single normal breath—is about 0.5 liters (0.0177 cubic feet) (Hallett 2022). You breathe more deeply when exercising: Anderson, studying a mixed-sex group, reports tidal volumes of 1.45 for light work (40 percent maximum aerobic capacity), 1.86 for moderate (60 percent), and 2.30 for heavy (80 percent). A maximum value of 3 has been suggested (USNDM 2016, 3–4.5).

The normal resting adult takes 12 to 20 breaths per minute (Chourpiliadis 2021). You also breathe more frequently when exercising. Anderson (2006) reports 21.9 breaths per minute for light work, 26.5 for medium, and 31.9 for heavy.

As for respiratory minute volume, a study of male car riders—presumably at rest—reported a minute volume of 11.9 liters (Zuurbier 2009). Male bicyclists had a minute volume of 22.0 liters. Anderson's values are 30.3 liters for light work, 47.4 for moderate, and 72.3 for heavy. The U.S. Navy suggests a broader range of "6 to 10 liters per minute at complete rest" to "over 100 … during severe work" (USNDM 2016). More particularly, 6 liters per minute sleeping, 7 sitting quietly, 9 standing still, and 16 walking (Fig. 3–6).

Oxygen consumption. If you are breathing fresh air (21 percent oxygen), the exhaled breath is only about 16 percent oxygen (Johnson 2018). The five percent difference corresponds to the consumed oxygen. The U.S. Navy assumes that the respiratory minute volume is 25 times the oxygen consumption rate, implying that the exhaled breath is 17 percent oxygen (3–4.6).

Oxygen consumption (as opposed to inhalation) varies from 0.25 to 4 liters/minute, depending on exertion and physical condition (USNDM 2016, Fig. 3–6).

Carbon dioxide production. Carbon dioxide is produced by human respiration. The exhaled breath is 4–5 percent CO_2 (Inglis-Arkell 2011).

When the body is "burning" fat, it produces 0.7 liters of carbon dioxide for every liter of oxygen thus consumed. When it is burning carbohydrate, it is one-for-one, and for protein metabolism, 0.9 (Patel 2022). Thus, for every liter of oxygen consumed, a person will produce and exhale 0.7–1 liter of carbon dioxide, depending on the work rate; 0.9 is a typical assumed value (USNDM 2016, 3–4.5). (This ratio is called the respiratory exchange rate or respiratory quotient.) An older edition of the diving manual said to figure maximum CO_2 production as one liter per minute for light work, two for moderate or heavy, and three for severe (USNDM 1979, 6–51). Persily (2017, 873–4) presents estimates of carbon dioxide production for, inter alia, 21–30-year-old males for a variety of activities; the range is from 0.24 liters per minute for sleeping to 1.92 for "vigorous calisthenics."

Carbon dioxide is also produced by combustion of fuel, whether in the wick of a candle or the furnace of a steam engine. A paraffin wax candle produces about 10 grams per hour (Chester 2017), as compared to 33–41 grams per hour exhaled by a person at rest (based on 37–46 mg per breath at NTP, Koennecke). How fast carbon dioxide is produced by a steamship will depend on the efficiency of its engines and the desired speed. One pound of carbon combines with 2.667 pounds of oxygen to produce 3.667 pounds carbon dioxide. If the coal used has a carbon content of 78 percent, complete combustion of a short ton produces 2.86 short tons carbon dioxide (Hong). Most of that, of course, should be exhausted by the smokestacks.

However, we will focus on human production of carbon dioxide.

Asphyxiation time calculation. We now have all we need in order to estimate how soon sailors would asphyxiate if they were limited to the initial air supply. Let's assume that the unventilated room in question is occupied by sailors at rest, with 200 cubic feet (5,663.37 liters) of initially fresh air per sailor. We assume they are breathing in 12 liters of air per minute, and thus 2.4 liters of oxygen. With a respiratory exchange ratio of 0.7:1, that means they are producing 1.68 liters of carbon dioxide per minute.

If we take the point of no return to be a 10 percent concentration (566 liters), and ignore the trivial amount of starting CO_2, that point would be reached in 336.9 minutes.

But would they have run out of oxygen earlier? The starting level of oxygen was about 1,189 liters. Let's follow the USNDM and assume that the critical concentration is 10 percent, or about 566 liters. To consume the difference (about 623 liters) will take our resting sailors about 260 minutes. However, if we followed the Air Force data and allowed oxygen to decline to 6 percent, that would take about 354 minutes.

Please bear in mind that there are significant individual variations in all of the component variables—oxygen consumption, carbon dioxide production, and reaction to low oxygen or high carbon dioxide levels, so the actual time to asphyxiation will vary. Also, chances are that the initial air supply isn't completely fresh, and that the rates will change as the oxygen declines and CO_2 increases.

Adequacy of Ventilation

While natural and artificial ventilation, by replacing old air with fresh, may inhibit disease transmission as well as keep oxygen and carbon dioxide levels within safe limits, the analysis in this section is focused on carbon dioxide.

Carbon dioxide limits. For obvious reasons, these limits are a lot more stringent than what is needed to prevent asphyxiation. In 1869, Roscoe stated that "when the air of a room contains 0.10 per cent of this gas, it is certainly unfit for continued respiration" (84). A more stringent standard of 0.7 per 1,000 was proposed for warships in 1912 (Ellis 1948). OSHA's current permissible exposure limit for eight-hour exposure is 0.5 per 1000 (and it assumes that fresh air has a CO_2 level of 0.4 per 1,000).

In the late 19th century, instances were reported of substantially higher carbon dioxide levels on shipboard. These include 1.03–3.21 per 1,000 between decks on the steam frigate *Doris*, 4.82 in the wardroom of the Arctic expedition (1875-6) steam sloop *Alert*, 5.57 on the lower deck of its companion ship, the steam barque *Discovery*, and 50 in the powder magazine of the armored corvette *Jackson* (Parkes 1883, 123).

Referring once again to Cope's British warship (1910), on the seamen's mess deck (11,900 cubic feet, 69 men present), CO_2 levels per 1,000 cubic feet air were as follows:

- no ventilation: 1.28
- only natural (seven 1.5-inch scuttles and a hatchway): 0.95

- only artificial (12.5-inch, 4,000 cfm fan): 0.85
- both: 0.63

Thus, the combination of natural and artificial ventilation cut carbon dioxide levels in half.

Ventilation rate. Alternatively, one could specify the fresh air ventilation rate that would, hopefully, achieve the desired carbon dioxide cap. This may be expressed in terms of the number of cubic feet inhaled per unit time per occupant, or the number of air changes per unit time. Cope's 13,000 cubic foot per hour porthole was said to be "sufficient for four men under good conditions" (443); thus, 3,250 per hour per occupant.

The 1911 *Encyclopaedia Brittanica*'s "Ventilation" suggests that fresh air be supplied and foul air extracted at a rate of 1,800 cubic feet per hour per person for barracks and 5,000 for workshops. Ellis (1948) says that meeting the 1912 carbon dioxide limit required a supply of 3,000 cubic feet (84,951 liters) per hour. However, by 1937 this was relaxed to 2,000 cubic feet (56,634 liters). (I believe that it was intended that these be interpreted as being on a per occupant basis.)

To place these recommendations in context, a contemporary Gunther five-bladed fan with a diameter of ten inches, turning 3,300–4,000 rpm, could exhaust 400–500 cubic feet of air per minute (cfm), while its 50-inch counterpart, running 550–700 rpm, exhausted 12,000–15,000 cfm (EB 1911, "Bellows").

In the 1910 British warship discussed by Cope, the stokers' and marines' mess decks had a 12.5-inch fan providing 4,000 cubic feet per minute, as well as scuttles and a hatchway. The space was occupied by 65 men, so just the fan ventilation works out to 61.54 cubic feet per minute (3,692 per hour) per occupant.

On World War II cargo ships, ventilation of berthing spaces and mess rooms was typically 20 to 50 cubic feet per minute per person (Markert 1944, 202). Ventilation requirements varied depending on the space, and Markert expressed it (206) in terms of the time for a complete air change. This ran from one minute for the gallery, bakery, pantry and laundry (where there would also be the issue of excessive heat), to six minutes for the "messing space" and 30 minutes for the cargo hold.

According to the 2018 International Mechanical Code, a ventilation system should supply outdoor air at a minimum rate of five cubic feet per minute per occupant for a dormitory sleeping area and 20 for a place where vigorous physical activity occurs (such as a gym or health club).

The American Society of Heating, Refrigerating and Air Conditioning Engineers (ASHRAE)'s 2019 standard has required rates of fresh air flow expressed on both a per-person and per-unit area basis for different spaces. Both criteria must be met. The ones most closely analogous (admittedly with a bit of a stretch) to some shipboard environments are:

Table 1–2: ASHRAE 2019 Ventilation Standards

	Rate Per Person cfm (L/s) Per Person	Rate Per Unit Area cfm/ft² (L/s-m2)	Assumed Occupant Density Per 1,000 ft² (Per 100 m²)
Break room (office)	5 (2.5)	0.12 (6)	50
Barracks sleeping room	5 (2.5)	0.06 (3)	20
Health club aerobics room	20 (10)	0.06 (3)	40
Manufacturing; school wood/metal shop	10 (5)	0.18 (9)	?

How do ventilation rates correspond to carbon dioxide concentrations? If the rate of carbon dioxide production is constant (i.e., people aren't moving in and out or changing activity level) and the ventilation rate is constant, the equilibrium carbon dioxide concentration C_{eq} will be

$$C_{eq} = Ng/aV + C_{amb}$$

with N (number of people in the room), g (average carbon dioxide production per person per unit time), a (number of complete air changes per unit time), V (volume of room), and C_{amb} (ambient, i.e. fresh air, concentration of carbon dioxide) (Luther 2014, eq. 5), of course with consistent units.

We may recast it in terms of the initial air supply and ventilation rate per person as follows:

$$C_{eq} = gV/ir + C_{amb}$$

where i is the initial air supply per person, and r is the ventilation rate per person. Note that this means that specifying a ventilation rate per person is not good enough to achieve a CO_2 target. That's because the bigger the room, the more air the ventilator must move to achieve the same rate of dilution of the stale air.

Ventilator sizing. Yet another approach to addressing the problem of carbon dioxide accumulation was to specify the number and size of ventilators. The Carriage of Passengers Act (1855) required that vessels used to carry passengers, with a legal capacity of a hundred or more, have at least two ventilators, one at the forward end of the passenger apartment, and the other at the aft end. Presumably, one was the uptake and the other the exhaust, but this is not specified. They had to have a capacity scaled in proportion to the legal capacity of the ship, and "for a capacity of 200, the capacity of such ventilators should be equal to a tube of twelve inches diameter in the clear." (It is not explicitly stated whether this refers to the intake or exhaust diameter.) The ventilators were required to "rise at least four feet six inches above the upper deck of the vessel," which leads me to believe that a cowl vent rather than a fan was contemplated. However, that could be the uptake tube for a fan. If these were fans, note that the speed of rotation wasn't specified, and that has a definite effect on its air-moving power.

Robin Sharma said that "to breathe properly is to live properly." We have a potent and urgent need to inhale oxygen and to avoid inhaling carbon dioxide. Brain cells begin to die after four minutes without oxygen, and carbon dioxide competes with oxygen for binding to hemoglobin.

But as Jacques Cousteau said, there are two essential fluids. Air is one, but water is the other. And water is the subject of chapter 2.

Chapter 2

Quenching Thirst

In August 1804, after a long blockade of Tripoli, Commodore Edward Preble ordered imposition on the U.S. Mediterranean Squadron of a short ration of just five pints a day of water, writing to the navy agent on Malta, "our water is nearly exhausted." Asking for water to be sent "immediately," he added, "if one is not here in 8 days we shall be ruined, as I have only 14 days water for the Squadron. Let no price stop you from chartering … and for God's Sake dispatch a Vessel in 24 hours" (Toll 2008, 242).

Drinking Water Demand

As physiological needs go, the need for water is second only to the need for fresh air. According to the Institute of Medicine and the World Health Organization, an average (70 kg) healthy adult male needs to drink a minimum of about 3 liters of fluid a day in a temperate climate, and in a tropical one, 4–6 liters (Grandjean 2004; Gleick 1996). Exercise also increases the water requirement. As we will see, the food served to sailors was frequently salted to preserve it, which would have increased demand.

The drinking water is to replenish fluid lost by sweating (possibly elevated by fever), urination, and, in case of sickness, vomiting and diarrhea. Thirst occurs when net fluid loss reaches 1 percent of body weight, and reduced work capacity at 4 percent. Collapse can occur at 7 percent (Grandjean).

Beverages

The body's need for fluid can be supplied not just by water, but also by other water-containing beverages, such as beer and wine. That is just as well, since in the premodern period, alcoholic beverages were likely to have a lower pathogen content than ordinary water. In the case of beer, the boiling during the brewing process rendered it free of bacteria. Subsequent bacterial infiltration was inhibited by various "antimicrobial hurdles," including not just ethanol but also hop bittering compounds, low pH, elevated carbon dioxide, and low oxygen (Menz 2011).

Wine, of course, was not boiled, but its greater alcohol content itself inhibited bacterial growth. Wine's high acidity and polyphenol content is also antithetical to long-term growth of most human pathogens. Certain preservatives (e.g., sulfur dioxide) were also used in winemaking in the premodern period (Azevedo 2016).

Brandy or rum of course would be even more potent (affecting both bacteria and sailors) and would have been initially sterilized by the distillation process.

A problem with alcohol beverages for thirst quenching is that alcohol is a diuretic—it stimulates urination. However, there is scientific literature indicating that the effect is short-lived and that drinking alcohol is not likely to cause you to be dehydrated (Abbott 2019).

Wine is presently defined as having an alcohol content of at least 7 percent. Rum typically has an alcohol concentration of at least 40 percent. Grog is diluted rum, but at a typical dilution of 4:1, it would still have a higher alcohol content than wine.

At least as early as the 10th century, Indian Ocean seafarers made use of coconut water. The coconut itself acted as a sterile water container (Lanbourn 2018, 178).

Tea, cocoa and coffee, as served, are largely water. In the Georgian navy, tea and cocoa were considered substitutes for cheese, and were usually drunk at breakfast, when hot water was available (Macdonald 2006, 43–4).

Other high-water content foods include the aptly named watermelon, as well as strawberries, cantaloupe, peaches, oranges, cucumbers, lettuce, zucchini, celery, tomatoes, bell peppers, cauliflowers, cabbages, grapefruits, lemons and limes. But only the last two, in juice form, were ever significant parts of sailors' diets, as will be discussed in chapter 3.

Water Rations

Spain. In the Spanish navy, in 1568, the daily ration included a liter of water and a liter of wine (Perez-Mallaina 1998, 141). The sailors had to weigh whether to drink their wine, a source of additional calories and some solace from hardships, or save it for sale in America. The French likewise took their liquor ration as wine (Spalding 2015, 70).

In 1588, prior to the departure of the Armada, the Duke of Medina Sidonia instructed shipmasters that "sufficient water must be given to each man for drinking and cooking purposes, but the ordinary water ration must not exceed three pints a day for all purposes." The normal wine ration was a third of an *azumbre* (one *azumbre* equaled nearly half a gallon), and the wines were to be used in a particular order based on their longevity at sea (Hume 1899, 269–70).

Britain. Premodern British sailors received a daily ration of one "gallon" (Childs 2009, 87); at least by the 18th century, this was measured as 14 fluid ounces because the remainder was the purser's profit of beer (or occasionally ale or cider). If they were in the benighted Mediterranean, they might instead receive one pint of wine. Obviously, if they were relying on that alone, they would be dehydrated.

The capture of Jamaica in 1655 meant that rum became the beverage of choice. Not only was it cheap, it was more potent than beer (whose brewing for the Royal Navy was discontinued in 1832 [Phillips 1970]) and thus kept for long periods in wooden barrels. Initially, the sailors could drink their daily half pints straight (if the captain permitted). In 1740, Admiral Vernon required that one part rum be mixed with four parts water, creating the famous "grog." An especially dilute grog (1:6 to 1:8) might be issued as a punishment (Blake and Lawrence 2005, 105). The men received two quarts of grog each day (Swinburne 1996, 309–10; Pope 2013, 150, 153; militaryhistorynow.com 2013). After 1810, lemon or lime juice was routinely added to the grog (Spalding 2015, 70).

The obvious problem with consumption of alcoholic beverages was that over-indulgence could impair crew performance or discipline. Less obviously, rum, at least, presented a fire hazard; in 1779 on the *Glasgow*, the purser's steward, while stealing rum from the aft hold, accidentally dropped a light into the cask and started a fire (a bad thing to do on any ship, but the *Glasgow* was shipping gunpowder to Jamaica) (Sugden 2012, 139).

The "tot" of half a pint (two gills) a day was reduced in 1825 to a quarter-pint, and in 1850 to an eighth-pint. The rum ration was abolished altogether in 1970 (Phillips 1970), but sailors were allowed to buy 1.5 pints of beer a day (Williams 2006, 243).

"The purser's steward then began to call the grog list." From E. Shippen, *Thirty Years at Sea: The Story of a Sailor's Life* (1879), page 67. It appears the grog is being taken from a tub rather than a barrel. Note the marine sentry.

There wasn't a formal water ration, but under normal circumstances the men were permitted to drink (but not take water away from) a water barrel on deck, the "scuttle butt" (Macdonald 2014, 43). Exceptions were made when the water was to be taken "for the express purpose of cooking, or for the use of the sick" (Johnson 1807, 2). A "butt" is a type of cask (strictly speaking, equaling two hogsheads), and a "scuttle" is a hole. According to Dana it was cut into the bilge of the cask (Dana 1841, 101).

In 1707, the volume of a "butt" (pipe) was standardized by section 17 of the Act 5 Anne c. 27, as 126 wine gallons, and the wine gallon itself was set at 231 cubic inches (Chisholm 1873, 35–36). However, in "ale measure," a butt was 108 gallons (Admiralty 1883, 337) and that was the volume of a water butt in Nelson's navy (Blake 2005, 249n10). The 1869 instructions to surgeon-superintendents of Queensland immigrant ships called for them to carry three scuttlebutts, containing 100 gallons each (Queensland 1869, Appendix 73).

A common location for the scuttlebutt was "under the topgallant forecastle" (Stayton 1895, 76). Melville's *Moby Dick* (170) refers to two scuttlebutts, one in the waist, and the other near the taffrail.

The men drank from a "parish pot," a common drinking cup. (For sanitation issues, see chapter 5.) According to Captain Griffiths (1828, 279–280) of the Royal Navy, it commonly held a quart (0.94 liters) of water. He thought this too much, arguing that they would drink it all, and this was excessive, and so they should be limited to a "half pint pot," and if they wanted more than one draft, they should be obliged to go back to the end of the line. A similar suggestion was made by Gihon (1871, 93), an American naval surgeon, who recommended use of a "small tin drinking cup, of the capacity of a gill [0.12 liters] … attached by a chain to the faucet of the scuttle-butt, and allowed to be filled but once at each drinking." He thought that the sailors would take "ten to fifteen full drafts a day."

The concept of limiting the rate of water consumption has some modern medical basis. There is a condition known as "water intoxication"; it is the result of drinking water faster than the kidneys can process it (about 0.8–1 liter per hour) and thereby upsetting the body's sodium balance (hyponatremia) (Semeco 2023). However, in practice, I suspect that the result would have been that the total daily water consumption was inadequate, and the sailors partially dehydrated.

America. United States Navy regulations in 1802 required that ships be provisioned with at least one-half gallon of water per man per day for foreign service (Brenckle 2018, 21).

American ships, like British ones, had a scuttlebutt. However, Herman Melville, who served as an ordinary seaman on an American frigate in 1843, reports that the cask stood on end, with its upper head removed (1922, 357) . Melville adds, "The scuttle-butt is the only fountain in the ship; and here alone can you drink, unless at your meals. Night and day an armed sentry passes before it, bayonet in hand, to see that no water is taken away, except according to law. I wonder that they station no sentries at the port-holes, to see that no air is breathed, except according to Navy regulations" (358).

American sailors additionally received a half-pint of spirits each day (Brenckle, 3). This was diluted 3:1 (19). In 1806, the U.S. Navy switched from West India rum to native whiskey (18).

In 1842, the U.S. Navy reduced the spirit ration to a quarter-pint and forbade it being given to a commissioned officer, a midshipman, or a sailor under age 21 (they got the cash value as a consolation). It also permitted substituting half a pint of wine for the spirits. However, it also added to the daily ration "one quarter of an ounce of tea, or ounce of coffee, or one ounce of cocoa" (Navy CyberSpace 1842). Given the quantities, these were flavorings to be added to the water ration. The spirit ration was abolished by the U.S. Navy in 1862 and by the British navy in 1970.

Coffee later became the beverage of choice. In the 20th century, Samuel Eliot Morison (1947, 252) wrote, "Although the United States Navy might win a war without coffee, it hopes never to be forced to make the experiment." Daily coffee consumption (based on the 1945 USN cookbook) has been estimated as 3.3 cups per sailor (Bean 2020).

The Ship's Water Allotment

The water required for a warship on an extended voyage was considerable. The rate of usage would vary, depending on the air temperature, the number of persons on board (itself subject to attrition due to combat, disease and desertion, and increase due to impressment), and rationing orders by the captain.

In 1636, the Amsterdam Admiralty declared that for each month at sea, a ship carrying 100 men had to carry "thirty-five tuns of ale in winter (forty-two in summer)" (Schama 1997, 176).

In the Napoleonic-era British navy, "a sloop with a crew of 135 usually carried about forty tons of fresh water and expected it to last three months, using it at a rate of about half a ton a day." (Pope 2013, 171). (The Royal Navy used the 28-day lunar month [Macdonald 2006, 78].)

The HMS *Pearl* (1762), a 32-gun frigate with a complement of about 220, carried 43.434 tonnes water (11,460 gallons) as three months' stores (Braithwaite 2009, 17).

In 1813, the USS *Constitution*, expecting to be six months at sea, sailed with 47,265 gallons of water and 5,074.7 gallons of spirits (Brenckle, 2). Brenckle estimates that a crew of 450 would consume 250 to 340 gallons drinking water a day, plus as much as another 150 gallons per day for cooking (21). Toll gives an even higher estimate for the *Constitution*'s consumption rate in the Mediterranean in 1804: 600 gallons a day (Toll 2008, 242).

Passenger ships. In America, the Steerage Act of 1819 required that all ships departing the United States carry at least 60 gallons of water per passenger. It seems strange that this was expressed on a per passage rather than a per diem basis, but perhaps the expectation was that the vessel owners would underestimate the duration of the journey. If we assume that the most typical route was from the United States to Europe, and that it could be traversed in 60 days, then the requirement would work out as one gallon per passenger per day. The 1882 Carriage of Passengers Act, section 4, required that each passenger receive not less than four quarts of fresh water per day.

Watering a Ship

Transferring water to a ship had its own difficulties. In the premodern period, water was likely to be in casks, either carried by boats or rafted over. Rafting was necessary on a coast with a surf, as a heavily laden boat could otherwise be swamped. The casks could be towed broadside on, in single file, or end on, in pairs. Note that fresh water is less dense than salt water, so the casks could float.

Later, canvas or leather bags were supplied to boats, and they would be filled by hoses from a shore pump. Typically, the bag had two halves that saddled the boat (Henderson 1907, 183).

Water Storage

Earthenware vessels. In ancient times, earthenware vessels, such as amphorae, were used for shipping liquids, such as olive oil or wine, and it is likely that they were also used to carry drinking water for the crew. Amphorae were heavy relative to their capacity. For example, an amphora with a capacity of 26 liters might weigh 26 kilograms, empty (De Grauuw 2017). They were also breakable, and therefore were protected with straw.

"By the Medieval period, at least in the central and western Mediterranean, the barrel had largely replaced the amphora as the preferred transport container for liquids ... having a better volume to weight ratio, more efficient stacking capability, and greater manoeuverability on land." And in the western Mediterranean, the switch was completed during the second half of the first millennium CE (Wilson 2011).

According to a late 17th-century report, on the Manila galleons, the initial supply of water was carried in two to four thousand earthenware jars. Some of these were hung in the rigging and others kept below deck (Schurz 1939, 269). The jars were of Philippine, Chinese, Vietnamese and Thai manufacture, and each jar held about 5.5 gallons (Lugar 2011, 37).

Wooden casks. In the early modern period, water was stowed in wooden casks. Casks were of a roughly cylindrical shape with a bulge at the middle. This shape made it easy to roll them and also to change the direction of the roll. They were formed of wood (typically oak) staves bound with iron hoops. They can be divided into wet casks for liquids and dry casks for solids, the former being made to narrower tolerances and therefore being more expensive. The seven standard English liquid (originally, wine) cask sizes, and their relationships, were as follows: 1 tun = 2 butts (pipes) = 3 puncheons = 4 hogsheads (quarters) = 6 tierces = 8 barrels = 14 rundlets. At least from 1587 to 1850, the tun was usually 252 wine gallons, and a tun full of wine weighed about one long ton (2,240 pounds). However, while the beer barrel was defined as 34 gallons in 1688 (and 34 × 8=252), it was increased to 36 (the pre-1688 value) in 1803 (Zupko 1985, 25–6, 186, 330, 357, 411, 425–6).

The price of oak staves climbed from 105 pounds per thousand in 1793 to 156 in 1805, leading to the substitution of beech and white oak for dry casks (Knight 2010, 62).

Typically, a ship would carry an assortment of water casks of different capacity. The 74-gun USS *Franklin*, in 1821, carried 10 casks each of 250, 200 and 50-gallon capacity, 130 of 100, 30 each of 40, 20 and 15 gallons, and 70 of 8 gallons—a total of 33,860 gallons (Williams 2012, 308). Most of the water barrels would be placed deep in the hold, serving as a form of ballast. Hence, as they were emptied, they would be refilled with seawater for the sake of stability.

The wooden barrels used to store water were not necessarily pristine. During the 19th-century Atlantic migrations, drinking "water was stored in old sugar hogsheads, in oil casks which had never been cleaned, in vinegar, molasses and turpentine barrels" (Wagner 2006, 62).

Water tanks. Square wooden water tanks were apparently used in ships built at Surat in the 18th century (Layman 1813, 17). In fact, fixed wooden water tanks (*fintas*) were still being used on the Arabian Gulf in the 20th century (Agius 2012, 142).

Metal water tanks were less prone to leakage and fouling, but of course were heavier and more expensive. In addition, since they were larger, it was harder to shift them around in order to adjust the trim of the ship as supplies were consumed and cargo loaded or unloaded. (Being able to pump water from one tank to another would help.)

The tank being larger, if it were only partially full, it would create a greater "free surface effect." In essence, when the ship heels over, the center of gravity of the water inside shifts, reducing the vessel's stability. The effect can be reduced by 75 percent by dividing the athwartships width of the surface area of the tank into two parts by means of a watertight bulkhead (International Maritime Organization 2008, 125). (The effect was small in the old barrels because they were so small and were used only a few at a time.)

In 1828, Captain Anselm Griffiths complained about having once taken command of a frigate whose water tank was lined with lead. Referring to evidence of lead and its compounds being poisonous, he had the tank taken to pieces and replaced it with a scuttlebutt (Griffiths, 300–2).

The 600-ton British sloops-of-war *Arrow* and *Dart*, built by Samuel Bentham (1757–1831) in 1795-7, had eight tinned copper water tanks, each holding 40 tons of water, which proved successful in preserving the sweetness of the water, and had double the capacity of casks taking up the same amount of deck space (Chalmers 2012; Bentham 1802, 238; Macdonald 2006, 85) Reportedly, they were too thin-walled to stand without support, and therefore were placed in wood casings (*Mechanics Magazine* 1849, 39). There was also apparently some problem with leakage, as there is an 1804 letter from Nelson saying, "if the tanks cannot be repaired, water casks must be substituted in their room" (Macdonald 2006, 85).

Despite the widespread use of copper pots for cooking on shipboard, copper tanks didn't catch on. I suspect that copper was used for cooking because its higher heat conductivity justified its use despite a higher density and cost than iron, but iron was better for cold water storage. Still, the *Leonora* (lost in 1874) had "several copper-clad wooden water tanks" (Lenihan 2010, 164).

Beyer (1848) asserts that substitution of iron for wood was proposed by a French captain, Sibon, in 1739. The mining and steam engineer Richard Trevithick proposed use of iron tanks in an 1808 patent application, which pointed out that iron walls could be thinner than wooden ones (Trevithick 1872, 1:285). The next year, his company approached the British Victualling Board, pointing out that the iron tanks were cheaper than their wooden equivalents of the same capacity. The Admiralty had the tanks fitted into five vessels and after five years' experience concluded that the iron tanks were superior (Macdonald 2010, 102). Captain Truscott's pump was officially adopted in 1812 and could be used to pump water from the water tanks directly to the cooking coppers, via leather tubes (Macdonald 2006, 85).

The iron tanks, typically of rectangular or square plan, could be close fitted and overall occupied half the space of an equal capacity of casks. However, a danger in close fitting is that water can leak if the tank is overfilled and, draining down through the cracks between the tanks, can cause rusting of the tanks and rotting of the wood underneath (Stevens 1871, 683). In 1871, Admiralty tanks ranged from 100 to 600 gallons (17–101 cubic feet) capacity, and weighed 364–1,190 pounds empty. Note that large tanks are more weight-efficient than small ones. The actual water weighs 8.34 pounds per U.S. liquid gallon (3.785 L, almost the same as the wine gallon of 3.79 L).

While it is certainly possible to fill or empty a large iron tank (too large or heavy to move) a bucket at a time, it is preferable if water can be pumped in or out with a hose (for pumps, see chapter 6).

Iron water tanks should be placed well away from the ship's magnetic compass, if possible, or the deviation caused by the tank should be noted for future reference. Commander Walker reported that in 1818, he set a WNW compass course, but found that the ship, equipped with new iron water tanks, actually bore NNE. After sailing 21 miles, it was at least eight leagues further south than it should have been (Walker 1863, 45).

Another problem with iron is its susceptibility to corrosion. Nor does the rust simply stay on the inner surface of the tank; it is mechanically weak, and the vibrations of the ship under way send particles into the water, imparting a metallic taste. Hence, experiments were made with various linings. Beyer (1848) declared that the least objectionable material was Portland cement. In the modern U.S. Navy, the insides of steel water tanks have an epoxy coating (*Manual of Naval Preventive Medicine*, 6–15).

The introduction of the water tank also changed somewhat the design and operation of the scuttlebutt. According to Herman Melville's *White-Jacket* (1850), within the latter there is an iron pump that brings water up from "the immense water tanks in the hold" (355). United States Navy 1864 "allowances" suggest a semipermanent installation, as they called for "a pipe to lead from three inches below the top of the scuttle-butt, down through its bottom," to connect with a pipe leading from a force-pump in the hold (USN 1865, 80).

It was not necessarily an "ever-full" receptacle. Gihon urged that "the whole daily allowance should not be pumped into the scuttle-butt at one time, but at intervals, during the day." He noted that it would be supplemented by "tea and coffee at intermediate times" (1871, 93).

By the early 20th century, the scuttlebutt was itself "a closed cylindrical iron tank with a net capacity of 75 to 100 gallons ... located forward on the gun deck." There were "faucets around its base," so it could be used by several sailors simultaneously, and it even could contain a refrigerating coil connected to an ice machine. But the water was still often drawn into "cups or mess bowls often attached by chains" (Gatewood 1909, 375), save in the case of men in the firerooms, who obtained their drinking water, often iced, in buckets (378).

Supplemental Water Sources

Seawater. If you run out of the stored beverage, you're in trouble, because drinking seawater in significant quantities leads ultimately to dehydration. Seawater is 3 percent salt, and the kidneys can't make urine from water saltier than 2 percent, so it take the water it needs from the tissues (Cerezo 2012). This is obviously self-defeating. As Coleridge's Ancient Mariner put it, "Water, water, everywhere, Nor any drop to drink." (Methods of desalination are discussed later in this chapter.)

Shore collection. If a ship was near shore, it might send a landing party to look for fresh water, hoping to find a spring or stream. There was of course the danger of encountering hostile inhabitants, or merely of unwittingly collecting water from a contaminated source.

Rainwater harvesting. One method of catching rain was simply to plug the scuppers and collect the water from the main deck. Of course, the water is likely to pick up all sorts of nasty stuff that was on the deck. In a similar way, you could catch it from the roof of the superstructure, assuming you could ensure that it wouldn't just run off the roof. You could set out buckets, or spread canvas horizontally, attaching it to the mast and rigging, and empty the harvested rain into a cask.

According to Lord Anson, the galleons on the Manila–Acapulco run took with them "a great number of mats, which, whenever the rain descends, they range slopingly against the gunwale from one end of the ship to the other, their lower edges resting on a large split bamboo, whence all the water which falls on the mats drawn into the bamboo, and by this, as a trough, is conveyed into a jar." Also, they hung heavy mats in the "upper works," and water could be squeezed from them (Schurz 1939, 269–270).

Some British naval captains also took advantage of rainfall, and not just when desperate. Captain Pasley, commanding the 28-gun *Sybil* in 1780, while off the Guinea Coast, "Filled several Tons of Water which, altho' we are not in immediate want, enables me to give a larger allowance of this necessary Element to the Seamen, than which nothing contributes more to their health in long Southern Voyages" (Society for Naval Research 2015).

There were several problems. First, whether it would rain or not was unpredictable. Second, the collection methods were likely to lose water if the ship were rocking (and rain tends to be associated with rough water). Third, the rainwater could be contaminated with salt from sea spray, especially if the wind was strong. Finally, how much rain could be collected would depend on how much surface area could be presented to the skies above, and on a ship, space was at a premium.

Melted ice. Ships traveling in high latitudes in winter may encounter icebergs, sea ice and snow. New sea ice (formed by freezing of seawater) is actually very salty, but as ice ages the brine is expelled by various mechanisms. Icebergs are made of freshwater ice.

Melting ice or snow requires much less heat than boiling seawater. Icebergs thus provide opportunity as well as risk: on January 9, 1773, the second lieutenant of the *Adventure*, James Burney, wrote, "being very fine Weather we brought too by an Island of Ice & hoisted our Boats out to pick up the loose pieces to water the Ship—we got 6 Boat Loads which when melted in the Coppers gave us 7 Tons of Excellent fresh water" (Boreham 2023). This was apparently resorted to at least as early as 1671 (Goethe 2012, 8)—I suspect this is a reference to a whaling expedition to Spitzbergen, memorialized by Friderich Martens in 1675.

Fish. While you do not want to eat high-protein food when you are thirsty and short of water, you can drink the aqueous fluid from the eyes and spine bones, which is almost free of salt (Kamler 2004, 107).

Desalinated seawater. The separation of salt from water goes back to ancient times, but generally it was the salt, not the water, that was sought. Salt production of course was simpler. The sea would rush over a dam at high tide, some of the seawater would remain behind when the tide receded, and the confined seawater would be heated and evaporated by the sun, leaving the salt behind.

A thirsty mariner, however, does not want the water vapor to escape, but rather to trap it and allow it to cool and condense. Methods of desalination are discussed in the following sections.

Thermal Desalination (Distillation)

Distillation is a process of purifying a liquid by first vaporizing the volatile components (usually by heating) and then condensing the vapor (usually by cooling).

If liquid water is in contact with a gaseous medium (like air), some of the water molecules will enter the vapor state. The greater the temperature, the more water molecules do this, and this vapor contributes part of the pressure in the medium. At 15°C (59°F), under normal atmospheric pressure, it is about 0.0168 atmospheres (Engineering Toolbox 2023).

When the vapor pressure of water equals the environmental pressure, the water is said to be a saturated liquid, that is, one on the verge of boiling. The temperature at which this occurs is called the saturation temperature (boiling point) and it increases with the environmental pressure. Likewise, for a given temperature, the pressure at which the water is on the verge of boiling is the saturation pressure.

If water is in the saturated state, it can be vaporized either by further increasing the water temperature or by further decreasing the environmental pressure.

In order to distill seawater, the distiller must first supply the energy needed to raise its temperature from its starting value (which will be higher in the tropics than at higher latitudes) to the boiling point (at the environmental pressure). Actual vaporization further requires providing energy to break the intermolecular "hydrogen bonds" that hold water molecules together in the liquid state. The necessary energy is called the (latent) heat (or enthalpy) of vaporization, and it decreases with increasing environmental pressure. The temperature of the water doesn't change during vaporization.

At normal atmospheric pressure, if the water is tropical water, at 80°F, then to raise it to 212°F requires 132 BTU/pound and the heat of vaporization is 970 BTU/pound, for a total of 1,102 BTU/lb (Telkes 1953, 1,108). (If the water were instead at 60°F, add another 20 BTU/lb.) One pound corresponds to roughly one pint of water, and 1,102 BTU to 0.323 kilowatt-hours.

For effective distillation you must be able to collect the vapor and then condense it (return it to liquid). Condensation requires removing the heat of vaporization, and that is usually done by circulating it over cold-water pipes.

Early thermal desalination. There was both classical and ecclesiastical authority for desalination by distillation. Aristotle (*Meteorology* lib. ii. ch. ii.) proclaimed,

"Sea-water can be rendered potable by distillation: wine and other liquids can be submitted to the same process. After they have been converted into humid vapors they return to liquids." In his fourth homily, Saint Basil said that having been shipwrecked on an island without drinking water, he and his companions heated saltwater in an iron basin and condensed the vapor on sponges, squeezing out the fresh water (Stevenson 1892, 2:569 n2). Note that Basil's sponges had to be cold for this to work. The medieval physician Gilbertus Anglicus suggested distilling seawater "though an alembic" and filtering it through "clean sand" (Keevil 1957, 1:18).

Distillation on European sailing ships. Sir Richard Hawkins wrote that in his South Sea voyage of 1593, "with an invention I had in my ship, I easily drew out of the water of the sea, sufficient quantities of fresh water to sustain my people with little expense of fuel; for with four billets I distilled a hogshead of water…. The water so distilled, we found to be wholesome and nourishing" (Hawkins 1847, 82). In 1606, a Spanish captain, finding that he had been shorted water barrels, likewise made "sweet water" by distillation using "a copper instrument he had with him"—that is, not an improvised device (De Queiros 1904, 196).

The early 17-century Dutch, notably Jan Huygen van Linschoten, Aegidius Snoeck, and Cornelius Drebbel, promoted shipboard distillation technology (Delyannis 1974, 6). Nonetheless, it appears that at least the Dutch East India Company crews were prejudiced against it for some reason (Beekman 1988, 23), and Snoeck's 1620 device was expensive (Torck 2009, 214).

There was also, unfortunately, a strange belief that distilled water was not pure, but rather also included "a bituminous substance, and a spirit of sea salt" (Lind 1788, 336). This in turn led those experimenting with distillation to add various neutralizing substances that probably did more harm than good. For example, in the time of Charles II it was proposed to add lime (Colomb 1898, 12). In 1753 Appleby tried *lapis infernalis* (silver nitrate) and calcined bones (Lind 1788, 334–5), and about the same time Alston promoted limestone, and Hales, powdered chalk.

Soon thereafter, Doctor James Lind conducted a controlled experiment to compare the various proposed additives, and was surprised to discover that the control (distilled seawater with nothing added) was equal in quality to distilled rainwater. He publicly demonstrated this in 1761 (1788, 329), and proposed that for distilling water, the ship's copper pots, used for boiling victuals, be fitted with "still-head covers" and a pipe used to carry the steam from the pot to a cask of cold water (1788, 336–344). A trial was conducted in 1768 on the *Dolphin*: "56 gallons of sea water were put into a still, and 42 gallons of fresh water drawn off in the space of five hours thirteen minutes, with the expense of nine pounds of wood, and of sixty-nine pounds weight of coals; this was upward of a quart of water for each man on board" (345). Note that the fresh water content of 56 gallons of seawater (specific gravity 1.026) would be about 54.6 gallons, so about 12.6 gallons were lost in the form of unrecovered water vapor.

Lind even described a method of improvising the still using the pot, a tea kettle (or a wooden hand pump), a musket barrel, and a cask. In the *Dorsetshire*, this kludge converted 22 quarts of seawater to 19 quarts of fresh water in four hours, expending 10 pounds of wood (Clarke 1838, 130).

On November 21, 1791, in order to expand use of distillation methods among American seafarers, Thomas Jefferson suggested to Congress that an account of the methods for obtaining fresh water from salt water be printed on the back of the clearance papers for all ships leaving United States ports.

In the 18th-century British navy, the ship's kettle was divided in half by a partition, and peas and oatmeal were cooked on only one side, but with water kept on the other. Doctor Charles Irving showed that the spare half of the kettle could be filled with seawater and distilled by a method similar to Lind's, while the peas or oatmeal were boiling, supposedly without any additional fuel consumption (Falconer 1815, 428) (presumably because of heat transfer across the partition).

One famous 18th-century user of the still was Captain Cook, on HMS *Resolution*. In 1775, fitting the still to a 64-gallon "copper," the crew lit the fire at four in the morning and water production began at six. In 12 hours, 32 gallons of fresh water were collected, at the expense of 1.5 bushels of coal, which was twice what was needed for boiling that day's victuals. Cook deemed it a "useful invention," but warned against trusting wholly to it, especially in hot climates where water demand was greatest (Cook 1813, 251–2). A still was also used by Louis-Antoine de Bougainville in his own circumnavigation (Dunmore 2022, 30).

In 1806, John Lamb received a British patent (2952) for a cooking and distilling apparatus. Lamb was a New Yorker living in London at the time, and a merchant by occupation (Repertory 1807, 407). It was tested on HMS *Trusty* in 1807 and, besides producing fresh water, it also used one-quarter less fuel than the one it replaced. At least 28 ships, including 74-gun warships and East India men, were equipped in 1809 with the "Lamb Patent Fire-Hearth." It had three boilers. With saltwater in one of them, it would produce eight gallons of freshwater an hour, without extra fuel being consumed. And of course you could use all three boilers at once in an emergency to produce 24 gallons per hour (*Naval Economy* 1811, 16–22).

Distillation on Asian sailing ships. Nineteenth-century Japanese mariners resorted to distillation of seawater in emergencies. When the Japanese merchant ship *Tokuju-Maru* was left adrift in 1813 as a result of a storm, they had plenty of soybeans but little drinking water. So Captain Jukichi rigged a makeshift still: "[S]eawater is boiled in a big kettle. A pipe is poked through a hole in the bottom of a big pot which is then placed on top of the kettle. As the steam passes through the pipe and cools, it forms into drops of water, which are then collected in the pot for drinking water. By using this *ranbiki* they were able to make about 7 or 8 shi [about 14 liters] of water per day" (Torck 2009, 221; Plummer 1991, 80).

A study of 19th-century Japanese castaways identified seven incidents, Jukichi's included, in which seawater was desalinated by distillation. Five of them claimed to have either designed the device themself, or to have been inspired by a dream (Jukichi included). But the frequency of adoption of this expedient suggests that "it was common knowledge among both maritime communities and alcohol distilleries" (Wood 2009, 110).

Jukichi's device was a makeshift *ranbiki*; the purpose-built one, used by Japanese apothecaries, was more elaborate. It was made of ceramic or copper, and had three chambers: one holding a liquid, to be heated by a charcoal fire; a top chamber to receive the evaporate; and a middle chamber that could be filled with herbs from which the steam could extract herbal oils. The evaporate condensed in the top chamber and the condensate descended through a side pipe, dripping into a receptacle. The top chamber was equipped with an annular cooling tube (Michel 2004).

Some think the Portuguese introduced the *ranbiki* (the name is thought to come from "alembic") to Japanese medical practice. If that's correct, then it occurred before 1639, when the Portuguese were evicted. On the other hand, it is possible that it was part of the "Dutch Learning," and if so, might have come later; a distillery for essential oil extraction was established by the Dutch on Deshima in 1671 (Michel 2004).

It may even have come from the Ryukyu Islands (conquered by the Japanese in 1609) rather than from the Europeans. On Okinawa, they drank *awamori*, made by distilling alcohol from fermented ice. It is sometimes called "island sake," but sake is not distilled; the Japanese equivalent is *shochu*. *Awamori* was distilled and sent as tribute to China at least as early as the 15th century and presumably was sent to Japan after the conquest (Esparza 2017).

Distillation on steamships. The development of steamships gave additional impetus to distillation technology. Seawater is about 0.15 percent calcium sulfate. The latter is only sparingly soluble in water and its solubility decreases with increasing temperature. If the temperature exceeds the solubility limit, the calcium sulfate precipitates out, forming scale on the inside of the boiler. This scale is a poor conductor of heat and therefore reduces the efficiency of the boiler. Worse, its deposition is uneven, and the boiler plate can be overheated in spots, with occasionally catastrophic results (Yeo 1894, 163).

Early ship boilers operated at relatively low temperatures and pressures and could therefore use seawater as the source of steam, provided that boiler water was periodically "blown out" (discarded) before the salt concentration reached the point of precipitation. This did have the consequence of wasting the heat associated with the blown-out water.

However, the power developed by a steam engine increases with increasing boiler pressure, and at higher pressures, the boiling point is also higher. So with high-pressure boilers, seawater couldn't be tolerated at all, as any calcium sulfate in the feedwater would be precipitated (Yeo, 164).

Hence, marine steam engines had to be designed as closed systems, with freshwater boiled into steam and the steam condensed back to liquid form to be returned to the boiler. (In contrast, say, to the typical steam locomotive, in which the steam was exhausted into the atmosphere.) The condensers used seawater as the coolant, but it was not permitted to mix with the boiler water. With efficient "surface" condensers, the initial charge of freshwater could be mostly recycled, but the system needed to take in additional feedwater ("makeup water") to make up for losses from leakage. Yeo estimated the daily waste as 2 to 3 percent of the water used (166).

In consequence, steamships were equipped with evaporators dedicated to processing seawater. The seawater was admitted into a vessel penetrated by tubes containing "primary steam" generated by the ship's steam power plant, heating and thus evaporating the seawater. Only some of the seawater was evaporated, leaving salt behind, and thus the remaining seawater (the "brine") was saltier as a result (Normandy 1909, 67).

The "secondary" steam produced by the evaporator passed to a condenser that supplied the makeup water to the boiler. Scale would form in the seawater evaporator, but the latter was designed for ease of opening and cleaning (Yeo, 167).

This setup was, somewhat confusingly, called a "double" distillation, because the first distillation was in the boiler and the second in the evaporator (Normandy, 172). In some instances, two evaporators were connected in series so the heat of the steam from the upstream evaporator was used to heat the seawater of the downstream one. If so, then Normandy would call it a "treble" distillation. The number of stages could be further increased, but with diminishing returns, and Normandy thought that quadruple distillation was the economic limit.

Thus, in single distillation, Normandy expected that burning one pound of "good Welsh steam coal (10,650 BTU)" would provide enough heat to produce 10 pounds of distilled water. With each stage added, he calculated that production would be increased by 80 percent from that of the preceding stage. So for quadruple distillation, a total of 29.52 pounds of water would be produced, although the fourth stage contributed only 5.12 pounds. However, with a multiple distillation, either the initial pressure must be increased or the final pressure decreased so as to have a sufficient temperature difference between "the steam inside the coils of each evaporator and the water outside such coils" (176–81). And of course, all of the constituent evaporators were the same size and cost regardless of how much water they produced.

Water production was typically about 5–5.5 lbs. for 7.5–8 lbs. of primary steam used, and thus for a pound of coal burned (Yeo, 168).

There was a choice as to when the primary steam was removed from the ship's power plant. It could be a portion of the "live" (high-pressure) steam generated by the boiler. Or it could be the exhaust (low-pressure) steam leaving the engine after it expanded against a piston or turbine blade. The main disadvantage of using the latter is that its pressure varies (Normandy, 68–9).

The steam could be introduced into the evaporator in a number of ways. First

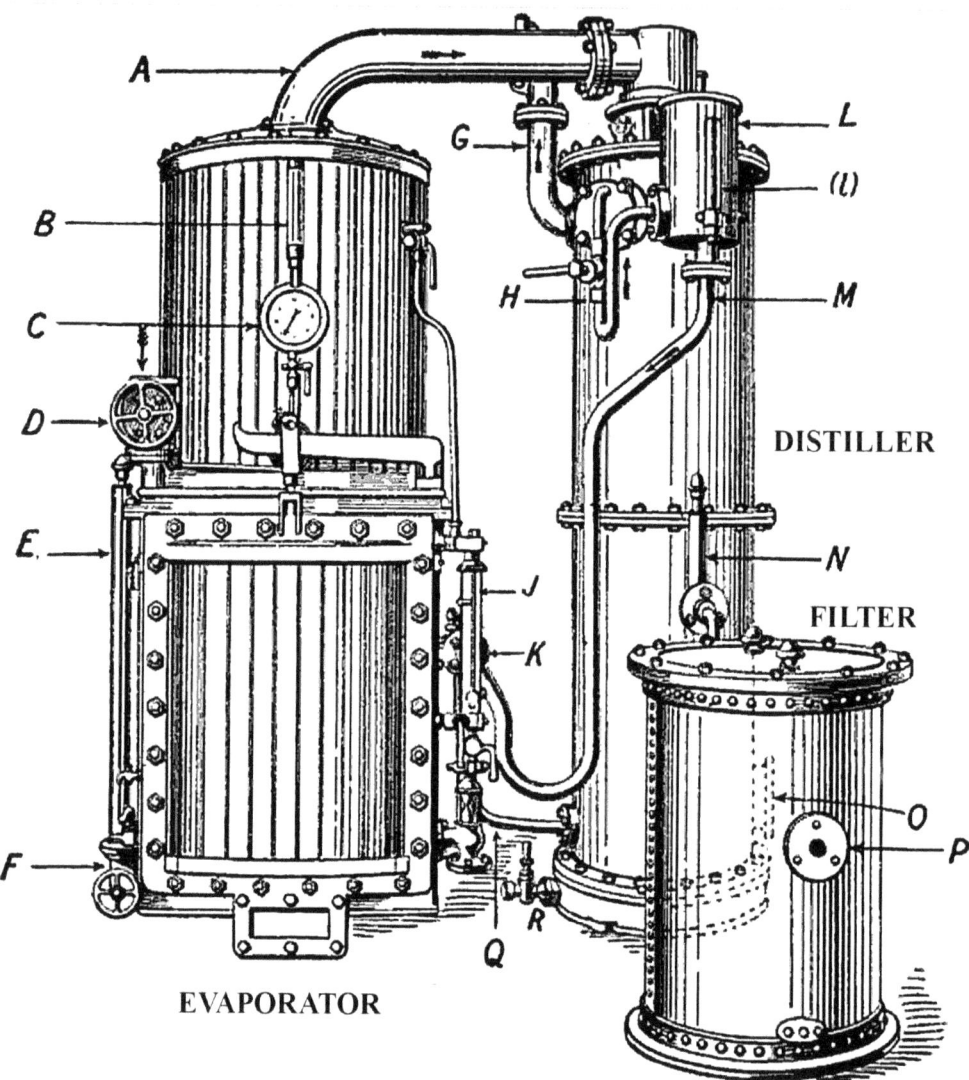

Normandy Double Distiller (from Frank Normandy, *A Practical Manual on Sea Water Distillation, etc.* (1909), Fig. 20). Parts are: A steam input pipe, B water pressure gauge, C steam pressure gauge, D steam valve, E brine discharge pipe, F brine blow-off valve, G circulation discharge, H feed pipe and cock, J water level gauge, K inlet from feed box to evaporator, L feed box, (l) gauge glass on feed box, M feed pipe to evaporator, N fresh water pipe from distiller to filter, O circulation inlet (from pump, not shown), P fresh water outlet from filter, Q pipe for filling evaporator with sea water, R drain cock.

straight vertical tubes were used, then horizontal U-shaped tubes, and finally (in the 1884 Weir evaporator) spiral tubes with a vertical axis (Normandy, 85–86). The straight tubes had the advantage that the water condensing from the primary steam could escape easily, but the pressure distribution inside the spiral tubes tended to flex the tube slightly, cracking the outside scale (90).

With regard to the choice of material for the coils, there were three considerations: how well it conducted heat, how resistant it was to corrosion by salt water, and what thickness of metal was needed to withstand the steam pressure. In the 19th century, copper was favored (Normandy, 97). In the 20th century, copper-nickel alloys became available (U.S. Bureau Naval Personnel 1956, 33–4).

For the distillation apparatus to work, seawater must be pumped in, both to feed the evaporator and to serve as a heat sink for condensing the distillate. The brine must be pumped away and the drinking water pumped to a water tank. This was all accomplished with steam pumps (Normandy, 186–7). A float-type feedback control could be used to regulate the flow of seawater to the evaporator (110).

The vapor could be routed to different condensers, depending on whether it was to be used to make up the loss in boiler water, or for drinking (Normandy, 8). In the latter case, it was cooled, and filtered and aerated if the best water quality was desired (9, 137, 150).

Assuming the seawater is not evaporated to dryness, the residue in the evaporator is an extremely salty brine. At one time, this was directed into the bilge and then pumped by the ship's bilge pump back to the sea. Later, it became customary to cool and dilute it, and then pump it away without passing it through the bilge (Normandy, 7, 9, 132). The brine could be discharged manually and intermittently (perhaps at the commencement of each watch) or automatically and continuously (130).

Scale could be loosened by thermal shock—that is, steam and cold water would be run alternately through the coil (Normandy, 156). Cleaning was otherwise mechanical, with scaling tools (159).

On passenger ships, there was a choice of putting the desalination apparatus above or below deck. The advantage of the former was that gravity could be used to move brine and fresh water out of the apparatus. But on warships, "it would not be expedient to place the distilling apparatus on an upper deck, where such important machinery might be shot away in action." Hence, it was placed in the armored engine room. It was also expedient for the warships to have duplicate sets of the apparatus, with interchangeable parts (Normandy, 9).

In theory, distillation should completely separate the water from the salt. In practice, there is a phenomenon called priming in which liquid droplets are entrained by the rising vapor and carried away, thus contaminating the condensate (Normandy, 12). Also note that if the water contains any substances more volatile than water, such as light hydrocarbons, they will vaporize and come over with the steam (110).

Flash Desalination

In flash desalination, evaporation occurs as a result of a reduction of pressure below saturation pressure rather than an increase of temperature over saturation temperature. The seawater is kept under pressure while it is heated to close to its boiling point. When the pressure is reduced, a portion of the seawater vaporizes (flashes) and the temperature of the seawater (liquid and vapor) drops to the temperature corresponding to saturation at the reduced pressure.

In a variation on the theme, the seawater is heated but left at atmospheric pressure. It is delivered to a flash drum that is at a partial vacuum (i.e., less than atmospheric pressure). The seawater need only be heated sufficiently that it will be above saturation point at the pressure encountered in the flash drum. However, for producing drinking water, the heat must still be great enough so that the water is sterilized, which requires a temperature of at least 180°F. The corresponding saturation pressure is about 7.52 pounds per square inch absolute, which is a little more than half atmospheric pressure.

In 1900, Addison Goodyear Waterhouse received what I believe to be the first American patent (643702) on a flash evaporator. His evaporator design also used the incoming seawater to cool down and thus condense the flashed steam. Moreover, he recognized that flash evaporators could be connected in series so that at each stage, an additional fraction of the seawater would be flashed. Each stage must be at a lower pressure than the one before.

Nonetheless, there was little interest in flash desalination until the 1950s. In 1954, the aircraft carrier USS *Independence* was equipped with four five-stage multistage flash (MSF) units, each producing 50,000 gallons per day (Al-Gobiasi 2010, 162). MSF is a popular system for desalination on both naval vessels and cruise ships, although now possibly in second place after reverse osmosis (see below).

Vapor-Compression Distillation (VCD)

Distillation in these systems is initiated by heating the seawater to the boiling point using an electric heating coil (preheater). The evaporate is then compressed by an electrically powered pump, raising its temperature as well as its pressure. The compressor output is passed into a heat exchanger. Heat is transferred from the compressor output (causing it to condense) to the incoming seawater (causing it to evaporate). Curiously, the pioneer in this field, as in MFD, was Addison Waterhouse (1897), although his compressor was steam-powered.

In 1937–40, a VCD apparatus was developed for use on U.S. Navy submarines. Previously, they were equipped with the Nelseco-Clarkson exhaust evaporator (1916), so called because it used the exhaust heat from the main engines to heat the water. However, the heat value of that exhaust heat varied, and was at a maximum when the ship was running on the surface at high speed. The submarine fleet needed a desalination unit that could be operated while submerged. The Model S VCD unit produced 750 gallons per day, and the Model-X (1943) up to 2,000 (U.S. Bureau Naval Personnel 1955, 1).

Sometime later, VCD technology was adopted for surface ships, too. Motorships,

ships propelled by internal combustion engines, first appeared on the world's oceans at the beginning of the 20th century, and they usually burned diesel fuel. It was more convenient to equip them with generators or dynamos to power the VCD preheater and compressor than with an auxiliary steam engine. The legislative history of the Saline Water Conversion Act includes a letter asserting in 1950 that "in general, diesel-driven vessels are equipped with Badger motor-driven vapor compression plants of either 1,000 or 2,000 gallons per day capacity" (Boehlert 1971, 113). In contrast, steam turbine-driven vessels used steam as the heat source.

Direct Solar Desalination

One of the disadvantages of the "fire still" is its demand for fuel. If the water is heated by the sun, no fuel is needed.

Early proposals. The first suggestion I have found of the use of solar heating for fresh water recovery at sea was in Arnott's *Elements of Physics* (1831). He proposed (75–6) that two airtight tanks, connected by a tube with a stopcock, be filled with salt water. One of these tanks would be blackened so it would heated by the sun. The other tank would be cooled by wetted coverings and a current of air. A vacuum would be created in the cool tank by pumping the water out from the bottom. The stopcock would be opened and the water vapor would pass from the hot tank to the cold tank, condensing in the latter. Arnott asserted that knowledge of this setup "would have saved shipwrecked crews from perishing by thirst," although the proposed apparatus doesn't seem particularly amenable to being assembled after a shipwreck.

An article in *Frank Leslie's Popular Monthly* (1880) depicts and describes a "simple apparatus for obtaining fresh water from the salt sea." It is a shallow box, say six inches deep, whose top is a slightly inclined sheet of glass. Seawater is added, to the depth of an inch, and the box is exposed to the sun. "The water now evaporates, condenses on the under side of the glass, flows down into the channel C, and from thence into a vessel D set to receive it." The staff writer comments, "Thus the burning sun, which in other circumstances would add to the torments of the shipwrecked, is made to minister to their relief." Moreover, it is asserted—with somewhat more justice than in Arnott's case—that "the various portions of the contrivance could easily be got together before abandoning the ship. The necessary piece of glass might be obtained from the cabin-windows" (254).

While *FLPM* didn't name the inventor, it appears from an 1875 publication that this was "C. Wilson." I believe this was Charles Wilson, the Swedish engineer who erected the "first large installation of solar passive desalination" in northern Chile in 1872 (Zaragoza 2012, 80). It is asserted that "a square meter of glass will condense daily two gallons of pure water" (Vincent 1875, 60–1).

The Wilson apparatus was a single-basin, fixed inclination solar collector. Since it has a transparent window, if the sun's rays aren't perpendicular to the window (zero angle of incidence), there is loss of energy due to reflection at the air-glass interface, which increases nonlinearly with the angle of incidence.

Emergency equipment. A small solar still for life raft use (primarily by naval aviators) was developed by Maria Telkes during World War II, and "more than 200,000 units were eventually produced as standard equipment for life rafts on aircraft and ships" (Talbert 1970, 4). While some sources describe it as spherical, it had actually had a "flying saucer" (oblate spheroid) shape, with a conical structure on the underside. It was about 20 inches in diameter and 15 inches high when inflated, and weighed one pound (233). The top was transparent Vinylite (a copolymer of vinyl chloride and vinyl acetate), specially treated so water would condense on it. A "black absorbent pad of cellulose sponge material," a quarter-inch thick and with an area (one side) of two square feet, was positioned horizontally at the "equator." This was saturated with seawater and the unit inflated by a blow tube so it could float. It came with a tow line so it could be dragged behind the raft rather than take up space inside. The conical bottom was weighted to keep the unit upright. Sunlight would pass through the transparent top and heat up the seawater in the sponge, causing it to evaporate. The water vapor would rise, leaving the salt behind, and condense on the relatively cool underside of the top. It would run down the sides, to the bottom of the spheroid, where there was a small hole, leading to the conical fresh water trap. This had a drinking nozzle at the bottom. In the tropics, it was expected to produce up to a quart a day. The calculated energy efficiency was 80 percent (Telkes 1953, 1,112).

Limitations on shipboard solar desalination. While the sun is a "free" source of energy, its use on shipboard is limited by the space available. From the discussion of fire stills, we know that 0.323 kilowatt-hours is sufficient energy to raise one pound water from 80°F to the boiling point, and then evaporate it. Recalling that in 1768, Lind produced 42 gallons of fresh water in 5 hours, 13 minutes with a "fire" still, let's calculate what the minimum area of a solar still must be to achieve this.

The higher the latitude, the greater the thickness of atmosphere that solar radiation must pass through, and the greater its attenuation. The sun's elevation and azimuth change during the course of the day and, if the solar collector is not directly facing the sun, there is a further diminution in energy capture per unit surface area because the solar energy is spread over a wider area. The daily energy collection also depends on the length of the day, which is also dependent on latitude and time of year, and of course on cloud cover.

Residential solar panels typically face south (in the northern hemisphere) and may have a fixed or adjustable inclination. This doesn't work as well for shipboard use, because unlike a house, a ship doesn't have a fixed orientation. Marine solar panels are usually installed in a more or less horizontal position. Hence, they directly face the sun only if the sun is directly overhead.

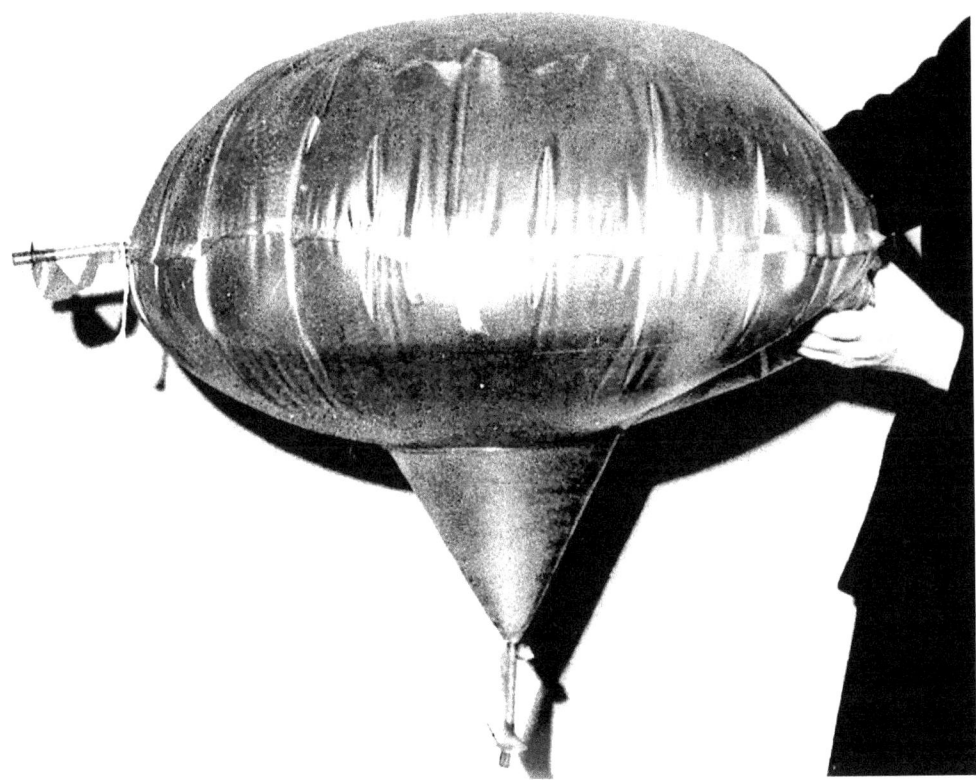

Telkes solar still, as photographed for the *Air Sea Rescue Bulletin*, published by the U.S. Coast Guard for the Air Sea Rescue Agency, June 1944.

Taking all of these factors into account, with clear skies, a horizontal solar collector at the equator would receive 8–9 kilowatt-hours of direct solar radiation per square meter collecting area per day. In contrast, at 30 degrees north or south latitude, it would range from 10 kilowatt-hours per day in summer to under 4 in winter (Honsberg 2019).

In a multistage ("multiple-effect") still, the "heat from the condensation of the first stage is used to evaporate the distill and of the second stage," and so on.

The efficiency of the solar still depends on a number of factors. There can be energy losses due to reflection by the transparent cover or the film of water on the underside of the cover, failure of the black evaporator to absorb all the light energy, and heat loss from the heated water by conduction through the floor and walls of the evaporator, or by reradiation to the cooler cover. Telkes says that early single-stage inclined plate stills had a maximum efficiency of 35 percent because of heat losses through the foundation. She calculated that an insulated single-stage still would have an efficiency of 74 percent and achieved 65–70 percent in practice (1953, 1,110–2). For Cambridge, Massachusetts, on a clear summer day, this translated to a yield of "nearly 1 quart a day per square foot of evaporator pan" (1,113).

The best performance I have seen reported for a solar desalination device was for a prototype tiltable, 10-stage evaporator, designed at MIT, that when irradiated perpendicularly by a solar simulator (constant 1,000 watts per square meter) had a steady state vapor production rate of 5.78 liters per square meter per hour. The solar-to-vapor efficiency was 385 percent (because of the energy reuse at later stages), and water collection efficiency was 75 percent (there was some leakage of water vapor and some small droplets that were not recovered from the condenser layers). In real-life performance—a partly sunny July day with scattered clouds, in Cambridge, Massachusetts, with solar irradiance of 200–800 watts per square meter, and a 30° tilt angle—water production was 2.6 liters per kilowatt-hour of solar energy (Xu 2020).

Indirect (solar-electric) desalination. An alternative to direct solar desalination—using the heat of the sun to evaporate the water—is an indirect approach, in which the light energy is converted by photovoltaic panels to electrical energy, which in turn powers some kind of desalination apparatus (as well as other electrical needs of the ship).

The first solar cells date back to 1883, but they used selenium wafers and were pathetically inefficient. Silicon cells were developed in 1954. In 1964, the U.S. Army experimented with a collapsible boat equipped with solar panels that could either directly power the motor or recharge the batteries on board (Prof. J. 1964, 100). By the 1980s, articles could be found in *Cruising World* about boats with solar panels mounted above the cabin (Naranjo 1986, 51).

Desalination by Reverse Osmosis

Osmosis is the process by which a liquid (such as water) passes through a semipermeable membrane so as to equalize the concentration of a solute (such as sea salts) on both sides of the membrane. The pores of the membrane are small enough so that the solute can't pass through itself, even though the solvent molecule can.

In reverse osmosis, the high-solute side is at higher pressure than the low-solute side, forcing the liquid to flow in the opposite direction, making the salty side saltier, and the dilute side more dilute.

The membrane is the critical component. It must be selective (passing water but not any of the sea salts), but porous enough so that the flow rate would be high at a reasonable overpressure. It also has to have a long life, bearing in mind that the higher the pressure needed to assure a reasonable flow rate, the shorter the life will be (Hearings 1965, 194).

Desalination of seawater by reverse osmosis was first demonstrated at the University of Florida in the 1950s (193). A decade later, membrane development had improved water flow rate by as much as a hundred-fold relative to the original experiment (195).

Theoretically speaking, the energy required for evaporative desalination is independent of the method used. All such methods must raise the temperature of the water, and then supply the "heat of evaporation" of water. However, in the 1960s, "the best evaporative conversion plants in operation consume approximately 30 times the minimum theoretical energy required for the separation of salt from water, whereas the reverse osmosis process consumes approximately 7 to 9 times the theoretical energy" (196).

The U.S. Navy became interested in the shipboard implementation of reverse osmosis in 1970. A major concern during the development process was the durability of the membrane. In the mid–1980s, prototype systems were installed on the *Stump* (DD 978) and the *Fletcher* (DD 992). The final version occupied less than a cubic foot but could produce more than 30 gallons of fresh water per day (Carlisle 1998, 387–9). Lifeboat models were in production by 1991 (448). Reverse osmosis is now one of the leading shipboard desalination technologies.

Control of Scaling

Scaling is an issue for all desalination methods, as the salts remain in the desalinator and their concentrations may exceed their solubilities in water. Desalinator feedwater may be pretreated with chemicals so that the scale is easier to remove. Sodium phosphate may be used so that non-adherent calcium phosphate is precipitated instead of calcium sulfate (Milton 2013, 420). Other scaling inhibitors have been developed. Chemical methods (typically acids) have also been devised for removing existing scale.

The scientist Albert Szent-Gyorgi wrote, "there is no life without water," and Captain Cook remarked, "I am well convinced, that nothing contributes more to the health of seamen, than having plenty of water" (Cook 1809, 252). But we need food—the body's fuel—as well. And that brings us to chapter 3.

Chapter 3

Filling the Stomach

In September 1773, Samuel Johnson wrote, "A man in a jail has more room, better food, and commonly better company" than a sailor (Boswell, *Life of Johnson*, 112).

Insofar as shipboard food is concerned, the shorter the passage, the less food needed to be provided, and the less likely it was that vitamin deficiencies would develop during the course of the passage. But in the Age of Sail, few passages were short enough that the ship didn't need to carry food. And there was always the risk that calms or storms, or navigational errors, would make the passage longer than anticipated.

Caloric Needs

The human respiratory process is essentially a combustion process in which a fuel (carbohydrate, fat or protein) is oxidized, breaking molecular bonds and releasing energy. The energy released by metabolism of food is usually reported in Calories. (Note the capitalization; it indicates that a kilogram-calorie, kcal, is intended. It is equivalent to 1,000 gram calories, each of which is 4.184 joules.)

Your daily energy requirement depends on age, sex, body weight, environmental conditions, and activity. The 2020–2025 U.S. Health and Human Services guidelines say that an active 16–18-year-old male needs 3,200 calories daily, and a 19–35-year-old, 3,000. "Active" is defined as "a lifestyle that includes physical activity equivalent to walking more than 3 miles per day at 3 to 4 miles per hour, in addition to the activities of independent living."

It is likely that this understates the energy requirements for sailors prior to the introduction of power-assisted appliances. (In 2015, the average U.K. construction worker consumed 4,050 calories daily—pbctoday.) However, we can get some idea of what those requirements might have been in two ways. First, we can look for analogous modern activities for which metabolic data exists (or has been estimated). Second, we can estimate the caloric value of known sailors' rations (on the theory that the sailors were neither starving nor getting fat with those).

Physiologists have rated the metabolic cost of various activities in METs (metabolic equivalents). It is defined as one kcal (Cal) per kilogram body weight per hour, so the underlying assumption is that the energy consumption is proportional to body weight (an average adult nowadays would be considered to weigh 70 kilograms, or 155 pounds) . Originally, it was thought that sitting quietly consumed one MET, but the current experimental value is 1.3. Sleeping is 0.95 METs.

Based on metabolic data, various activities have been given mean MET ratings, and compiled in the 2011 Adult Compendium of Physical Activities (Ainsworth). Note that the compendium does not account for differences in age, sex, or environmental conditions. Some of the values I thought interesting were sailing in competition (4.5 METs), rowing in competition (12), rock climbing by rappelling (5), shoveling coal (6.3), outside construction (4), walking 2.5 mph carrying an object weighing more than 25 pounds (3.8), pulling a rickshaw (6.3), pushing a plane in and out of a hangar (6), moving furniture or carrying boxes (5.8), running (6–23, for speeds of 4–14 mph), and mopping, light (2.5) or moderate (3.5) effort. Thus a 155-pound competition sailor would be estimated to be burning 70 times 4.5, or 315 kcal, per hour.

Compare these to climbing and manipulating the rigging, stowing cargo, heaving on a capstan, traversing cannon, rowing the ship's boat, manning a pump, mending sails, and swabbing the deck.

The intensity of physical activity is only half the story; duration is the other half. In battle, or clawing upwind, the sailors would have been working strenuously for hours. With a good wind on the beam, in the open sea, in peacetime, the sailors were probably given make-work to keep them busy.

In 2010, NATO's Research and Technology Organization accepted estimated energy expenditures by soldiers as being about 3,600 kcal daily for normal operations and 4,900 for combat operations (NATO 2010, 4–2).

MacDonald (2006, 10) has pointed out that "Georgian sailors did not have reliably waterproof clothing or warm, dry sleeping quarters," and hence probably needed to "expend calories to keep warm."

Humans are homeotherms, that is, we have metabolic control mechanisms to keep our core temperatures within a fairly narrow range. We therefore experience "cold-induced thermogenesis" (Brychta), which in turn may be fueled by increased food consumption. A 2017 study found that men in an outdoor school averaged burning 4,787 calories daily hiking in winter (15–23°F) and 3,822 hiking in spring (mid–50s) (Begley 2017). But there is considerable variation in the data.

Food Rations

Spanish practice. In the Spanish navy, as of 1568, the staple daily food was 1.5 pounds of "biscuit" (*galleta*), a double-cooked unleavened bread. To eat it, it had to be soaked in water or wine. To this add, four days a week, 150 grams of *menestra,* a mixture of horse beans and chick peas and one-third pound of salted fish;

two days a week, one pound of salted meat and two ounces of cheese; and one day, a half pound of salt pork and one-tenth pound of mixed rice and oil. There was also a monthly ration of one liter of oil and somewhat more than a half liter of vinegar. It is estimated that the caloric content of the daily meal was 3,500–4,200 calories (some from the wine), with a protein content of 13 percent. The principal deficiency was the lack of vitamins, because of the failure to include fresh fruits and vegetables (Perez-Mallaina 1998, 141–3).

Dutch practice. The premodern Dutch mariner's fare was similar; to feed one hundred men, the ship had to carry for each month at sea: "2250 pounds of hard bread, 40 sacks of soft bread, 450 pounds of cheese, 5 tuns of meat, 400 pounds of herring, 1.5 vats of butter, 5.5 pipes of barley, 4.5 narrow casks of white or green peas, 2.5 narrow casks of bean or yellow peas, and 0.5 narrow casks of white salt" (Hoving 2012, 198).

French practice. The earliest data I have for the French navy is from the *Ordonnance* of 1689; the daily ration was "one and a half pounds of biscuit, a midday meal of bacon, salt beef, fish or cheese, and a supper of dried peas or beans, prepared with oil and vinegar. The fish might be herring or sardines. There was a monthly ration of mustard seed" (Spalding 2015, 70).

British navy and merchant marine. In the Tudor navy, at one point the weekly rations were seven pounds biscuits, eight pounds salt beef, three-quarter pounds each stock fish and cheese, and three-eighths of butter. By 1588, there were three salt beef days (totaling six pounds), three fish days, and one day on which the sailor was served a pound of bacon and two pints of peas. The calorific value of the diet, including beer, was estimated at 4,265–5,132 kilocalories (Childs 2009, 87–88). The fish day on Friday was technically a half-ration, but Childs suggests that it was the day that leftovers from earlier in the week were thrown into the stew.

Between 1677 and 1733, the fish was replaced by oatmeal (Macdonald 2006, 9). In Nelson's navy, British sailors received just a pound of bread (weevil-enriched hard biscuit) daily. Twice a week they were fed a pound of salt beef, twice again a pound of salt pork, four days a week a half-pint of "pease," and three times a week, a pint of oatmeal, two of butter, and four of cheese. However, these were nominal weights, because in purser weights, each pound of most items was required to weigh only 14 ounces, butter only 12, and cheese only nine (Pope 2013, 151–5). (In 1797, the Nore mutineers demanded that "our provisions be raised to the weight of sixteen ounces to the pound" [Brenckle 2018, 1]). The calorific value is estimated as 4,888, but substitutions could change this (Macdonald 2006, 177).

There was a list of official substitutes, which included flour, raisins, currants, beef suet, and mutton for the beef and pork; navy beans, chick peas and lentils for peas; wheat, pot barley and molasses for oatmeal; and rice, sugar and oil for several standards (Macdonald 2006, 176). The practical significance was that this was what the Victualling Board would pay for, and a captain who bought off the list might get in trouble if caught (and lacking justification of the absence of all official substitutes or an admiral's order) (Macdonald 2006, 10). A purchase ordered by the captain would be charged against the captain's salary until the captain persuaded the board that the purchase was appropriate.

That said, there were authorized (or tolerated) purchases of lemons, oranges, and various vegetables (notably cabbages, onions, leeks, pumpkins, kale, collard greens, carrots, turnips and, rarely, potatoes). Officers generally brought additional food purchases at their own expense. This could be fruits and vegetables, or even livestock for subsequent slaughter.

In theory, crew could bring personal supplies too, but in practice usually couldn't afford to (Bown 2005, 24). When they could, this created its own problems. On the 28-gun *Sibyl* in 1780, at the Cape Verde Islands most of the messes (groups of four to eight crewmen who ate together) bought "three or four pigs, as many goats and half a dozen fowls," leading the captain to order the pigs to be killed first, as the goats made less of a mess (Macdonald 2006, 19).

There was no regulation of food and drink served in the British merchant marine until the mid–19th century (Macdonald 2006, 12).

American navy. In the United States, the Naval Act of 1794 and the 1801 Act "providing for a naval peace establishment" both specified the rations the sailors were to receive on each day of the week.

In 1794, there were two 1.5-pound beef days, three one-pound pork days, and a one-pound fish day. A pound of bread was served every day, and the weekly diet also included rice, peas or beans, cheese, potatoes or turnips, pudding, and butter or molasses (Navy CyberSpace 1794). The 1801 legislation reduced the rations: bread to 14 ounces, beef to 1 or 1.25-pounds—and Friday fish was replaced with four ounces cheese and a half pint of rice. Tuesday pudding was replaced by Sunday and Thursday suet (Navy CyberSpace 1801).

The following table shows both the weekly allotment per sailor in the U.S. Navy in 1813, and the provisioning of the USS *Constitution* when it embarked on a projected six-month cruise that year (Brenckle 2018, 2–3):

Table 3-1: 1813 U.S. Navy Weekly Allotment and USS *Constitution* Loadout

	Weekly Allotment		Loadout on Sailing	
Beef	3.5	lb.	51,969	lb.
Pork	3	lb.	39,840	lb.
Flour	1	lb.	12,544	lb.
Suet	0.5	lb.		
Raisins			360	lb.
Bread	98	oz	76,234	oz

	Weekly Allotment		Loadout on Sailing	
Cheese	6	oz	2,174	oz
Butter	2	oz	1,765.5	oz
Peas, Beans	1	pt	1,286.4	gal
Rice	1	pt	1,316.9	gal
Molasses	0.5	pt	870	gal
Vinegar	0.5	pt	870	gal
Crout (sauerkraut)			800	gal

On the USS *Constitution*, the 1813 menu provided about 4,240 calories a day, mostly from fat. Three modern MREs (Meal, Ready-to-Eat) add up to 3,750 calories, 36 percent from fat (Biesty 2011).

While the *Constitution* sailed without suet, in 1816 a 44-gun frigate needed 11,700 pounds of suet annually (Brentle 13).

In 1842, Congress amended the navy rations. It no longer specified the rations by day of the week but gave the captains a choice among several options for the main course. It also permitted the daily ration to include a "quarter pound raisins, dried apples, or other dried fruits" and gave a weekly allotment of "half pound pickles or cranberries." For the sailors' sweet tooth, there was a daily ration of two ounces sugar and a weekly one of half a pint of molasses. There was also a weekly half pint of vinegar. Lastly bread was replaced with biscuit (Navy CyberSpace 1842).

The law also made it lawful to substitute "vegetables or sour-crout for the other articles usually issued with the salted meats," as long as the quantity issued cost as much as the articles they replaced. Also, "the articles of butter, cheese, raisins, dried apples or other dried fruits, pickles and molasses, may be substituted for each other and for spirits."

If sailors were placed on short rations, they were entitled to compensation for the cash value of the food they failed to receive.

The 1902 Act provided that the meat could be smoked rather than salted, included dried or canned fruit and canned or desiccated vegetables, and doubled the daily sugar to four ounces.

The weekly allowance specifically included macaroni, cheese, tomatoes, vinegar, pickles, molasses, salt, pepper and dry mustard. It also provided that "an extra allowance of one ounce of coffee or cocoa, two ounces of sugar, four ounces of hard bread or its equivalent, and four ounces of preserved meat or its equivalent shall be allowed to enlisted men of the engineer and dynamo force when standing night watches between eight o'clock postmeridian and eight o'clock antemeridian under steam" (Navy CyberSpace 1902).

In 1945, the battleship USS *Iowa*, with a complement of 2,800, consumed "about 7 tons of food per day.... 1.5 tons were fresh, 2 tons frozen, and 3.5 tons dry. The ship's storerooms could carry 100 tons of fruit and vegetables, 84 tons of frozen meat, and

Chapter 3. Filling the Stomach 53

Mess hall, USS *San Francisco*, 1898 (Naval History and Heritage Command. Photo catalog NH 119202).

650 tons of dry stores." In 1941, the supplies the battleship USS *Washington* took on board included lemons, cucumber, lettuce, sweet potatoes, tomatoes, asparagus, celery, carrots, oranges and rhubarb (Bean 2020).

In 2007, the U.S. Navy introduced the Navy Standard Core Menu, which has a 21-day cycle of program menus. There are five "platform-specific menus" (core, submarine, LCS, carrier, and ashore), rather than individual menus for each ship (Smith 2007; Person 2008).

The U.S. Navy's current food service takes into account the 2015–2020 Dietary Guidelines for Americans. Menu items are green-, yellow- or red-coded depending on their total and saturated fat content. Menus also offer good sources of vitamin A, vitamin C, folate and calcium daily (Naval Supply Systems Command 2016, 1–3, 1–5).

Food Quality

Despite Johnson's quips, according to Macdonald 2006 (11–12), at least by the 1790s, the food issued in the British navy (as opposed to the merchant marine) was generally good as well as plentiful, and was at least comparable to what unskilled laborers on land were eating.

The British Victualling Board tried to control food quantity and quality. Some food was purchased from outside contractors, with contracts awarded based on competitive bidding, but over time, more and more was produced at the board's own depots and yards (Macdonald 2006, 46, 52). The board issued instructions with respect to how the food was prepared, packed, transported, issued and, if need be, condemned. In the case of cheese and butter, if a batch didn't last for six months, the supplier wouldn't be paid for any of it, and the purser had to issue them within three months of receipt (31). Documentation was needed for purchases (and the board wanted to see originals), and ships made weekly reports of provisions on board (72ff). Inspections were made at certain points in the chain (22, 75ff), and the board was notoriously concerned about the fate of every penny (49).

The board had to make some difficult choices. Fresh-baked bread tasted better than biscuit, but wouldn't keep longer than ten days, whereas biscuit lasted for "many months" (Macdonald 2006, 16). Suffolk cheese (thrice skimmed of cream) had a long shelf life, but was said to be so hard as to be fit only for making wheels for wheelbarrows, or buttons for jackets (and when old, it was infested with red worms). In 1758, after many complaints, it was replaced with Cheshire or Gloucester cheese. They didn't last for long but were far more palatable (31).

Substituting food of inferior quality definitely occurred in all navies. For example, in Spain, there was an instance in 1566 of a steward mixing two jugs of water with one of vinegar, and characterizing the mixture as three jugs of wine. In Britain there were cases of collusion between Victualling Board clerks and contractors, and between contractors and pursers, to supply food in quantity or quality inferior to what was supposedly supplied (Macdonald 2006, 49). But there were cases of adulteration and mislabeling on land, too; bakers used alum to disguise the texture and flavor of bread made from inferior flour (16).

Even without fraud, food that was wholesome when delivered would deteriorate under shipboard conditions. There was a case of 41,440 pounds of biscuit gradually powdering, over the course of ten weeks, to produce 2,420 pounds of dust (Swinburne 1996, 313). Also, the sailors involuntarily shared their provisions with rats and insects, and frequently these vermin got to the table first.

Fresh meat. On 16th-century Spanish ships, passengers took on board chickens, pigs, sheep and goats so that they could be slaughtered as desired to supplement the normal fare (Perez-Mallaina 1998, 130).

In 1746, a British navy ship sailed with "a goat, sheep, a sow in pig, six and half dozen hens and 13 ducks on board." The hens of course could also supply eggs, and the nanny goat milk. Cows were also sometimes taken on board (McKay 2011, 38).

If the ship was tied up at a jetty, the animals could be driven up an enclosed

Engraving from *Harper's Weekly*, February 12, 1876, showing sled dogs, casks of provisions, and sheep on the deck of a vessel on its way to the far north (Naval History and Heritage Command, Photo catalog NH 95128).

gangway. If not, they would have to be slung on board or, in the case of the smaller animals, hoisted up in a net or carried by hand.

In Nelson's navy, pigs were placed in a sty constructed under the forecastle until

1801 and in the waist afterward. Sheep pens were built between the capstan and the main hatch, and if there was a goat, it probably slept with the sheep. Before 1815, poultry (chickens, ducks, geese, turkeys) were kept in movable coops, placed on deck in the daytime and below at night. Later, the coops were fixtures of the waist. Cattle were tied between the guns, heads facing the ship's side (Macdonald 2006, 86ff).

Macdonald 2006 (130) has expressed surprise that she found no record of rabbit-raising, even though they are "very efficient converters of food into meat." But Cooke (2014, 29) claims that "in the eighteenth and nineteenth centuries English sailing ships commonly carried rabbits as an extra source of meat." When the *Guardian* transported convicts to Port Jackson (Sydney), it had pens for a "dozen rabbits" (Schotte 2019, 160), although I do not know whether they were intended for consumption on board or by the convicts after they came ashore. In 1765, the *Purisima Concepcion* was "shipwrecked near the coast of Tierra del Fuego and the crew introduced rabbits in the island with the intention of being supplied with food" (Jaksic 2021, 189). Obviously, they didn't plan on being shipwrecked, so they had live rabbits on board for some other reason. In 1740, HMS *Victory* "was supplied with one 'rabbit coop,' … and the *Unite* carried breeding pairs in summer 1808" (Blake 2005, 39). Blake concedes that "examples are rare, presumably because of the difficulty of carrying green food for them."

In keeping livestock, there was a risk that the keeper, the crew or the officers would become fond of a specimen and turn it into a pampered pet. One such pig got so fat that she couldn't walk, but grunted to have food brought to her (Macdonald 2006, 131). Granting mercy to the animal not only kept it out of the food supply, it also meant that it competed with the sailors for drinking water (Brenckle 2018, 23).

It should be noted that sailors turned adversity into fortune by eating rats ("full as good as rabbits, although not so large") (Bown 2005, 20; Macdonald 2006, 97).

Food Preservation

Food can be preserved by drying, cooling, freezing, salting, sugaring, smoking, fermenting, canning, etc. In the Age of Sail, the foods were preserved mostly by drying (peas) or salting (beef, pork). Salted meat had to be soaked in freshwater, typically overnight, before it was boiled (Brenckle 2018, 9).

The term "preserved" comes with some caveats. The environment below decks was humid and ideal for mold growth. Even food that hadn't yet putrefied was likely to be hard. If softened, it was with seawater (Bown 2005, 20).

In the 1660s, the aristocrat-chemist Robert Boyle demonstrated that cooked meat could be preserved in butter for over six months. This was more or less the same concept as the confit of southern France and was later known as "potted meat." (Macdonald 2006, 29).

"Portable soups"—essentially what we would call soup mix—were introduced in the 18th century.

"The French Navy in the 1860s complained of rancid butter on long journeys and requested a cheap replacement. In 1869, a French food chemist named Hippolyte Mege-Mouries patented a method for the creation of margarine" (Mollahan 2012, 3).

Canning. In 1806, the French navy tested confectioner-chef Nicholas Appert's canning methods—he sealed food in airtight glass jars, and boiled them to kill the bacteria inside. Tin cans, as an alternative to glass jars, were introduced by English entrepreneurs in 1813, but did not make it into British sailors' rations until 1847 (Macdonald 2006, 171–2).

In theory, even in the Age of Sail, it was possible to keep food cold for an extended period. You needed to load ice on board just before you left port and put it in a well-insulated chamber. Frederick Tudor's 19th-century ice trade demonstrated that ice can be shipped a long distance if it is appropriately stored. Of course, opening up the ice chamber to remove ice, or food if it is already stored with the ice, is going to speed the melting.

Refrigeration. In the mid–1870s, meat was refrigerated experimentally on ships using "ice containers filled with natural ice, circulating a current of air through the ice by means of a fan" or pipes carrying brine cooled with an ice-salt mixture (Critchell 1912, 25, 191). In 1878–80, shipboard mechanical refrigeration was used successfully by Carré (SS *Paraguay*) and Bell (SS *Strahleven*) (28–32). But it was initially used just for cargo. That said, the cruiser USS *Olympia* (commissioned 1895) had a steam-powered icemaker (Bean 2020).

Provisioning en Route

The typical pattern was that the food options narrowed the longer one was at sea. There were three ways of producing food between ports. First, livestock could produce milk (usually from goats, not cows) and progeny.

Secondly, the crew could fish when their duties and sea conditions permitted. Fishing lines were among the items in Antonio Gonzalez's sea chest in 1571 (Perez-Mallainas 1998, 150). In Nelson's navy, all ships were supplied with fishing tackle. Trawl and seine nets might also be carried. The speed with which the vessel was traveling affected the practicality of fishing; optimal trawling speed is two to four knots (Brenckle 2018, 24). The sickbay had first dibs on what was caught (Macdonald 2006, 38; Blake 2005, 227n25). It was also possible to catch seabirds (Macdonald, 30).

Finally, there is the possibility of cultivating a vegetable garden on board. This was in fact attempted by the Dutch, but proved impracticable because of waves (and sea spray) coming over the bulwarks (Torck 2009, 24 n54; Carpenter 1986, 23; Bown 2005, 39). Curiously, even though the Wardian case (a small glass terrarium) was used to ship exotic plants by sea, I have found no reports of its use in sick bays to grow medicinal plants.

Wilson (1870, 93) reports use of an apparatus for growing vegetable seeds, so sailors can enjoy "a nice salad." This "consists of a series of earthen vessels, fitting on top of each other, perforated like colanders, except the lower [sic, lowest] one, and provided with a loose lid." Seeds are placed in all save the lowest one, and a cup of water poured into the uppermost vessel. This water percolates down through all the vessels, moistening all the seeds in the process, and any excess water collecting in the lowest vessel is returned every few days to the top. Wilson assures the reader that "in the course of a week or more, … the whole apparatus is found to be packed full of tender sprouts."

Provisioning on Station

The British navy ultimately developed a complex set of practices for keeping blockading or patrolling warships provisioned. Obviously, sending the warship home was not desirable. Rather, victualling yards were established abroad, merchant ships were hired for transport use, and purchases were made from local merchants at the nearest friendly or neutral port (or even from local "enemy" fishermen). Ships sometimes traded provisions, too. Of course, if a ship was sent home for repairs or to carry messages, it likely would be expected to return with fresh food (Macdonald 2006, 59ff, 72).

Cooking

On the British *Mary Rose* (sunk 1545), food was cooked "in two large cauldrons supported on iron bars over a fire box." The cauldrons were made of a copper alloy, and were of 360 and 600 liters capacity. There were also small metal and ceramic cooking pots (Marsden 2019; Dobbs 2012).

For safety, the cauldrons would have been set inside a brick fire hearth with a chimney venting to the upper deck. The lighter iron fire hearths were introduced in 1728. The separate furnace and hearth were merged into a vented stove, the Brodie stove appearing in the 1780s, and the Lamb and Nicholson stove (with integral water distillation) in 1810 (Macdonald 2006, 105). Copper cooking implements were tinned to avoid poisoning by verdigris (copper acetate) (133).

The Brodie stove, up to six feet square and five feet high, was made primarily of wrought iron, with cast iron fire boxes and copper ventilator and hood. It was equipped with one or two ovens, one or two lidded boilers, and a range; a spit or pots on cranes (hinged arms with cutouts) could be swung over the latter. The spit was turned by a chain and pulley. A provision in the patent specification that you won't see in a Home Depot description for a modern stove is that "there are double dish'd screwed plates to mend the boilers in case of accidents by shot."

The Lamb and Nicholson stove was larger and had three boilers instead of two, but lacked the open range. On these stoves, temperature was controlled by moving the pots closer to or further away from the heat source, rather than adjusting the fire.

There was no cooking when loading powder in port, or when the ship was heeling markedly because of wind or wave (this problem could arise even on a calm sea when in the trade winds). British captains did favor giving the crew a warm meal before a battle, and that could be done when they were confident that there was sufficient time before the ships closed to serve the food and then clear for action.

Food (and Drink) Storage

In Nelson's navy, there were separate rooms for bread, fish and spirits. The bread room was starboard aft, high enough to be free of bilge water. The fish room was also aft, but mostly used as a coal store. The spirit room was aft, under the cockpit, locked and with a marine guard. Livestock were kept on the main deck, in the manger by the bow (Pope 2013, 59; Macdonald 2006, 78).

Within these rooms, the food was kept in canvas sacks or dry casks, but some victuallers economized on the latter. The casks might be of overly light construction, or green wood (Pope 155), or not properly bound with iron hoops (Jones 1996, 59). Such casks took a beating when the ship rolled and pitched, and some leaked or even collapsed. To keep biscuits dry and sweet, they need to be packed in airtight boxes, as was done by the Dutch in the 17th century, and by the Americans by the War of 1812, but not by the British (Macdonald 2006, 18; Brenckle 2018, 12).

Water (or beer), salted meat and other provisions would have been kept in casks, stowed deep in the hold (Macdonald, 78). How did they get there? Hatches gave access to the holds, and casks could be lowered in slings connected to pulleys attached to overhanging yards or booms (84). And the casks would be rolled whenever possible.

The basic British naval ration nominally weighed about 11 pounds per man per day, but Macdonald 2006 (78–9) says to allow 15 to account for the weight of brine in the meat casks and of cooking water. Ships destined for Channel service would usually get three or four months' provisions, and those in foreign service six months', and sometimes different terms were specified for different classes of provisions.

The food that spoils first should be consumed first, and therefore ideally is stowed so it is the most accessible. But that was not the only consideration. The frigate *Doris* in 1821 carried 141 tons of food and drink for 240 men for four months; that works out to 11.75 pounds/day (28 days/month). How is this enormous weight stowed to maintain the ship's stability?

There are several stability considerations. First, the ship's center of gravity should be as low as possible, so you want to put the largest, densest casks near the bottom of the hold, just above the ballast. Second, you want to evenly distribute the provisions between port and starboard to avoid a list, and between fore and aft to maintain the ship's trim. Third, you may want to adjust the position of the casks relative to the centerline of the ship, because that distribution of weight affects the ship's roll period, and a short period is productive of seasickness.

But you have the further problem that the provisions are consumed. So you want to stow them so that the effect on trim is small and averages out in the course of a few days. Still, the consumption of provisions would progressively lighten the ship. While the reduced draft was by itself desirable, a warship tended to be top-heavy (because of the positioning of the guns), and so this would mean that the center of gravity rose (which reduced stability). Hence, empty casks would be filled with seawater and thus serve as temporary ballast.

Macronutrients

In 1827, William Prout identified proteins, fats (lipids) and carbohydrates as the three principal constituents of food—the modern "macronutrients." All three are potential sources of energy, but they vary in energy density (4.1 kilocalories per gram for carbohydrate, 9.45 for fat, and 5.65 for protein). They also vary in digestibility; we digest about 98 percent of carbohydrate, 95 percent of fat, and 92 percent of protein. The available energy is therefore about four kilocalories per gram for carbohydrate and protein, and nine for fat (Stare 1984, 129). Alcohol, while not considered a macronutrient, was a prominent part of sailors' diets, and it provides about seven kilocalories per gram (Hayes 2019).

The human body continually loses proteins through metabolism, excretion, skin exfoliation, and growth of hair and nails. Dietary proteins are broken down to their component amino acids that are assembled into new proteins. Some amino acids are deemed "essential" because the human body cannot synthesize them itself. If the intake of carbohydrate and fat is too low, some dietary protein will be used to generate energy rather than in protein synthesis.

The macronutrient content of 16th-century European naval diets has been studied by Hayes (2019) (percentages in table show percent energy contribution from stated constituent):

Table 3–2: 16th-Century Naval Diets

Navy	Protein (g)	Fat (g)	Carbohydrate (g)	Alcohol (g)	Energy (kcal)
Denmark 1557	534 (24%)	125 (13%)	1125 (48%)	186 (15%)	8,858
France 1582	395 (33%)	35 (7%)	355 (28%)	217 (32%)	4,805

Navy	Protein (g)	Fat (g)	Carbohydrate (g)	Alcohol (g)	Energy (kcal)
Sweden 1546	111 (10%)	59 (11%)	812 (66%)	87 (13%)	4,637
Spain 1560	195 (19%)	46 (10%)	513 (48%)	130 (23%)	4,043
England 1565	216 (22%)	89 (20%)	401 (38%)	110 (20%)	3,962

In the United States, dietary studies by the navy date back to the late 19th century, although they seem oriented to finding cheaper sources of the macronutrients, rather than the healthiest proportions.

The following comparison was made in 1913 by Admiral Cowie (House Committee on Naval Affairs 1914, 139):

Table 3-3: Early 20th-Century Naval Diets

Navy	Protein (g)*	Fat (g)*	Carbohydrates (g)*	Energy (kcal)
U.S. (sea)	138	269	556	5,180
Japan	126	56	607	3,430
France	170	34	524	3,078
Britain	127	110	601	3,891

(* eaten)

The first United States Navy cookbook to refer to macronutrients was the one published in 1944. The cookbook stated that active men need 3,000 to 4,500 calories per day, and that these should be derived 10–15 percent from protein, 55–70 percent from carbohydrate, and 20–30 percent from fat (Bureau of Supplies and Accounts 1944, 3).

In 2010, NATO general purpose combat rations were 9–13 percent protein, 21–36 percent fat, and 49–69 percent carbohydrate (NATO 2010, 3-3).

Vitamins and Minerals

The recognition that certain diseases can be prevented or cured by a change in diet came centuries, or even millennia, before the identification of the causative factor. The most extreme example of this is that Eber's papyrus (circa 1550 BCE) describes the treatment of night blindness with liver (Combs 2022, 12).

The formal proof of the dietary dimension came through deprivation/reinstatement studies. For example, in 1897 Eijkman fed polished rice to chickens, who developed beriberi-like symptoms and were restored to health by providing the rice kernel or unpolished rice. This and other studies led Hopkins and Funk to independently propose that some foods contained "accessory factors" (Hopkins) or "vitamines"

(Funk)—in addition to the macronutrients—that were necessary for health (McCollum 1918, 18–19).

The 1932 U.S. Navy cookbook identified "butter, milk, eggs, leafy vegetables … and fresh fruits and vegetables" as "protective foods" because "their use is necessary to prevent certain diseases" (168–69). However, the underlying basis for this protective effect was not spelled out.

The 1944 edition of the cookbook was the first to make explicit reference to vitamins (A, B1, "G" [riboflavin; B2], niacin, C and D) and minerals (calcium and iron) as necessary to "growth and the maintenance of normal body functions."

Vitamin C (Ascorbic Acid) Deficiency (Scurvy)

Scurvy was the scourge of sailors; indeed, it was also called *mal de mer*. The *Mayflower* lost 50 out of 102 on board, mostly to scurvy, during its 56-day 1620 voyage (Baron 2009, 317).

A vitamin C deficiency results in a failure of collagen synthesis, leading initially to bleeding gums and loose teeth, and later to muscle pain and degeneration, fatigue, lethargy, internal hemorrhaging and consequent anemia, susceptibility to infection, and skin lesions.

The healthy human body contains a reserve of 900–1,500 mg of vitamin C (the ability to store it declines with age), and uses about 30–60 mg daily (Bown 2005, 42–3). In 2021, the U.S. National Institutes of Health's recommended dietary allowance for males was 75 (age 14–18) or 90 (age 19+) mg, and 65 and 75 mg, respectively, for females (NIH 2021).

There were many theories as to the cause of scurvy, and many attempted remedies. I survey below the reported pre-20th century uses of both effective and ineffective agents. Unfortunately, the use of the former was haphazard. A potentially effective agent might be prepared or stored in an inactivating manner, discrediting it. Or given too little, too late. Or mixed with other agents that in turn were given the lion's share of the credit. So there was no consensus in premodern times as to what worked.

France and Spain. It's been asserted that the French and Spanish had a lower rate of scurvy than the English because they ate onions and garlic (Goethe 2012, 7). The U.S. Department of Agriculture states that modern raw onions have an average vitamin C content of 8.1 mg (red, USDA 2020, NDB 100252) or 8.2 mg (yellow, USDA 2020, NDB 100253) per 100 grams. For raw garlic it is 10 mg (USDA 2020, NDB 11215). Unfortunately, I have not been able to determine when these foods were introduced into standard rations. They are not in Perez-Mallaina's (1998, 67) 16th-century Spanish provisions list.

In the winter of 1535–36, the French explorer Cartier wintered by the Iroquois village near what is now Montreal, and his men developed scurvy. The Indians successfully treated them with an extract made from a tree called *annedda*, now believed to have been spruce or hemlock. But it was not a sure thing; in 1743–4 at Churchill, 11 men died of scurvy despite having drunk spruce beer (Erichsen-Brown 2013, 10–1).

Netherlands. In 1598 the Dutch East Indies fleet took lemon juice and grew horseradish (*Armoracia rusticana*, syn. *Cochlearia armoracia*) and scurvy grass (*Cochlearia* spp., spoonwort) on board, suffering the loss of only 15 men (whereas the 1595 fleet lost 88) (McDowell 2013). The USDA lists "prepared horseradish" as containing 24.9 mg vitamin C per 100 grams (USDA 2018, NDB 2055).

Those were not the only Dutch antiscorbutics. In 1564, shipwrecked sailors ate oranges (Goethe 2012, 15). There was also sauerkraut, a pickled fermented cabbage. The Dutch navy reportedly served it as early as the 16th century (Bloch-Dano 2012, 47). However, another source says that sauerkraut was not supplied to Dutch ships until the end of the 18th century (Beekman 1988, 22). The truth probably lies somewhere in between. Of course, getting sailors not accustomed to it to eat sauerkraut may be another matter; the British adopted it two centuries later on the recommendation of Doctor Lind, but Captain Cook had to order his men to eat it (Carpenter 1986, 77).

Raw cabbage itself has a vitamin C content of 40.3 (green) (USDA 2022, NDB 11109) or 53.9 (red) (USDA 2022, NDB 1112) mg per 100 grams. For "Sauerkraut, canned, solids and liquids" (USDA 2018, NDB 11439), it was 14.7 mg. For "Sauerkraut" in "Survey Foods" (USDA 2022, Food Code 75230000), it was 12.8 mg. There are 280 branded versions of sauerkraut listed as of May 7, 2023, and the first five I looked at quoted values of 12.5, 8, 8, 4 and 8 mg, respectively.

Clow and Marlatt (1929), reacting to modern advertising of sauerkraut as a vitamin source, questioned whether the vitamin C content would survive the processing of the cabbage. They made sauerkraut from the All Seasons variety of cabbage and found that guinea pigs, fed five grams daily of this sauerkraut, were protected from scurvy. However, they noted that two prior studies, using commercial sauerkraut, had mixed results.

Britain. Gilbertus Anglicus, in *Compendium Medicinae* (ca 1230, published 1510), recommended that seafarers take with them "apples, pears, pomegranates, cucumbers, citrons, lemons, muscatels, and vegetables pickled in brine" (Keevil 1957, 1:18). This somewhat scattershot advice does embrace one good antiscorbutic.

Nonetheless, the English (and others) drank beer and, at least in the 18th century, believed that beer was a scurvy preventative (Stubbs 2003; Bown 2005, 120). The beer they had in mind was produced by fermenting malted barley or wheat, and flavoring the brew with hops. I see no reason to expect that this beer contained significant vitamin C (in some modern beers, vitamin C is added as an antioxidant).

However, more effective agents were used in the late 16th century. Both Francis

Drake and Richard Hawkins sought out oranges and lemons at their tropical landfalls (Bown, 74). In 1601, James Lancaster, in the *Red Dragon*, led four English ships to the East Indies. By the time they arrived at South Africa, three and a half months into the voyage, 80 of the 480 sailors had died of scurvy. After rounding the Cape, the *Red Dragon* picked up oranges and lemons at Madagascar, and Lancaster gave three spoonfuls of lemon juice to his sailors each morning, as long as it lasted. It has long been claimed that by this means Lancaster saved many of his men, but actually, the mortality rate on the *Red Dragon* was only slightly less than on the other three ships (33 percent vs. 34, 38 and 45 percent). Nonetheless, Lancaster persuaded the British East India Company to use lemon juice on the voyages of 1604 and 1607 (Baron 2009, 316).

Interestingly, Sir Hugh Platt in 1607 recommended covering it with an olive oil supernatant—this reduced the loss of activity by slow oxygenation. The idea didn't catch until about 180 years later (Baron 2009).

James Woodall, in *The Surgeon's Mate* (1617), promoted the use of lemons, limes, tamarinds and oranges (Reiss 2005, 130) So, too did John Smith in *An Accidence* (1626) (Baron n39). By the early 17th century, the British East India Company had settled on tamarinds (which lack vitamin C) and oil of vitriol (sulfuric acid) as the "answer" to scurvy (Bown 2005, 75).

Sweden. In 1628, scurvy ravaged the Swedish squadron commanded by Admiral Fleming, patrolling the Polish coast, leaving only 19 out of 115 men fit for work. Subsequently, he obtained 200 lemons for them (Baron 2009, 317). This showed that he knew the remedy even though he wasn't willing to include lemons in daily rations as a preventative.

Asia. It has been argued that the Asian habit of eating fruits and vegetables, including those with vitamin C, helps explain the relative scarcity of references to scurvy in East Asian sources (Torck 2009, 250). Foods in the Japanese diet with significant levels of vitamin C include *kabu* (turnip) roots, *komatsuna* (mustard spinach) and of course certain seaweeds.

The vitamin content of seaweed varies by species, season, water temperature and salinity, light exposure, etc. Red seaweeds (*Palmaria, Porphyra*) are rich in provitamin A and brown seaweeds (*Undaria, Laminaria*) in vitamin C (Škrovánková 2011, 360ff). At least one seaweed provided, in a single cup, 15 percent RDA when fresh and 8 percent when pickled.

After the early 17th century, we enter the "dark age" of scurvy treatment, with scurvy attributed to a host of non-dietary causes, and numerous ineffectual or even hazardous treatments proposed.

The first insight was in 1747, when the physician James Lind conducted a comparative trial of cider, vitriol, vinegar, seawater, oranges and lemons, and a medicinal paste. The oranges and lemon treatment resulted in a complete recovery, and the cider provided some relief (Bown, 96ff). But others claimed to have refuted Lind's findings (Bown, 167), and Lind later advised that beer could prevent scurvy (1788, 140).

James Cook used a smorgasbord of reputed antiscorbutics in his voyages, both effective (rob [a lemon or orange juice concentrate], sauerkraut, spruce beer) and otherwise (wort of malt). Even Blane, in the 1780s, combined citrus juice and wort of malt, most likely for political reasons. It was not until 1795 that he persuaded the Admiralty to require a daily ration of lemon juice (0.75 oz/day) (174–82). (Or did he? Macdonald 2006 [160ff] says that the Admiralty just sent out lemon juice to the Channel Fleet, with instructions that it be issued at the discretion of the ship's surgeon. And outbreaks of scurvy continued.)

If lemon juice was known to be efficacious, why did the British replace lemons with limes? The Royal Navy initially imported lemons from Spain, and later from Portugal, Malta and Sicily. In 1869, it switched to lime juice from the West Indies. The cost was higher but the profits went to British plantation owners in Montserrat. Also, they thought that the greater acidity of the lime juice implied that it was more potent. (In fact, the potency of the West Indian lime was about a third that of lemon [Bown, 212].)

An 1864 *Times* exposé revealed that some "lemon" juice was manufactured in England from tartaric and other acids, with essence of lemon added to give it a lemony flavor. Such juice would have been completely ineffectual. Nonetheless, after scurvy forced abandonment of the Nares Arctic expedition in 1877, the navy decided that "lemon" juice didn't prevent scurvy after all (Baron 2009).

In the 20th century, there was and is still some dispute as to the relative potency of lemon and lime juice. A 1918 study asserted that fresh lime juice was only one-quarter the potency of fresh lemon juice, but that doesn't jibe with modern data as to the relative ascorbic acid content; the USDA cites 38.7/100g for raw lemon juice (USDA 2018, NDB 9152) vs. 30 for lime (USDA 2018, NDB 9160, species/variety not stated, it could be the Mediterranean lime). Carpenter (1986, 237) says that by modern analysis, sour lime juice is 23–59 mg/100g with an average of 30, and lemon juice is 31–61, with an average of 45. Carpenter notes that the "fresh lime juice" of the 1918 study was two months old when given to guinea pigs, so this may be another illustration of handling pitfalls (see below).

Handling pitfalls. Even if one identifies a vitamin C-rich food, the manner of preparation and storage can drastically affect its potency. Vitamin C is sensitive to oxidation and begins degrading immediately after harvest (Njoku 2011). Vitamin C is water-soluble, and thus easily leached into water and inactivated by heat. Thirty minutes at 60°C reduced the vitamin C content of peppers by 64.71 percent (Igwemmar 2013). Heating, followed by a month's storage, destroys the efficacy of gooseberries and spruce beer (Bown, 120).

Cabbage soup cooked in a copper pot for 40 minutes loses 75 percent of its vitamin C; if cooked in an iron pot, only 50 percent. The reason is that copper catalyzes the aerobic breakdown of the vitamin (Reiss 2005, 130). Lemon and lime juice were sometimes run through copper pipes or exposed to prolonged heat (Bown, 212).

Dried peas and beans do not contain significant vitamin C, and dried "scurvy grass" is seriously impaired (46, 76).

Lind created a lemon juice concentrate ("rob") by evaporating lemon juice. The evaporation process concentrated the lemon juice ten-fold but lost about half the original vitamin C. He nonetheless hoped it would stay potent for years; it didn't. After a month, it lost 87 percent of its activity (Bown, 120). Later, Lind and Trotter proposed straining the juice rather than boiling it, and then using olive oil as an air barrier (Baron 2009). This worked.

Baron observes that the risk of scurvy was reduced by the introduction of steam power, which "shortened all sea journeys."

Vitamin A Deficiency

Vitamin A is found in food in the form of retinol (from animals) and various carotenes (provitamins, from plants). One of the first manifestations of vitamin A deficiency is night blindness, and it can be followed by total blindness.

In 2022, the recommended daily allowances of vitamin A were 900 µg RAE ("retinol activity equivalent") for males and 700 for females (NIH 2022).

In the premodern sailors' traditional diet, the foods offering the most vitamin A were butter, cheese, and peas. If there are chickens on board, note that eggs are also respectable sources. Overall, the richest sources of vitamin A are the livers of various animals (or their oils, notably cod liver oil). Among plant sources, dandelion greens, sweet potato, carrot, broccoli leaf, kale, spinach, pumpkin and cantaloupe score high.

Despite its presence in the sailor's diet, vitamin A deficiency did arise if the ship were at sea long enough. One notable instance involved the French warship *La Cornelie* in 1862. The first case was a topman who complained that he could no longer work at night. A few months later, other sailors experienced the same problem. The French frigate *L'Andromede*, attempting a Pacific exploration, reported that three-quarters of the crew were impaired, and aborted the mission. Nor did the disease single out the French; there was an outbreak of night blindness on the Prussian ship *Arcona* in 1861, and in 1851, 44 percent of the men on the British brigantine *Griffon* had to be led about after the sun went down (Koletzko 2012, 3–8).

Night blindness takes longer to develop than scurvy in part because vitamin A is stored in the liver. If a healthy person with a history of adequate intake of vitamin A is suddenly deprived of it, he may have 8–12 months' worth in reserve. On the *Novara*, which circumnavigated the world in 1857–9, scurvy appeared on several of the long legs between ports where citrus fruits and potatoes were available, whereas night blindness only arose near the end of the voyage (9).

Vitamin B1 (Thiamine) Deficiency (Beriberi)

Beriberi became a problem for the Japanese navy in the late 19th century. It is attributable to thiamine deficiency. The body does not store thiamine, and symptoms arise in 1–2 months if it is not supplied. In dry beriberi, there is long nerve and muscle degeneration, leading to paralysis. The first signs are tingling toes and painful "burning" or cold feet. In wet beriberi, there is a fluid overload of the circulatory system, injuring the heart. A patient may develop both types (Roman 2013).

Unpolished rice is rich in the B-group vitamins, with 38.8 percent of the B-1 (thiamine) in the bran and 51.7 percent in the germ (Carpenter 2000). Traditional methods of polishing (pounding) removed only part of the bran, leaving some thiamine behind. With considerable labor, all of the bran could be removed, resulting in rice "as white as snow," an expensive prestige food. By the late 17th century, such rice was available in Edo, the capital of Japan, and beriberi became known as the "Edo affliction."

Machine milling made white rice more economical to produce. Since white rice lasted longer than brown rice, it became a staple of the rations for both soldiers and sailors in Japan. In 1878–1881, the beriberi incidence rate in the Japanese navy ranged from 25.06 percent to 38.92 percent. In 1882, it was 40.45 percent and the death rate 2.64 percent (Bay 2008, 117).

The navy doctor Takaki Kanehiro conducted surveys and found that the beriberi rate in various naval units was inversely correlated with the protein-to-carbohydrate ratio, which on average was less than the European standard of 1:15 (Bay, 116). The rate was also higher among officers than sailors, and beriberi was not a problem in Western navies (Ewbank 2018). He therefore concluded that beriberi was a protein deficiency disease—he was half-right, as it is really attributable to thiamine (an amino acid found in protein) deficiency.

In 1883, a training ship returned from a transpacific voyage with 169 cases of beriberi (and 25 deaths) among the 370 cadets and crew. Takaki arranged for the next training ship to sail the same route but carry bread and meat rather than just white rice. It returned, and only 14 crew members—the ones who hadn't eaten the Western food—had contracted beriberi.

Takaki first tried to prevent beriberi by having the navy sailors eat meat and bread, but they refused to do so. He therefore resorted in 1884 to including barley in the diet. Barley, which fortuitously is a rich source of thiamine as well as protein, was part of the traditional treatment for beriberi (Bay, 114). In 1884, the incidence rate was 12.74 percent; in 1885, 0.59 percent; and in 1886, 0.04 percent (117).

Unfortunately, Tokyo Imperial University and the Army Medical Bureau were convinced that beriberi was of bacteriological origin and the army declined to follow suit. (It didn't help that there was an academic prejudice against traditional medicine, and barley fell in that category.) During the Sino-Japanese War (1894–95), there were more than 30,000 cases of beriberi in the army, but none in the navy (121). It was not until the 1920s that there was consensus in the Japanese scientific community that beriberi was a dietary deficiency disease.

The late 19th-century Dutch East Indian navy also experienced problems with beriberi. Prior to 1874, the diet of the native crews was 77 percent rice, whereas that of the European crews was only 29 percent rice. And in that year, the incidence rate of beriberi was 7 percent in the native crews and 0.07 percent among the Europeans. Van Leent concluded that rice consumption per se caused beriberi, and the natives were placed on a European diet at the end of 1877. The incidence rates fell to 5.4 percent (1878) and 2.3 percent (1879) (Braddon 1907, 240ff). Of course, we now know that the problem was not the consumption of rice, but rather the consumption of polished rice without also consuming some adequate source of vitamin B1. Even Eijkman erroneously thought that there was a "poison" in the white rice and an "antidote" in the polishings (357), rather than an essential nutrient in the latter that was absent in the former.

The anti-beriberi factor, later called vitamin B1, was isolated in purified form from rice bran in 1926 and its structure determined in 1936 (Stare 1984, 195).

Chapter 4

Alert Above, Asleep Below

The crew of a ship at sea must be vigilant 24 hours a day, seven days a week. In the Age of Sail, they had to respond to changes in the strength and direction of the wind or, given the uncertainties of period navigation, unexpected proximity to the coast or a shoal. Even when ships were equipped with engines and more reliable navigational tools, they could encounter a squall, an iceberg or another ship. Since the crew could not all remain awake, they had to be put on schedules so the crewmembers took turns being on watch. If the turns on watch were too long, there was a danger of loss of efficiency due to fatigue. If the turns off watch were too short, then there would be difficulty getting enough good sleep.

Traditional Watchkeeping Schedules

In the Anglo-American navy and merchant marine, the traditional watchkeeping schedule, which survived well into the 20th century, divided the crew into two teams, the "port watch" and "starboard watch," and called for a change of watchkeepers seven times during the course of a 24-hour day. (The modern U.S. Navy uses the term "section.")

The watch periods were also called "watches," and were as follows (expressed in modern military time):

first watch, 2000–0000;
midwatch, 0000–0400;
morning watch, 0400–0800;
forenoon watch, 0800–1200;
afternoon watch, 1200–1600;
first dogwatch, 1600–1800; and
second dogwatch, 1800–2000.

Under ordinary conditions, the two "teams" alternated who was on duty. Since there were seven timekeeping watches, that meant that there was a daily rotation between who had the first and morning watches and who had the midwatch. Moreover, if on one day you had only four nighttime hours for sleeping, the next day you would have eight.

The Lookout (1895–6), by Winslow Homer (American, 1836–1910). Note that the sailor is dressed for bad weather (see chapter 5). (Gift of Charles Savage Homer, Jr. Cooper Hewitt, Smithsonian Design Museum. Accession 1912-12-32.)

Of course, if there was need for the entire crew to be at work—to haul up a stuck anchor, adjust the sails in the face of a storm or in a wartime chase, or to man the guns—then the watch below would be called up and no one would getting any sleep for a while.

The traditional two-watch schedule was not universal. There were variations both in the division of the day and in the number of sections.

The late 19th-century Imperial Russian navy had the peculiar arrangement of dividing the day into five periods for officers (changing at 0800, 1300, 1900, 0000, and 0400) and four for crew (changing at 0800, 1200, 1800, and 0000). Nonetheless, there were just two watch sections (Jane 1899, 480).

Captain James Cook, on HMS *Resolution*, organized his crew into "three watches, except on some extraordinary occasions. By this means they were not so much exposed to weather as if they had been at watch and watch: and they had generally dry cloth to shift themselves when they happened to get wet" (1776, 404). No particulars are given, but the simplest three-watch schedule is four hours on, eight off.

In 1797, six of the 11 ships in St. Vincent's fleet were on three watches, rather than two (Coats 2011, 201). And in 1804, a "Royal Navy Captain" (John Davie per Lavery 2020, 245), published a book setting forth a detailed watch bill for a ship's company of a "first rate" (a ship with at least 100 guns), organized into three watches, with 225 men per watch (Captain 1804, Appendix). He remarked that this system would "see their health and vigor preserved, unimpaired by fatigue, or the disastrous consequences of wet clothes, which men at two watches, have neither the requisite time, nor means of getting dried."

Nonetheless, the two-watch system remained dominant for at least another century.

A watchkeeping system must balance the sailors' need for sleep against the ship's need for a sufficient number of sailors to be on duty and alert at all times. So let's look more closely at sleep.

Sleep

During sleep, your body repairs tissues, fights infections, and stores memories. We all need sleep, but our sleep needs vary. Most people need 4.5 to 10.5 hours a day, with 65 percent of adults falling in the 6.5 to 8.5 hour range (Lavie 1996, 113). The FAA assumes that "it takes 8 hours of sleep to balance a normal day of wakefulness."

Sleep deprivation can result in fatigue, drowsiness, irritability and hallucination, although the thresholds for the onset of these phenomena vary depending on the individual. In general, it is possible to function for several days without sleep—in combat conditions, motivation, stress and physical activity are all factors that help fend off sleepiness (Lavie, 120).

That said, in 1872, Hutchinson reported "there are instances related of sailors falling asleep on the gun-deck of their ships while in action" (39 40).

A 2023 survey of French merchant marine officers found that 29 percent admitted to having fallen asleep at least once while on navigation watch duty (Giot). For fishermen, Gander reported in 2008 that "23% of days at sea permitted less than four hours sleep"; Allen in 2010 that "60% ... believed that their personal safety had been at risk due to fatigue at work, and 16% had been involved in a fatigue related accident" (Jepsen 2015, 108).

Sleep deprivation was a factor in the grounding of the *Exxon Valdez* oil tanker on Bligh Reef in Alaska in 1989; the crew had barely slept during the previous 16 hours. And the National Transportation Safety Board concluded that sleep deprivation on the part of the bridge watchstanders was a contributing factor in the 2017 collisions of the USS *Fitzgerald* and USS *John McCain* (Fuentes 2021).

Even if sleep-deprived sailors don't fall asleep on duty, their decision-making may be impaired. In 2013, the Royal Norwegian Navy conducted a study of the effect of long-term partial sleep deprivation on decision-making. This was a crossover study; half the subjects were sleep-deprived in the first half of the study and rested in the second half, and this sequence was reversed for the other subjects. The test subjects were naval cadets in their early twenties, and when sleep-deprived, they slept an average of 2.5 hours a day for five consecutive days. They were given mission briefings and "instructed to anticipate individually all potential tactical and moral problems they found to be associated with their present assignment" within a ten-minute period. The ability to anticipate these problems was found to have been severely impaired by the sleep deprivation (Olsen 2013).

There are both homeostatic (self-regulation) and circadian (time-of-day) processes affecting alertness and sleepiness. The homeostatic processes are the "sleep drive" (the slow increase in sleepiness that occurs the longer you have been awake and on alert) and its opposite (the slow increase in wakefulness the longer you have been asleep).

The longer the duty shift, the greater the risk of sleepiness, especially in the midnight to sunrise period. Thus, in marine simulator studies, the incidence of sleepiness was higher in the six-on/six-off system than in four-on/eight-off (Jepsen 2015, 109). As compared to eight-hour shifts, working 10-hour shifts increases the risk for accidents and errors by 13 percent, and 12-hour shifts by 28 percent (NIOSH 2020).

At the other extreme, on the Japanese warship *Asahi* in 1905, the 10-hour period 2000–0600 was divided into five two-hour watches, covered by four sections (Japan Bureau of Medical Affairs 1911, 362).

Sleeping reduces the sleep drive, and failing to get enough sleep each day results in the accumulation of a "sleep debt" that puts you at greater risk for fatigue.

In 1978, an international convention on Standards of Training, Certification and Watchkeeping for Seafarers (STCW) was adopted by the International Maritime Organization, and it entered into force in 1984. There were major revisions of the STCW in 1995 and 2010.

The STCW applies to civilian vessels. It requires "flag States to establish and enforce rest periods for watchkeeping personnel, and to require that watch systems are so arranged that watchkeeping personnel are not impaired by fatigue." More specifically, it requires that except in an emergency, "watchkeeping personnel ... shall be provided a minimum of 10 hours of rest in any 24-hour period. The period of rest may be divided into two periods, one of which must be at least 6 hours." Moreover, it warns that the minimum rest requirement "should not be interpreted as implying that all other hours may be devoted to watchkeeping or other duties."

Speaking of emergencies, Van Leeuwen 2013 reported that a "free watch disturbance" during an "eight-on/eight-off" bridge simulation, resulting in a 16-hour duty shift and a missed sleep opportunity, profoundly increased the participants' subjective scores on the Karolinska Sleepiness Scale. NIOSH 2020 contends that being awake for 17 hours impairs performance to the same extent as a blood alcohol concentration of 0.05 percent; being awake for 24 hours, 0.10 percent.

If the watch below is called up to help, sleep inertia (grogginess after awakening) may impair their initial effectiveness. Sleep inertia is more likely to occur and to last longer in individuals who were previously sleep-deprived, or if awakened from a deep (stage 3) sleep or during their "circadian low" (Hilditch 2019).

Humans (and other animals) have a biological "sleep clock," which establishes the circadian rhythm: the normal interval between wake times in the monophasic cycle. The normal circadian rhythm is actually slightly longer than 24 hours, but exposure to light (especially blue wavelengths) and morning activities help to reset it and keep it in sync with the normal 24-hour day.

Exposure to light at a markedly unexpected time (e.g., as a result of flight between time zones, or artificial lighting) resets the clock in a way that can result in reduced sleep duration or quality, and thus in subsequent sleepiness. (In 2020, the USS *Hamilton* [DDG-60] began a study on whether thicker curtains on berthing racks can improve the quality of sleep. Mittleider 2020 [36] found that "half of all participants with standard rack curtains were judged to have moderate or severe clinical insomnia. In contrast, no participants with enhanced rack curtains" did.)

Langer and Launert (2018, 8) propose that work areas needing high alertness be lit with "activating and alertness enhancing light" (over 1000 lux and over 5000°K color temperature—"cold" or "blue" light) They also propose temporal variation of lighting, with "activating" pulses of bright, "cold" light at several points during a day shift and at the beginning of a night shift, and dim, "warm" lighting before bedtime (6). However, they do not directly address the need for dark adaptation on the navigation bridge (see chapter 8).

Light, of course, is not the only phenomenon that can disturb sleep. There's also noise (ship crews have been given earplugs), erratic ship motion, and stress. Seafarers have combated sleepiness during work shifts with caffeine consumption, social interaction, physical activity, music, cool air, and unscheduled rest breaks and naps. It is rare, even in the merchant marine, for sailors to ask to be excused from a shift because of severe fatigue (Giot).

The primary circadian low—the period of lowest alertness and performance—for the average person following a monophasic sleep pattern with sleep at night has been variously reported as 4 to 7 a.m. (Valdez 2019) and 2 to 6 a.m. (FAA 2012; NASEM 2011, 80). There is also a secondary low, at about 2 to 6 p.m. (Valdez) or 3 to 5 p.m. (FAA), but it is not as low as the broad nighttime low.

Nowadays, the most common human sleep pattern is monophasic: sleeping once a day. This of course matches the day-night cycle. In the tropics, a biphasic pattern is seen; there is an afternoon siesta. A polyphasic (three or more sleep periods per day) pattern may be adopted by individuals who perceive a need to increase their work or study hours.

Ekirch has assembled literary evidence that in preindustrial Europe, a different biphasic pattern ("segmented sleep") was common. Individuals went to bed "between 9 and 10 p.m." This was followed by "two major intervals of sleep, 'first sleep' and 'second sleep,' bridged shortly past midnight by up to an hour or more of wakefulness" (Ekirch 2015, 152; Ekirch 2001, 364). This was not, he argued, limited to the "longer winter nights in higher latitudes," but rather occurred in the summer, and even in southern Europe (Ekirch 2016).

Ekirch regarded segmented sleep as the natural human condition, citing experiments by Thomas Wehr in which subjects "deprived at night of artificial light over a span of several weeks, eventually exhibited a pattern of broken slumber.... Wehr's subjects first lay awake in bed for two hours, slept for four, awakened again for two to three hours ... then fell back asleep for four more hours" (2001, 367–8; 2015, 177). He also cited anthropological research on three African villages (367) in which adults awoke after midnight. However, some recent studies have identified several present-day non-Western, preindustrial societies whose members do not typically experience segmented sleep (Yetish 2015; Smit 2019).

The switch from "segmented" to "seamless" sleep occurred gradually, most likely driven by the dissemination of and improvement in artificial lighting, which made it possible to work or play well past sunset (Ekirch 2015, 175).

This author wondered whether the prevalence of segmented sleep in the Age of Sail societies might have made it easier for their sailors to adjust to the Royal Navy's traditional watchkeeping schedule, in which one shift would sleep 8–12 p.m. and 4–8 a.m., and be awake from midnight to 4 a.m. Danielski (2023) has gone a step further, asserting that "the 18th Century Royal Navy based its watch schedule to reflect the wider sleep habits of the era." However, the length of the midwatch is longer than the typical hiatus of segmented sleep, and that hiatus was often spent in quiet meditation (Ekirch 2001) rather than physically and mentally challenging activity.

In 1804, Davie contended that with two watches, "it is scarcely possible to prevent" the seamen from "lying down on the damp deck, or in wet places" (Captain 1804, 58). Did he mean that the men on duty become so fatigued that they lie down for part of the time? Or that the men going off duty collapse on deck rather than make it to their hammocks below? Either way, it does not appear that the commonality of "segmented sleep" did much to accustom the men to the Royal Navy's traditional seven-period, two-section schedule.

Fatigue is a progressive deterioration in mental (attention, cognitive function) and physical (speed, strength, balance, reaction time) abilities. It is a complex phenomenon that depends on more than just how long you have been awake or your sleep debt. Intense or prolonged physical exercise and mental stress may keep one awake in the short term but increase the need for rest in the long term. Eating may make one sleepy in the short term, but in the long term it also provides the energy needed to function. A source of fatigue unique to ships under way is that sailors must expend energy "simply from continuously balancing and adjusting to the ship's degrees of pitch and roll" (Brown 2012, 7).

Modern Watchkeeping Schedules

In general, the greater the number of teams sharing the watch duty, the more off-watch (and hopefully, sleep) time each team has (Van Leeuwen 2021, "Conclusion"). On modern ships with large enough crews, there may be more than three sections. The 1964 *Admiralty Manual of Seamanship* declares that "in the Royal Navy there are two types of watch organization, known respectively as the two-watch and three-watch systems" (334).However, each watch is divided into two parts. Thus, these can also be operated as four- or six-section systems, respectively.

A "circadian" watchbill is one whose work/rest cycle adds up to a twenty-four-hour day, and the cycle is divided such that, for a given team, work and sleep occur at the same time every day. The disadvantage of the circadian schedule is that one or more teams will endure, day after day, a night shift. Consequently, it is customary with this and other schedules that do not have an intrinsic daily rotation to share the burden by intermittently but periodically (weekly, biweekly, monthly, etc.) rotating the shift assignments.

Some sailors will need to be awake at night, which means that there will be a conflict between their need to stay alert and their circadian rhythm. This is true even if they are placed on a circadian watchkeeping schedule, where you might expect them to become acclimated to the night shift.

Those sailors who are required to be on a "reversed" schedule—on duty at night rather than during the day—are likely to suffer more fatigue even if their nominal sleep period is the same length as that of their day shift counterparts. The increased fatigue was shown in a 2002 study on sailors aboard the USS *John C Stennis* (CVN 74), who shifted from day to night air operations (Fuentes 2021). It is interesting to note that the effect was most pronounced on those who "spent any time outside," since we know that light exposure affects the circadian clock.

This is seen in other occupations. Night shift "workers try [to] sleep during the day and invariably experience shorter (less than 5–6 h in every 24 h) and more disrupted sleep.... Irrespective of the years spent on a permanent night shift, nearly all (approx. 97%) of night shift workers do not adjust to the nocturnal regime but remain synchronized to daytime" (Foster 2020). The light experienced in the workplace is relatively dim—this would be especially true on a ship's bridge at night, where light is limited to preserve night vision—and "after leaving the night shift, an individual will usually experience bright natural light during the day and the circadian system will always lock onto the brighter light signal as daytime and align internal biology to the diurnal state."

Analysis of International Maritime Organization statistics have shown a correlation between the frequency of ship collisions and local time (to which ship time is usually adjusted). The three most dangerous four-hour periods are (most to least) 0000–0400, 0400–0800, and 2000–0000 (Vinagre-Rios 2021). However, it is difficult to parse out how much of this is attributable to the circadian rhythm and how much to reduced visibility at night. These things are not independent, as the low navigation bridge light level at night encourages sleepiness.

The simplest modification of the traditional watchbill was to merge the first and second dogwatches into a single watch. This is called "four-on/four-off," and it is a circadian schedule. A "six on, six off" circadian schedule was proposed for East India Company ships by Alexander Stewart (1798, 60–61). He urged that the night watches be 2000–0200 and 0200–0800, arguing that the crew would benefit from having "six hours of uninterrupted sleep instead of four."

Other two-section circadian schedules include "eight-on/eight-off" and "12-on/12-off." These provide longer periods for sleep at the price of also spending a longer period on watch. Stewart was perhaps the first writer to acknowledge this tradeoff (62); it has been reiterated by van Leeuwen 2021.

In the modern merchant marine, the most common watchkeeping system is "four-on/eight-off," which is a three-section circadian system. But "ships engaged in coastal navigation or short voyages usually arrange a six-on/six-off watch system … to reduce manning costs" (Vinagre-Rios 2021).

Assuring that a schedule is circadian does not guarantee that it fosters alertness on duty. Van Leeuwen 2021, using a mathematical model of sleep, predicted that with the 12-on/12-off schedule, the risk of sleepiness on watch would be almost twice that as with six-on/six-off or eight-on/eight-off. This is because of the intensification of the sleep drive during long wakefulness (combined with the nighttime circadian low).

The "U.S. submarine" and "five-and-dime" are popular modern non-circadian (daily rotation), three-team schedules. In general, three-team schedules provide more sleep than two-team ones (Van Leeuwen 2021).

On the "U.S. submarine" schedule, one team takes a twelve-hour shift (1730–0530) and the other two each take a six-hour shift (1130–1730 or 0530–1130). On the "five-and-dime," two teams each take two five-hour rotations (the "dime": 2200–0200 and 1200–1700 for one; 0200–0700 and 1700–2200 for the other) and the third takes the easy five-hour 0700–1200) shift. It seems odd to this author that on both schedules, the shifts with longer on-duty hours are also given watches that embrace circadian lows.

While van Leeuwen 2021 reported that two of the daily rotation systems—the "Royal Navy" and "five-and-dime" schedules—had the lowest theoretical risk of sleepiness while on watch, they acknowledged that "having to work (and hence being forced to sleep) at different time points every day is unprecedented in shift work ashore."

In June 1987, the USS *Reeves*, deployed to the Persian Gulf after the attack on the USS *Stark*, used a six-on/six-off ("Blue"/"Gold") circadian rhythm schedule. The USS *Cushing* (DD-985) in 1992–93 tried out a 7/5/5/7 division of the day. And the USS *Lake Erie* (CG-70), in 1999, settled on 12-on/12-off. Its Captain Capello considered this to be successful and indeed, the Commander, Naval Surface Forces Pacific "endorsed the voluntary use of Blue/Gold in other Pacific Fleet ships." The same schedule was subsequently used by the USS *Oscar Austin* (DDG-79), but the watchbill was unpopular with the specialized technicians (Cordle 2013).

More recently, the USS *San Jacinto* (CG-56) adopted a German navy schedule—three-on/three-off/three-on/15-off, intended for four watch sections. The three-hour off component proved problematic and they shifted to a simpler three-on/nine-off for most divisions. Those divisions that lacked the depth to support four sections tried two sections in a six-on/six-off schedule. Captain Cordle opined that "the ability to preserve a circadian rhythm—working and sleeping the same hours each day—paid huge dividends in reducing fatigue." Moreover, the "short watch in the hot spaces ... and triple the time outside the hot area virtually eliminated the need for 'stay time' limitations and mitigated the effects of heat and humidity on engineering watchstanders."

Even with a circadian rhythm watch schedule, there is a need to rotate the sections periodically, so that the good and bad schedules are evenly distributed. Cordle rotated them once a week. The rotations of course disturbed the sailors' circadian rhythms, but not as much as the daily rotation of the traditional schedule would have. The Naval Postgraduate School (2017) actually recommends that the period between rotations be three weeks or more.

Anti-Fatigue Measures

In 2005, a Bridge Navigational Watch Alarm System (BNWAS) was proposed to the International Maritime Organization, and it was adopted and made effective in 2014. Essentially, once the autopilot is activated, the bridge officer "is required to signal his presence to the BNWAS every 3 to 12 minutes in response to a flashing light." If the officer doesn't respond within 15 seconds, alarms sound: first on the bridge, then in the captain's and first officer's cabins, and finally in other locations (Anund 2015, 29).

Bedding

Early modern sailors didn't get much time to sleep, and when they did, the bedding left much to be desired. On 16th-century Spanish ships, only the superior officers and distinguished passengers could sleep on a cot or bed. The rest of the voyagers had to lay a sleeping pad—usually a sack filled with straw—somewhere on one of the decks. The open decks were preferable because of the poor ventilation of the decks below. Blankets could be used in cold weather, but in hot weather one had to hope for a bit of wind. A tarred awning could be used to keep off the rain, if any (Perez-Mallainas 1998, 136–7).

Hammocks. Columbus first saw a native hammock—a bed slung between two trees—on "Crooked Island" in the Bahamas in 1492 (Barratt 2004, 98). Authorities are divided as to when the Spanish adopted them for shipboard use. Perez-Mallainas could only find one 16th-century reference to them, as opposed to many to pallets, but in 1595 Sir Walter Raleigh reported that "in hot countries all the Spaniards use [them] to lie commonly" (Dean 2014). It is possible that this referred to terrestrial use.

The British gave them official status in 1597, when "Sir Edmund Carey paid Roger Langford 300 pounds for 300 bolts of canvas 'to make hanging cabins or beds' for the Earl of Essex' fleet." (Stockwin 2011, 101).

Hammocks were hung "fore-and-aft from the beams" (Dana 1907, 193), that is, parallel to the keel. Since ships are longer than they are wide, they have greater pitch stability than roll stability. The longitudinal suspension was presumably adopted because rolling is more frequent and severe than pitching in most sea states. As Dana observes, hammocks do not rock like a cradle; "it is the ship that rocks, while they always hang vertically from the beams" (1907, 240).

If the men were sleeping on a deck pad (or a bunk), and the seas were rough, they could be thrown about and injured. This was a problem with a hammock only in a particularly violent storm, such that the hammocks were "pitched against each other and the bulwarks" (Panico 2018, 69).

While the longitudinal orientation was the norm, Stedman (1813, 8) reports that on a Dutch troopship in 1773, "by way of experiment we had slung the hammocks athwart ships, and not as usual fore and aft; which method we found however to be both so roomy and convenient, that it has been since adopted by several other vessels." Louise Rouppe, also referring to Dutch seamen, said that their hammocks "are hung on both sides of the ship transversely, so that the rope of one end of the hammock is fastened to the side, and the other end fastened to the deck toward the middle of the shippe" (Roupe 1772, vi–vii).

Examining a typical hammock in more detail, 8 to 12 "kettles" ("knittles") were threaded through grommets on the canvas at each end of the hammock and secured to it. The far ends of these short lines were gathered into clews, woven to form an eyelet, or fastened to a ring. Depending on the length of the kettles, either the eyelet or the ring would be looped directly over a hook in the beams overhead, or it would be connected to the hook via a rope ("hammock stop") (Panico 2018, 43–44, 52–53). The hook used would be the one on the far side of the beam, so the hammock was less likely to slip off (56). The head and foot hooks were likely to be about nine feet apart (Fincham 1825, xviii; Admiralty 1901, 6), although it was reportedly eight feet on the USS *Constitution* (Magoun 1928, 103).

The hammock could instead be hung from battens (wooden beams with slots for receiving the hammock stop) or iron bars (Panico, 51–55). These offered better handholds than the hooks for swinging oneself into and out of the hammock. It does not appear that there was a simple progression from hooks to battens (Glascock 1848, 73, speaks of battens being replaced by screwed hooks).

Some 19th-century sailors definitely preferred hammocks to bunks. Dana said that "hammocks … are the best things in the world to sleep in during a storm," and William Don thought them "distinctly preferable to a hard ship's bunk" (Panico, 70).

I have not found any study comparing the sleep quality provided by hammocks versus bunks at sea. There are two studies in which a bed or hammock rocked, but shipboard hammocks are actually stationary and it's the environment that rocks.

In 2011, Bayer reported that "lying on a slowly rocking bed (0.25 Hz) facilitates the transition from waking to sleep, and increases the duration of stage N2 sleep." (The "0.25 Hz" means that a single rock back and forth occurs every four seconds.) In 2021, Estrella compared sleep quality in regular bed and hammock users in southern Mexico (hammock use remains common in the neotropics). They reported that "hammock users show poorer objective sleep quality (higher index of activity and shorter sleep episodes) than bed-users," but the body mass index of the hammock users was higher and this might be the actual cause, via respiratory disturbance, of the impaired sleep quality.

Neither study, obviously, was of sleeping on a ship at sea. But to be blunt, the adoption of the hammock was less about ensuring restful sleep and more about efficient use of space. While the 1597 Admiralty warrant to make the "hanging beds" was said to be for the better preservation of the sailors' health, in 1627, John Burgh wrote to the Duke of Buckingham, "you cannot transport so many men in a shippe without hamackes as you may with them" (Blomfield 1911a, 144–5).

Hammocks were needed primarily by warships, which needed to carry large crews in order to man guns, rapidly adjust sails, and conduct (or defend against) boarding actions (all despite possibly having taken casualties earlier). Those crews were too large to be housed just in the forecastle and had to go below deck. Hammocks not only could be hung close together, they were removable, allowing the deck below them to still be used to man the guns.

In 1629, hammocks were provided "for overseas service in the proportion of

1 for every two men" (Blomfield 1911a, 145). The implication is that hot-bedding was practiced; that is, the hammocks remained below and the port and starboard watches took turns sleeping in the same set of hammocks. It would still have been necessary to remove the hammocks from the gun deck if the ship were going into action, so the guns could be worked, and the Duke of Buckingham's 1628 fighting instructions made reference to this requirement.

By 1798, each sailor was issued two hammocks (Panico 2018, 82). Thus, one could be washed while the other was in use. These were "made of canvas 1.8 m by 0.9 m [71 by 35 inches], each containing a mattress made from flock or rags, a blanket and a coverlet" (Stockwin, 101). Glascock (1848, 73) indicates that a new hammock was only 64 to 66 inches long, and should be made to measure 62 inches. This was, I imagine, a problem for taller sailors.

On ships of the line, the hammocks were slung above the guns, on the two (or three) gun decks. Hence, they were subject to two constraints, the distance between floor and ceiling, and the height of the guns.

Sovereign of the Sea, launched in 1637, had three gun decks. "Amidships ... the deck-to-deck height was 7ft 0in on the second deck and 8ft 0in on the gun deck"; the height of the average man at the time was about five feet six inches (McKay 2020, 38). (The term "gun deck" here refers to just the lowest deck bearing guns.)

For the American "North Carolina" class ships of the line—74-gun two-deckers authorized by Congress in 1819—the height from the lower gun deck to upper gun deck was seven feet two inches to seven feet four inches, and from the latter to the spar deck, about two inches less (Navy 1969, 602–3).

As previously noted, another practical limitation on hammock height was that when they were hung on a true gun deck their bottoms had to be higher than the guns.

The heaviest armament typically mounted on the lower gun deck was a 32-pounder. USS *Constitution* Museum Plan 14933 (1907) depicts a 32-pounder gun on a naval carriage. The center of the trunnion (the elevation axis) stands two feet eight inches above the floor, and the top of the horizontal barrel would be several inches above that.

Of course, you wouldn't want the bottom of the hammocks to actually touch the gun barrels. Judging from an 1851 illustration (Keck 1851, Plates 2:VI: 21), the "sag" of the hammock was about a foot above the guns. Wynter (1870, 98) reports that hammocks are hung at least four feet five inches in height from the deck. Mannix (2014), speaking of the U.S. Navy of the late 19th and early 20th centuries, says they were suspended "about the level of a man's shoulders."

The higher they were suspended, the more difficult they were to enter and leave, and the less ventilation space there was between the hammock and the ceiling.

On frigates, beginning in the mid–18th century, there was just one enclosed full deck carrying guns. The designers reduced the height of the deck below, relative to

Chapter 4. Alert Above, Asleep Below

Hammocks slung on gun deck. From Heck, *Iconographic Encyclopedia* (1851), Plates, Volume 2, Division VI, plate 21, figure 4.

that of a two-decker ship-of-the-line, and thus reduced the ship's profile and center of gravity (Lardas 2012, 9). The Royal Navy, unfortunately, persisted in calling this "gunless" lower deck a "gun deck," whereas the U.S. Navy called it, more accurately, the "berth" deck (Gardiner 2011, 43).

In the 1794 American specifications for building a 44-gun frigate, the height between gun (deck) and lower deck was six feet four inches (American State Papers 1831, 11). This is consistent with Lieutenant Lord's plan of the USS *Constitution*'s midship section (Magoun 1928, 86). Given the absence of guns on the lower (berth) deck, that provided considerably more room for suspending hammocks there than would have been available on its gun deck, despite the latter's ceiling height of seven feet. However, the berth deck did not have portholes, so ventilation was limited to what came through the hatches.

Commodore David Porter alluded to a difference of opinion in the American navy as to whether to permit "the crew to sleep on the gun-deck, with the ports open." He thought this contributed "not a little ... to the preservation of their health." The principal objection was that the hammocks would be "in the way of the guns," but Porter had two counterarguments. First, that only the men quartered at the guns were allowed to sleep on the gun deck, and it was advantageous to have them near the guns so they could respond more quickly to a call to action. Second, "should circumstances make it necessary to pipe up the hammocks on seeing a strange sail at night, they can be lashed up much sooner and with less confusion on a roomy gun-deck, than from a dark and crowded birth-deck." He was confident that all "preparation for action" could be completed in 15 minutes. And if, somehow, the enemy was able to make a close approach unseen, the sailors could cut the cords and throw the hammocks below or clear of the guns (Porter 1815, 29).

Luce (1884, 294) reports use of the gun deck as well as the berth deck on a frigate, with the forecastlemen and topmen on the gun deck and servants, most "idlers" (anyone who didn't need to serve a night watch because they worked all day), the engineer's force and marines on the berth deck.

The orlop deck was the lowest deck on a sailing ship, and thus was generally below the waterline. It was primarily used for storage. Freemantle (1904, 37) reports that on an old three-decker, the *Camperdown*, the ship's hammocks were stowed on the orlop deck since it didn't have hammock cloths (coverings used to protect hammocks stored on the rails of an open deck).

Kingston (1861, 43, 134) asserts that on Royal Navy line-of-battle ships, "midshipmen and other junior officers sleep in hammocks" in the "after cockpit" section of the orlop deck. Likewise, in America, in 1841, when William Parker (1883, 3), then a "green" midshipman, came aboard the 74-gun *North Carolina*, he found that he had to sleep in a hammock on the orlop deck.

Platt (2019, 20) comments, "though there was no natural light and little air, the orlop deck was spacious and quiet compared to the gun decks above."

Hammocks were hung closely spaced. The earliest statement I have found is that the hammock hooks are spaced "at distances not exceeding eighteen inches for petty officers berths, nor less than sixteen inches for common men" (Blake 1758, 3). But a few decades later, we are told that the separation is "fourteen inches" (Moore 1784, 244). The *Encyclopædia Britannica* of the day was more equivocal, allowing "14 to 20 inches," and explaining that it "must in some measure depend on the number of the crew, etc. in proportion to the room of the vessel" (1780, 3,505). Contemporary Dutch practice may have been more restrictive; allowing "ten, or at most twelve Rhineland inches" (Rouppe 1772, vi). An unusually liberal 22-inch allotment was said to have been provided on the USS *Constitution* (Magoun 1928, 103).

Masefield states that in Nelson's time, the "rule was 'fourteen inches to a man,' but some ships afforded sixteen inches, and one or two as much as eighteen. The petty officers, who slept by the ship's side, had each about two feet of space, as 'they are not to be pinched'" (1905, 131).

Proceeding to the mid–19th century, we are told with respect to the Royal Navy that it was "seldom exceeding sixteen inches" (Lukin 1852, 90), and "fourteen inches for common sailors and twenty-eight for petty officers" (Glascock 1849, 73).In the Union navy, the spacing was "only seventeen inches" (Porter, 108).

Turning to the beginning of the 20th century, on British transports it was "18 inches apart for seamen and marines, 16 inches apart for troops, and 20 inches for petty officers and sergeants" (Admiralty 1901, 6). On the British cruiser *Leander* in the mid–1930s, the crew were allowed 18 inches between hooks (McKee 2002, 94).

Even in 1937, the average spacing between the suspension points of hammocks in the U.S. Navy was just 18–20 inches. To increase head-to-head separation, and thus improve effective air quality, a committee recommended that the men "sling alternately head to foot" (Ellis 1948).

How does this compare to the average breadth of an adult male's shoulders? In the early 1960s, adult American males had an average shoulder width of 15.6 inches (Watson 2018). In the late 19th century, Seggel reported that soldiers enlisted at Munich had an average shoulder width of 41.1 cm (16.2 inches), and further suggested that the width of the shoulder in recruits should be at least 22–25 percent of the man's height (Notter 1896, 918).

Plainly, unless the sailors had a lighter frame than Seggel's soldiers, they couldn't sleep prone or supine if everyone were in hammocks at the same time. Fortunately, at sea, only one of the two watches slept at a time. And from side to side, the hammock berths were assigned alternately to the port and starboard watches. This "effectively doubled the amount of sleeping space allotted to each man since the hammock on either side of a sleeping sailor was empty due to his neighbor being on deck" (Panico, 56–7).

Captain Glascock (1848, 72) reported that a lower-deck mate made the egregious error of assigning all of the starboard watch to the starboard side and all of the port watch to the port side. However logical that seemed, it meant that "the weight of the watch below is to be all on one side," and he had to repaint all the berthing marks.

Masefield warned (1905, 132) that in those ships that did not berth the watches alternately, "the watches slept on their respective sides, jammed together in great discomfort." Nor was discomfort the only concern. In 1860, the crew of the *St. Jean d'Acre* (part of the British Mediterranean fleet) became "very sickly," which Doctor Hill (1864, 47) attributed to the entire crew sleeping below deck when the ship was in harbor, and thus having a limited supply of fresh air.

A berthing plan for the lower deck of the HMS *Bedford*, a 74-gun third-rate two-decker launched in 1775, exists. Its maximum beam was 46.75 feet. The plan (Royal Museums Greenwich, ID ZAZ6793) shows that from the forehatch to the main hatch, the center was kept clear, with up to ten hammocks on each side. Forward of the forehatch, there was an athwartships row with 24 hammocks.

Daniel Ross proposed staggering the hammocks so they would "interlock like a honeycomb" (Blake 2005, 181). That is, the heads and feet of the hammocks were overlapped, which permitted them to be packed more tightly together. This is shown on the *Bedford* plan, and also in Steel's "plan of the upper deck of a seventy-four gun ship" (1794, 428).

The plan was still in use a century later, in an 1893 berthing plan for a transport (Admiralty 1901, 18). The overlap on transports was 18 inches at each end (Goodrich 1883, 169); this was called "locking in."

However, the overlap could be more pronounced; Fincham (1825, xviii) refers to "every alternate hammock each way locking in about 3 feet." Glascock (1848, 73) comments that "the locking-in will depend upon the carlins; but from eighteen inches to two feet will be found sufficient" (carlins are fore-and-aft supports for the deck beams).

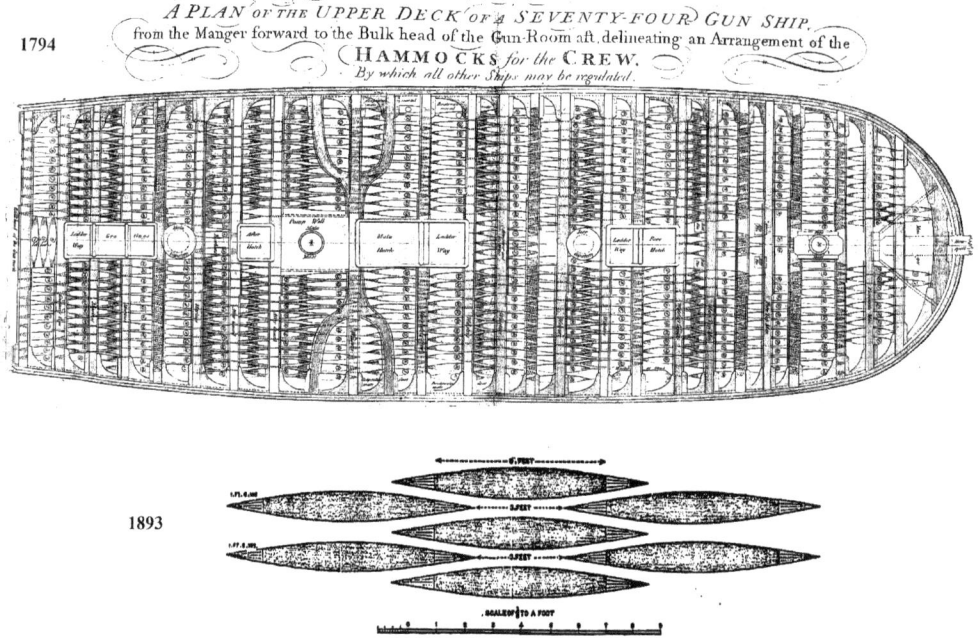

The upper illustration is the "Plan of the Upper Deck of a Seventy-Four Gun Ship, from the Manger forward to the Bulk Head of the Gun Room aft, delineating an Arrangement of the Hammocks of the Crew," from Steel, *The Elements and Practice of Rigging and Seamanship* (1794) (access courtesy of the Rare Book and Special Collections Division of the Library of Congress, Washington, D.C., photograph courtesy of Eric Frazier, curator). The plan was "drawn allowing 16 inches in width for every Hammock, but if the complement of men be full, only 14 inches for each Hammock can be allowed." The lower illustration is from the 1893 Admiralty berthing plan for a transport.

When the watch below was called up, their hammocks normally had to be stowed away (although Lukin [1852, 90] refers to empty hammocks being "compressed" by their neighbors). Curiously, they usually weren't normally stowed below deck, but rather were rolled and lashed up (tightly enough to be "put through the hoop," an iron gauge, by a bosun's mate), taken up to the main deck, and hoisted into trough-like netting on top of the bulwarks. The netting was supported by wooden or iron stanchions with forked ends, surmounted by rails (Panico 2018, 94–99; Wilson 1873, Fig. 53, 237–8) . Sometime in the 19th century the netting was replaced by a rigid trough (Luce 1884, 6). However, since this was light wood, there was a risk of it splintering in battle (Davis 1929, 109–110). The method of filling the netting or trough was odd; "the first hammock was doubled over … and then each succeeding hammock was laid at an angle of about forty-five degrees."

At least from 1746, the rolled-up Royal Navy hammocks were covered with painted or tarred canvas "hammock clothes" to shield the hammocks from sea spray and rain (Panico, 111).

However, the covers inhibited the drying out of the hammocks. Hence, the

Cropped image of USS *Ossipee* (1862–1891), a screw sloop-of-war, circa 1887–8. "Hammocks [are] stowed in hammock rails over the bulwarks." Note that they are laid on a slant. (U.S. Naval History and Heritage Command photograph. Photo catalog NH 42949.)

hammocks were intermittently hung on impromptu "hammock gant lines" to air and dry out.

Pope (2013, 165), Stockwin (2011) and Falconer (1780) say the hammocks served as a protection against small arms fire, but Hickox (2007, 16) says that they were up there to air out. Rolling them up would have slowed the airing-out process, so that appears to have been a secondary goal.

On the Laird ironclad rams (built 1865–1880), the hammocks were stowed "three deep around the upper edge of the turret roofs" to provide protection for marine riflemen lying there (English 2016, 98, 2).

Lavery (2021, 94) characterizes the hope that the stowed hammocks would "stop musket balls" as "vain," whereas Mondfeld (2005, 202) opined that it was "good protection against musket balls and caseshot fire." It seems to this author that their effectiveness would depend on the range, the motion of the two ships, etc.

On HMS *Excellent*, a marine's musket with a charge of 4.5 drams was fired from a distance of 40 yards at a target. When the target was "netting fitted with a single row of hammocks," it "did not stop a ball, but a double row resisted every one" (*Excellent* 1866, 20–21). The dimensions of the rolled-up hammock are not stated, but according to Ryder (1871, 58), it had a diameter of 9.25 inches (235 mm). At the same range, most balls passed "through 2 inches, but 3 inches of oak was musket proof" (*Excellent* 1866, 5). Hammocks, surprisingly, were more resistant than well-quilted woolen mattresses, placed together; at the same range, a musket ball was found to penetrate more than 40 inches into the latter (Talcott 1841, 296).

Cropped image of USS *Ossipee* again, in February 1887. This photo shows the hammock rails covered with tarps. (Naval History and Heritage Command. Photo catalog NH 42942.)

It is difficult to say at what range a single row of hammocks would provide protection from musket fire. That would depend on the rate of deceleration of the musket ball with distance (which is a nonlinear function of Mach number, see May et al. 2020) and on the relationship of striking velocity to penetration (there is disagreement whether penetration is proportional to momentum or to kinetic energy, or a combination thereof). However, experiments at West Point in 1837 showed that a ball from a musket charged with 134 grains penetrated 1.43 inches of seasoned white oak at 50 yards; 1 inch at 100 yards, and 0.66 inches at 150 yards (*Aide-Memoire* 1862, 3: 103). So if there were a similar range-to-penetration relationship for rolled-up hammocks, a single row might suffice at 80–100 yards. But this is highly speculative.

That said, there is anecdotal evidence that the hammocks provided some protection. According to Rear Admiral Daniel Ammen (1891, 364), while patrolling the St. John's River in Florida in 1863, the gunboat *Seneca* was fired upon, and "although the hammock-rail netting that protected us to above the hips was

Hammocks hung out to dry on a ⅓₃-scale model of the 74-gun *Achille* (built 1803) in her 1805 state. Model in collection of the Musée National de la Marine de Paris, inventory number 17 MG 5 (photo taken [2007] by "Rama," made available under the Creative Commons Attribution-ShareAlike 3.0 France, https://commons.wikimedia.org/wiki/File:Achille_mp3h9307.jpg). A similar procedure on a frigate was depicted by Jean-Jerome Baugean (1764–1819), reproduced in Charles Gerard Davis, *Ships of the Past* (1929), 110, and again in Naval History and Heritage Command photo NH 1017. Hammocks were still hung out to dry in this manner in 1920, on the Spanish battleship *España*.

well-riddled, … we escaped injury from the dozen or more guns that had been aimed at us."

Whether effective or not, it was possible that enemy fire would rip "most of a ship's hammocks to shreds. This resulted in men sleeping on bare decks until the ship was reprovisioned" (Panico 2018, 71, 103).

Admiral Sir Edmund Lyons, during the Crimean War, had the spare oars of the *Agamemnon* cut up and fitted to the spare hammocks, thus improvising stretchers for carrying the wounded (Blomfield 1911a, 147). During or after an action, hammocks can be used to temporarily plug below-the-line holes (Magoun 1928, 90). Hammocks could serve as life preservers (Stockwin, 102), as is discussed in more detail in chapter 7.

As for the timing of the stowing of hammocks, the captain's 1788 orders for HMS *Andromeda* typify the normal practice: "whenever it is fine weather, the hammocks are to be piped up at half past seven in the summer and at eight in the winter.... At sunset, the hammocks are to be piped down" (Martin 1803, 345). ("Some accounts specify eight or nine in the evening"—Panico 2018, 86.) When "piped up," they were carried up to the main deck and stowed. When "piped down," they were rehung.

"If the weather will not permit the hammocks to be piped up at the usual hour, they are then to be triced up." Steel (1805, 139) makes reference to "tricing battens," by which "the sailors trice up the middle of their hammocks out of the headway." ("Tricing" refers to lashing something up.) These battens were "about two inches thick and four inches broad, nailed up under the deck between the beams."

It is important to note that whether they were stowed above or triced below, the hammocks were not available for daytime use, even by off-duty sailors.

Van Leeuwen 2021 applied a standard mathematical model (the three-process model of alertness regulation, "TPMA"), used in aviation and other industries, to several common modern maritime watchkeeping schedules. For the traditional ("Royal Navy") system, it predicted that the two teams would average (over two days) five hours and 23 minutes of sleep daily, and that this would include some sleep time outside the three night watches. However, during the "hammock" era, if hammocks were stowed during the day, such daytime sleeping would have been possible only by lying on the deck (if permitted). Losing the daytime sleep opportunities would have reduced the total sleep by almost two hours.

It took time to learn how to sleep in a hammock. In the Civil War navy, "new sailors routinely spent [their] first night climbing in then roughly falling out of their hammocks" (Bennett 2005, 36). This was still a problem in World War II. A recruit, recalling his time at the Great Lake Naval Training Depot, said, "We were often woken [sic] by sound of one of [the recruits] falling out of the hammock onto the deck" (Hargis 2012, 11).

It was important for sailors inspect the cordage of the hammock from time to time. Willie Leonard, a sailor on the USS *Constitution* on the eve of the Civil War, admitted that his old hammock head clew "gave out last night and let me down head foremost. The fall stunned me; it was almost an hour before I could collect my senses" (Panico 2018, 44–5).

Each year, there were injuries attributable to sailors falling out a hammock, or hammock supports failing and taking the ensconced sailors down with them. For example, for 1931, in the U.S. Navy, there were 57 cases of the former and five of the latter (Navy 1932, 458), but no deaths. In the same year, there were 10 falls from bunks (460), but relative risk cannot be ascertained without knowing the number of person-hours that sailors slept in hammocks versus bunks.

These falls weren't always accidents. Hammocks could be sabotaged as a form of hazing of new recruits, or to punish another sailor for a real or imagined slight, or just to amuse the perpetrator and his buddies. In his essay, "Life in an Indiaman," Silas Webb declared, "the cutting down of hammocks is a common practical joke, but then is usually done upon the humane system of cutting the foot lanyard, which is not dangerous" (Webb 1856, 270).

Howe (1870, 370) refers to three forms of "cutting down" a midshipman who failed to timely relieve his fellows. The most benign was the one alluded to by Webb. Cutting the head suspension was, obviously, more serious. A third form was to "cut away both ends at once," often after placing some hard object, such as a chest, underneath the offender's rump. Or a line ("belly band") was tied around the sleeper's middle, and suspended from above, so when the ends were cut, he would sag at head and foot, suspended "ingloriously."

It was "usual to sleep with the head forward" (Blake 2005, 183) but there were "several … courts martial for [attempted] murder in which the victim is saved only by hanging the wrong way around" (263n9).

Kindleberger (1918) stated that "a certain number of men cannot sleep comfortably in the Navy hammock on account of the exaggerated curve, which causes morning stiffness of the back, due to overstretching of the lumbar muscles and ligaments for a number of hours." The extent of the curvature would depend on the spacing of the deck beams from which the hammock is suspended, but of course that is driven by structural strength considerations rather than crew comfort.

Inside a ship, hammocks interfere with the movement of air and thus make providing adequate ventilation more difficult. On the USS *Nipsic* in 1880, the berthing deck had 108 spaces for hammocks, with a 14-inch separation. The total air space was 8,643 cubic feet and fresh air was provided by 10 seven-inch diameter air ports, two hatches and two coal-holes (Hunt 1881, 561).

In addition, the hammocks are suspended so the sleepers are close to the beams above. Given that warm air rises, that means the air they are breathing is hot, humid, and probably high in carbon dioxide (Reid 1844, 357). Commodore Perry was presciently critical of this, saying that their "whole atmosphere" must be "polluted because the men exhaust more air than is supplied to them" (Perry 1875, 108).

Another comfort issue was dampness. The sailors might climb into their hammocks while still wearing wet clothes (despite regulations prohibiting this), and even if they removed their clothes first, water could enter the "between decks" space "through hatchways, gun ports, or any seam not tightly caulked," increasing the humidity or even leaking directly onto sleeping sailors (Panico 2018, 73, 84). The dampness of course fostered the growth of mold, leading to foul smells (74) and perhaps also fungal diseases. This was only partially alleviated by issuing two hammocks to each sailor and requiring that bedding be "aired, and exposed to the sun at least once a week" (84–5).

Hammocks were not used by all 19th-century navies. On the *Mahmoud*, a 140-gun Ottoman warship, "instead of hammocks, there are little raised platforms in the berth deck, for the men to lie down upon" (*Sailor's Magazine, and Naval Journal* 1835, 243).

Sleeping on an open deck. In 1828, George Jones reported that a master-at-arms told him that when sailing in a schooner in the West Indies, "the men were permitted to sling their hammocks on deck; the moon shone upon them, and a general inflammation of the eyes was the consequence, ending in a few cases in fevers and death" (65). He quoted scripture: "The sun shall not smite thee by day, nor the moon by night." James Boyle, a colonial surgeon in Sierra Leone, wrote in 1831 that seamen who "contrived to stretch themselves on deck, exposed to the beams of a brilliant moon, have been known, in numerous cases, to contract severe fever" (76).

In 1798, Alexander Stewart published, with the "approbation" of the British East India Company, a proposed "medical discipline," including preventing the men "sleeping on deck," which he considered "extremely prejudicial to their health, particularly when the deck is damp, or when they lie down in a draught of air" (64). The 1811 instructions to officers commanding British army troops traveling on transports for service in tropical countries warned, "The strictest attention must be paid to prevent the men from sleeping on the deck in the arm weather, which they are very apt to do—This practice is generally productive of fevers and fluxes" (British Army 1811, 244). In the 1841 U.S. Navy regulations, Article 818 directed that ship captains "shall not allow the men to sleep about the deck, or in situations where they will be exposed to night dews."

However, some questioned whether sleeping above deck was truly unhealthy (whether attributable to moonbeams, night dews, or drafts). In 1857, Tom Cringle noted that the seamen of the Indian navy "are the healthiest afloat," attributing this to their "privilege of sleeping on deck in a pure atmosphere" (1863, 59–60). In 1866, Parkes questioned the assertion that sleeping on the open deck led to more frequent "fevers and fluxes." He urged, "there is no harm in sleeping on deck when the weather permits.... I paid particular attention to this point in India.... The pure sea air is infinitely better than the hot foul atmosphere between decks" (601).

Nonetheless, as late as 1916, the Royal Navy declared, "a large number of men will request to sleep in the open air in summer; and if this is made possible, the sick list will soon show the effect" (Panico 2018, 109).

Civilians on passenger ships were of course free to sleep above deck, rather than in their cabins. But even if they did not fear contracting a fever as a result, there remained the chance of being unpleasantly awakened by a sudden tropical squall, and the morning dew could be heavy. Or, as mentioned by Glascock (1829, 80), of having a practical joker empty a bucket of water upon them.

Swinging cots. According to Falconer's 1780 dictionary, the officers slept in a "cott," not a hammock. But this was not a modern cot with fixed legs. Rather, it was a wooden frame "suspended from the beams of a ship." The frame provided a chest-like form "about six feet long, one foot deep, and from two to three feet wide," with a canvas bottom and sides, to sleep in. Falconer remarked that it was "much

Chapter 4. Alert Above, Asleep Below

Swinging cot in the captain's cabin on the "frigate" USS *Constellation* (actually the 1854 sloop-of-war, preserved as a museum ship in Baltimore, Maryland, and built using a small amount of material from the 1797–1853 frigate), photo taken October 18, 1969. Courtesy of Frank D. Scott, Maryland Constellation Chapter, Nautical Research Guild, Baltimore (Naval History and Heritage Command. Photo catalog NH 69818).

more convenient at sea than either the hammocks or fixed cabins." Unfortunately, he does not provide details of the suspension.

Sir William Crookes, writing of an 1870 expedition to observe a total solar eclipse, remarked, "[s]everal of the party, having found that the incessant rolling from side to side all night long interfered materially with their rest, gave up their fixed berths and decided to try hammocks and swing cots; they accordingly enjoyed a somewhat quiet night in spite of the liveliness of the ship. So long as the motion is confined to rolling, the fact of sleeping in a cot almost entirely neutralizes the annoyance, … but when the vessel pitches fore and aft a swing cot or hammock is of very little use" (D'Albe 1924, 144).

However, a means of counteracting pitching motion had been devised several decades earlier. On April 6, 1838, the astronomer Sir John Herschel, journeying from the Cape of Good Hope to England, noted in his diary, "devised a perfect Swing Cot

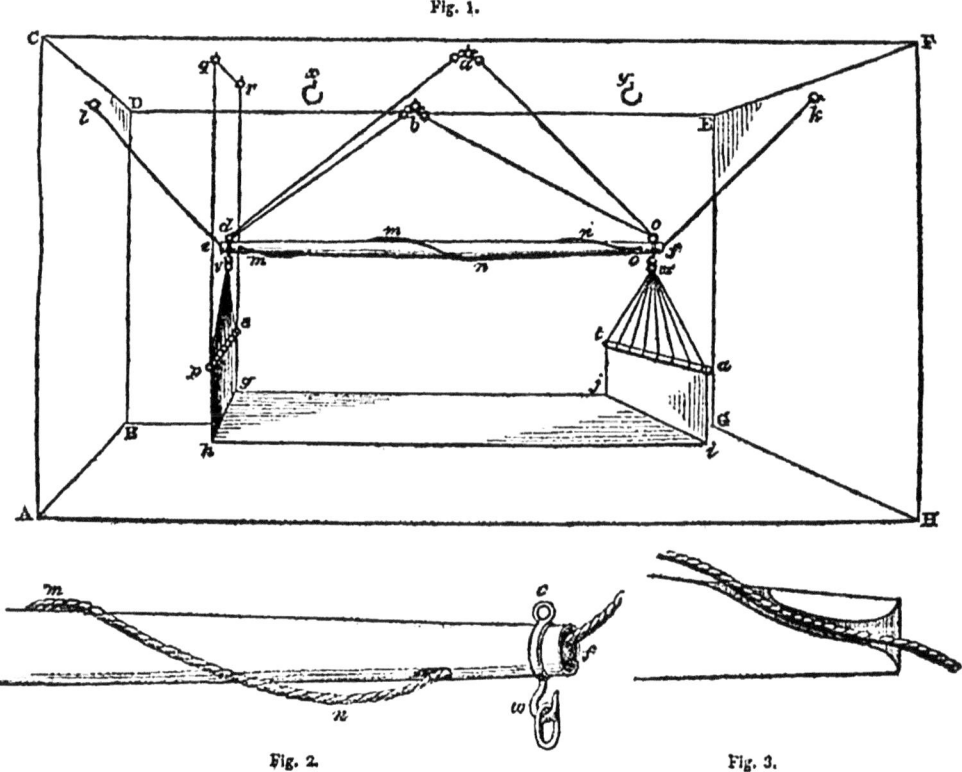

Herschel swinging cot, from his "Swing Cot for Sea Voyages," *Journal of the Society of Arts*, 9: 97–8 (January 4, 1861).

and proceeded to fit ours up on the principle. The rolling & pitching are both separately and independently corrected" (Evans 2014, 351). The accompanying illustration (350) shows the basic features of the cot suspension he described publicly in 1861.

That suspension, he wrote, "relies on friction as a means of deadening oscillation." Multiple cords fanned out from rings attached to both ends of a hollow cotpole to the headboard and footboard (there were no sideboards). The cotpole, in turn, had a four-point suspension. One cable ("friction band") ran between hooks high on the fore and aft walls, passing through a center hole of the cotpole, out a side hole, around the cotpole several times, and into a side hole and out the center hole at the far end. This would inhibit pitching. Two more cables ran between hooks on the port and starboard sides of the ceiling, and through rings on the ends of the cotpole. A final cable, intended to inhibit rolling, ran from one end of the headboard to the other but passed through two overhead hooks (Herschel 1861).

In 1918, Kindleberger designed a swinging cot for use on board ship. It was rectangular, and it was to be suspended from two tripods rather than from the beams. However, it lacked any feature for counteracting pitching motion.

Fixed beds. There was some use of fixed beds in the 19th century. Research on the issue is tricky, as sometimes the term "bed" or "bunk" was used to refer to a hammock, but fixed beds were mostly likely to be used by the navy in troopships, wardrooms and sickbays. Merchant ships carried fewer sailors relative to their size than did warships, and hence were more likely to house them "in the forecastle on permanent bunks" (Panico 2018, 68).

In the 1920s, the German regulations for merchant ships ordered, "each seaman must have a separate bunk. Double bunks without partitions are not permitted. The minimum size of a bunk is [6 ft × 2 ft]. The distance between the floor and the lowest bunk must not be under [10 inches].... The mutual separation of the bunks and that from the uppermost bunk to the ceiling must not be under [30 inches]" (Kari 1927, 131).

"In the year before Pearl Harbor, most navy ships replaced the hammock with cots, and eventually, bunks" (Rose 2007, 2:330). Double- or triple-stack bunks increased the number of people that could be accommodated on a given floor space, but deck height and ventilation could limit their use. Symondson (1876, 153) noted that top bunks are "greatly preferred," as "in bad weather, a sea breaking in through the doorway seldom wets them, whereas the lower ones are often drenched."

Bunks are rigidly attached to the deck. While that keeps them from sliding about, it means the occupant feels the pitch, roll, and heave of the ship. While traditionally oriented, like hammocks, fore and aft (longitudinal), there was some athwartship (transverse) bunking even in the 19th century.

On a sailing ship, which is heeled over its lee side if it is sailing other than directly downwind, this could be a problem when the ship changes tack, and thus the heeled-over side. After an 1841 transatlantic passage on a square-rigger, Robert Charles Leslie wrote, "all those below in athwartships bunks find themselves with their heels higher than their heads, unless they rouse out and shift end for end" (Leslie 1894, 24). There was a mix of longitudinal and athwartship bunks in a single berth on Symondson's first ship.

A 1974 *All Hands* article reported, citing the Naval Ship Engineering Center, that "athwartship berthing is used extensively aboard commercial passenger, cargo and fishing vessels.... Safety is increased ... since men are not being rolled out of their bunks in heavy seas ... by mixing the traditional longitudinal berthing and athwartship berthing, more space is made available." It also cited the Bureau of Medicine to the effect that there is "little difference between athwartship and longitudinal berthing in regard to safety, performance, and motion sickness."

In 1975, Martin stated, "Interviews with knowledgeable personnel resulted in a wide variety of strongly felt opinions about personal preferences for berth orientation, but no respondent was able to cite deleterious effects of this [athwartships] orientation" (Martin 2010).

Nonetheless, Matthews (2021) reported that sleep time is shorter, with more frequent awakenings and less REM sleep, when participants in athwartship bunks are exposed to head-to-foot motion (pitching), than when subjects in fore/aft bunks are exposed to side-to-side motion (rolling).

CHAPTER 5

More Creature Comforts

Air-Conditioning

Comfort is a function of both temperature and humidity. The interior of a ship is often characterized by both high temperature and high humidity.

Temperature. The temperature of exhaled air is on average 34.5°C (94.1°F) (Cowan 2010). Save in the tropics in the daytime, that is warmer than outside air, and the effect is that the temperature in the interior of even a sailing ship could be excessive.

In a 19th or early 20th-century steamship, a great deal of heat was generated, and most of that was radiated by exposed surfaces of boilers, steam connections and machinery. Not only did this warm the air in the rooms containing such apparatus, the heat was transferred by conduction through metal decks and bulkheads, and by convection of the air, into other compartments (Gatewood 1909, 175).

In a 20th-century motor ship (i.e., one propelled by an internal combustion engine), there is no steam, but the fuel efficiency is likely to be less than 50 percent, and the engines generate waste heat. If the engine is air-cooled, then that heat is going directly into the air.

Cope (1910) has described the ventilation of the engine rooms on an early 20th-century British warship. The supply fans provided a total of 155,000 cubic feet per minute, and the exhaust fans could move 162,000 cubic feet per minute. "This system is of extreme value in keeping under control the wild heat which naturally exists in large quantities in this vicinity" (444).

On warships, the problem of excessive temperature was aggravated by preparation for combat. "As soon as a ship enters dangerous waters it is battened down and hermetically sealed, to minimize the risk of rapid flooding due to enemy action.... Ventilation suffers drastically.... A steady rise of temperature follows, which in the Tropics may attain potentially dangerous levels" (Critchley 1945; *British Medical Journal* 1948). "During the Spanish-American War there were ships in the tropics whose sleeping quarters were never below 93°F, whose decks in compartments over boilers and engine-room felt hot to feet within shoes and whose firerooms gave air temperatures as high as 170°F" (Gatewood 176). In a hot fireroom, the sailors' body temperature typically rose 2°F during a four-hour watch, and could rise almost 4°F, despite "profuse perspiration."

Seawater and sea air also have an influence on interior temperatures. The strength of that influence depends on the material used to form the hull and weather deck. Wood is a good insulator, steel is not. This is one reason that some steel-hulled ships retained wooden decks (Ellis 1948).

Wood, unfortunately, is flammable, which is a problem, especially for a warship. Critchley proposed the use of non-flammable insulation, such as cork, or certain plastics. Insulation can be deployed on the sides of ships, or more selectively, such as around steam pipes.

The paint job is also relevant to the heat balance of a ship. A ship with a white-painted hull and superstructure will reflect more solar radiation than one painted a different color, and that would be advantageous in tropical waters. Of course, with a warship, there is the problem that white makes the ship more visible at a distance to enemy warships and merchantmen.

The human body's main cooling mechanism is perspiration, and if the humidity is high (see below), perspiration is inhibited. We also lose heat by radiation, but if the room temperature is higher than the body temperature, the net heat transfer is inward rather than outward.

The human body may suffer heat stress as a result of exposure to high ambient temperatures, or to even moderate temperatures when combined with intense exercise. The body temperature rises if the heat production plus inward heat transfer exceeds outward heat transfer. Exercise is associated with increased heat production by the body, a consequence of the associated metabolic reactions. Heat transfer occurs by four mechanisms: radiation, conduction, convection and change of state (e.g., evaporation), and for the first three mechanisms is from where the temperature is high to where it is low.

The body's principal thermoregulatory mechanism is evaporative cooling by perspiration.

High humidity exacerbates the effect of high temperature because it inhibits perspiration (hence, the meteorological "heat index" that takes both into account). Ventilation has a cooling effect by increasing the removal of heat from the body by convection. Clothing has a complex role; it may reduce the rate of absorption of radiant solar heat, but retard evaporative cooling via perspiration.

Unfortunately, perspiration reduces the body's salt and water levels, although the salt content of sweat can vary depending on circumstances. Heat cramps are brief, intermittent and often severe muscular cramps that frequently occur in muscles fatigued by heavy work or exercise. "They are believed to be caused by a rapid change in the extracellular fluid osmolarity resulting from sodium and water loss" (Weinmann 2003; Wexler 2002), although this has been disputed (Noakes 1998; 2008).

Heat exhaustion is more severe, and is "characterized by minor changes in mental status (poor judgment, irritability), dizziness, nausea and headache"

(Weinmann). "Heat exhaustion occurs both as water- and sodium-depleted types, with associated symptoms such as malaise, vomiting, and confusion" (Wexler).

Heatstroke (hyperthermia), a medical emergency, occurs when the body's thermoregulatory mechanisms fail. The body core temperature rises above 40°C (104°F) (Wexler), and there is "multisystem tissue damage and physiological collapse" (Weinmann). Risk factors (besides those already mentioned) that are likely to affect sailors include lack of acclimatization to the ambient conditions, dehydration, sleep deprivation, and various illnesses (Coris 2004).

Exertional heatstroke was the third-ranked cause of death among high school athletes in the United States, and they share with sailors a strong motivation to perform strenuous exercise in the heat (even though it may come from peer pressure or competitiveness rather than military discipline) (Hubbard 1990).

"Of the fifty-nine cases of heatstroke that occurred in United Sates navy in 1896, no less than forty of these were caused by heat in fire and engine rooms" (*Medical Record* 1898). During the steamship era, the firemen and stokers were given oatmeal water to drink, which they greatly preferred to water alone (Wilson 1879, 92). It appears that this retarded the development of nausea (Gatewood 1909, 379).

Oatmeal is a soluble fiber, rich in carbohydrate (beta-glucans). Insofar as electrolytes are concerned, oats are richer in potassium (380 mg/100g fresh weight) and magnesium (145 mg) than in calcium (54 mg) or sodium (9 mg) (Webster 2016, 101). So oatmeal water is inferior to modern sports drinks when it comes to restoring electrolyte balance. However, the caloric content of the oatmeal water replenishes the body's energy stores.

Humidity. The level of humidity inside a ship is dependent, to some degree, on conditions outside. If you are sailing for prolonged periods in the humid tropics, the humidity inside will be high, absent some mechanism for dehumidification.

Additionally, the level of humidity inside tends to be higher than that outside. Exhaled air contains water vapor at levels higher than atmospheric (5–6 percent vs. perhaps 1 percent). Thus, an occupied room, without ventilation, will experience a slow increase in humidity. If the humidity is already high, that may cause discomfort. Combustion of a hydrocarbon, as in a steamship, will also produce water vapor. If coal were pure carbon, its combustion would not produce water vapor, but coal typically contains some organic compounds and even water (2–15 percent for bituminous coal) (Bowen 2008).

We have already discussed shipboard ventilation; the fresh air provided is hopefully cooler and less humid than the air it displaces. True air-conditioning involves cooling and possibly also dehumidifying the air.

Cooling Systems

Cooling systems were first used to cool drinks and preserve perishables. One could place snow or ice, harvested in winter or from high altitudes or latitudes, in proximity to a food or drink container to be cooled. Heat would flow from the latter to the ice, melting it, but the concomitant loss of heat reduced the temperature inside the container. The melting process is isothermal; that is, the temperature of the water does not change as it changes state from liquid to solid. In 1873, frozen meat was shipped from Australia to England in a saltwater tank covered with ice and insulated with tanning bark. The tank design was faulty and the meat spoiled. However, later in that decade, naturally refrigerated meat was shipped in winter from North America to England.

Another approach was evaporative cooling. The evaporation of a liquid (such as water) removes heat from the container it is in contact with, cooling the latter. So here the heat changes a liquid to a gas, whereas in the first approach, it changes a solid to a liquid. Evaporative cooling was used in ancient Egypt to cool air; the air was fanned over clay jars filled with water.

It was common in the tropics to use wet mats to cool rooms by evaporation. In theory, the method could be used below deck for cooling, by wetting mats with seawater and then letting the water evaporate. However, if the air is already saturated with water vapor, evaporative cooling will not occur, and when it is possible, it increases humidity.

Paris (1825, 205) asserted that "at sea, wine and other liquors may be cooled by enveloping the bottles in wetted linen, and exposing them to a current of air in the sails."

A fundamentally different cooling strategy involved exposing the food or drink container to a mixture of chemicals that react endothermically. An endothermic reaction is one that needs heat to proceed. One such reaction is the dissolution of saltpeter in water. This was known in India, and was introduced to Europe by the 16th century.

I have not been able to document the carriage of saltpeter, or other salts, on shipboard for the purpose of making ice *in situ*. However, it does appear that in the mid-19th century there were instances in which salts were used to make a freezing mixture in the summer, and the ice produced was then used to preserve provisions on shipboard, rather than to carry livestock at sea (Tudor 1859, 100).

Modern refrigeration is achieved by a mechanical process that transfers heat from an area of low temperature to an area of high temperature, i.e., the direction opposite to that in which heat flows naturally.

In **ab**sorption refrigeration, two chemicals are used, a refrigerant and an a**b**sorbent. The refrigerant (for example, ammonia) is at a low partial pressure and evaporates, extracting heat from the refrigeration space. The refrigerant, in vapor form, dissolves into the liquid absorbent. The gas-liquid solution is heated, releasing the refrigerant. It passes into a condenser and is returned to liquid form. A related scheme, **ad**sorption ("d" not "b"!) refrigeration uses a solid a**d**sorbent rather than a liquid a**b**sorbent.

In vapor-compression refrigeration, a refrigerant, in the form of a saturated vapor, is compressed, raising its temperature. As a superheated vapor, it is passed to an air- or water-cooled condenser, where it loses heat, becoming a saturated liquid. Note that a saturated liquid is one on the verge of vaporization, and any increase in temperature, or decrease in pressure, will result in the change of state. The saturated liquid passes through an expansion valve, and the pressure is abruptly reduced, causing some of it to evaporate. The saturated liquid-gas mixture takes heat from the refrigeration space, evaporating the remainder of the liquid. And then the saturated vapor is passed to the compressor.

In 1878, the SS *Paraguay*, equipped with an ammonia compression refrigeration system, carried frozen meat from the Plate to Le Havre.

In 1879, the SS *Strahleven*, relying on a refrigerator employing the Bell-Coleman cycle, carried frozen meat from Australia to London (Farrer 2005, 52ff). The Bell-Coleman cycle is what a thermodynamicist would call a reverse Brayton cycle. The working fluid is air. The air is compressed, cooled, expanded and heated, but it remains in the vapor state (Cengel 2002, 584).

It was not until the 1930s that air-conditioning was installed on passenger ships (Cruise Arabia Online 2013).

Dehumidification

There are two basic approaches to dehumidification. The first is to pass the air through a removable filter containing a desiccant, which is a substance that has a high affinity for water. Possible desiccants include silica, activated charcoal, and calcium chloride and sulfate. Once the desiccant has reached its water-absorption capacity, it must be recharged by heating off the water.

The second approach is to pass the air over a cold surface, causing the water vapor to condense out and run off. Of course, providing a cold surface will typically mean providing a refrigeration system.

The first machines for large-scale dehumidification of cargo holds, using silica gel, date to 1939 (Duffy 1950).

Air Filtration

Air quality may be improved by filtration. Simple filters remove dust, and treated filters may kill microorganisms, neutralize odors, or decompose or absorb toxic substances. However, filters increase air resistance, too. By World War II, there were air filters on some merchant vessels (Markert 1944, 212).

If air must be recirculated for a long period of time—as on a submarine—methods of removing carbon dioxide become of interest. These involve chemical reaction of the carbon dioxide with strong bases (sodium, potassium, calcium or lithium hydroxide, or calcium oxide), or physical absorption on activated carbon. The system ultimately must be regenerated by driving off the CO_2 when fresh air is available for this purpose.

Clothing

Heating the body is easier than heating the air in a room, and so it would be logical to look at sailors' clothing before considering any more elaborate heating systems for ships.

There were no standardized (fleet-wide) sailor uniforms in the early 17th century, but function did dictate form to a degree. The sailor is exposed to heat and cold, to wind and wave. It would therefore seem to be important that the sailor's clothing be waterproof and have good insulating qualities. Clothing also had to be loose enough so one could climb rigging, and free of adornments that might catch on the ship's hardware.

Moreover, the supply system constrained what sailors wore. Contractors supplied clothing ("slops") to the British navy, and these were issued as needed to the sailors (who were charged for them). The clothing was made according to "sealed patterns" (Brenckle 2016). The captain of a particular vessel could order, and pay for, customized apparel for his crew.

The first regulations concerning British naval uniforms were issued in 1748 for officers, and in 1857 for the crew. The latter called for a "pea" jacket made of "Blue Flushing," another jacket made of "Navy Blue Cloth," a blue serge frock, a duck or white drill frock, navy blue and white duck trousers, a black silk handkerchief, a black or white hat ("according to climate"), a cap made of No. 1 cloth for use at night, and a dark blue woolen comforter (Osborne 1857). There were various stylistic specifications, such as the colors of the buttons, which I am not going to get into.

In the U.S. Navy, the first regulations for officers' uniforms were issued in September 1776. The first regulation to cover enlisted personnel was promulgated in 1841. However, before that date, as in Britain, clothing was bought by the navy on contract and sold to the crew by the purser. In 1797, the crew usually wore "a short jacket, shirt, vest, long trousers and a black low crowned hat." Preble's regulations for the USS *Constitution* called for each seaman to have "two blue jackets, waistcoats, pairs of trousers, both blue and white, black neckerchiefs and either a hat, or a hat and cap," and for warm weather, "a white jacket and vest" (Naval History and Heritage Command).

The 1841 regulations stated that for seamen, the outside clothing "shall consist of blue woollen frocks, with white linen or duck collars and cuffs, or blue cloth jacket and trousers, blue vest when vests are worn, black hat, black handkerchief and shoes, when the weather is cold; when the weather is warm, it shall consist of white frock and trousers, and black or white hats, as the commander may direct…, black handkerchiefs and shoes" (U.S. Navy 1841, 13).

The regulations in both navies appeared to be more concerned with aesthetics and distinctions of rank then with more mundane functions.

Bodywear. If you were able to examine the contents of the sea chest of a Spanish sailor in 1571, you would see shirts, trousers and jackets of linen and wool, and wool capes (Perez-Mallaina 1998, 150). The Vasa Museum states that 17th-century Swedish sailors wore "a linen shirt, a short, skirted woollen doublet (jacket), [and] wool trousers that ended below the knee." Wool has the advantage in that it is warm and the lanolin in its fibers provides some water resistance.

Headgear. In England, sailors wore the famous Monmouth knit wool cap (Long 2007, 187), also known as a "watch cap" because of its nautical associations. The Spanish had their own wool cap, the *boneta*. Felt caps were also used. On the *Vasa*, one might have found "broad-brimmed hats or conical caps." But looking at contemporary paintings, these weren't the only options.

In the 19th century, some straw or canvas hats were tarred, presumably to make them water resistant. Dana called this a "tarpaulin hat" (Dalton 2017).

Footwear. Some sailors undoubtedly worked barefoot, for better traction on wet decks or grip on the yards. However, shoes do provide some protection against the cold, dropped objects, sharp things on deck, etc. Shoes have been recovered from the wrecks of the *Mary Rose* and the *Vasa*.

Waterproofing. The waterproof fabrics available to 17th-century European sailors were leather and oilskin. Oilskin was made by soaking sail canvas in linseed or other oils. There is reference to oilskin in a Dutch fur trader's diary of his dealings with the Susquehanna Indians in the 17th century (Kent 2004, 307).

In the early 19th century, Charles Macintosh discovered that hardened latex can be resolubilized with naphtha and then applied to a fabric. Historically, the first widely available rubberized fabrics (1824) were not particularly successful because they were tacky when hot and stiff when cold. An 1865 fire on the steamship *Seine* was blamed on "the spontaneous combustion of the macintosh clothing shipped back from Havana" (Stevens 1871, 685).

The Goodyear method of vulcanizing rubber (with sulfur) was invented in the 1830s and commercialized in the 1840s. The vulcanized rubber is more tolerant of temperature extremes.

In 1879 Thomas Burberry invented gabardine, a finely woven cloth that was waterproof, but I have not been able to ascertain precisely how this was achieved.

Washing clothes. A day's wage of a Spanish armada sailor would buy 1.4–1.9 kg soap. Once clothes were soaped up, the bucket, with line attached, was tossed into the ocean "for the waves to make a primitive but effective washing machine" (Perez-Mallainas 1998, 116).

Chapter 5. More Creature Comforts

"Washing Day" on board U. S. Man O'War.

Washing day aboard unidentified U.S. Navy battleship, circa 1906 (Naval History & Heritage Command. Photo catalog NH 82779-KN).

In Nelson's navy, urine was saved in a tub by the bowsprit (which also served as the head), and thus served as an ammonia bleach. Soap was not provided until 1796 (Pope 1987, 148, 181).

Salt water was used for washing and rinsing. Unfortunately, when clothes wet with saltwater rub against the skin, it forces salt into the pores and can "lead to severe boils" (Fairfax 2015).

"In 1804, the Royal Navy Admiralty ordered hammocks cleaned once a month…. In 1884, the…. United States Navy created a set cleaning schedule…, advising mattress covers cleaned on the second and fourth Tuesday, blankets washed on the third Wednesday, and hammocks scrubbed on the second Thursday of every month" (Panico 2018, 87).

Drying clothes. A sailor's clothes will get wet, from sweat as well as waves, and thus the logical question is, how is are they to be dried? In Nelson's navy, once the hammocks were stowed, the order was given to hang up the clothes on lines (Pope 1987, 168). The lack of air circulation and high humidity below deck would have made the process very slow.

An alternative, employed in the Tudor navy, was to put them near the galley

The ship's laundry on the USS *Houston* (CA-30) during the 1930s (Naval History and Heritage Command. Photo catalog NH 53591).

(Childs 2009, 91). Or, on steamships, by the boiler room. However, the clothing presents a fire hazard once it dries out.

Washing machines. HMS *Warrior* (1861) was the first warship to have washing machines. It was a steam-powered warship, and the entire crew was involved in coaling operations. The stokers of course were exposed to coal dust throughout their work day, but wore white uniforms. The washing machines were manually powered. Muir writes: "The washing machines were filled with hot water. The clothes were put in, along with scrapings of soap. Turning the handles worked all the machines at the same time. Clothes were then put through the wooden mangles to squeeze out the excess water" (Muir 2012).

Heating Systems

Sailors will certainly be exposed to cold temperatures when voyaging across the North Atlantic in winter. Even in other waters and seasons, it can be cold, and it doesn't help that the occasional wave or rainstorm will get the sailors wet, leading ultimately to evaporative cooling. The cold is at its most insidious when the sailors are trying to sleep, and thus are not warmed by activity.

Chapter 3 noted that cooking in 17th-century sailing ships was done in copper cauldrons and ceramic pots set over a fire box inside a brick fire hearth. So one option is for sailors to sleep in or near the galley.

While it is generally assumed that fires were allowed only in the galley, in the British navy, "moveable iron stoves were issued from 1783" (Blake and Lawrence 2005, 93) and used in the crew's quarters (Hill, 1838). (Of course, any burner should be properly vented, so you don't have a buildup of carbon monoxide.)

In World War II, the U.S. Navy porcelain coffee mug, which lacked handles, served as an impromptu hand warmer (Mair 2014).

When ships added steam engines, whether for propulsion or for cargo handling, the option of sleeping near the boiler presented itself. In addition, there was the possibility of routing the exhaust steam through ducts to heat other parts of the vessel.

Human Waste Disposal

John Smith, in *The Sea-mans' Grammar* (1627), delicately refers to the beakhead in the bow as "a place for men to ease themselves in" (chapter 2, 10). By 1712, this informal onboard toilet was referred to as the "head." In the 15th century, all that was provided were slots or gratings, but in the 16th, simple seats (holed boards) and rails were offered (Simmons 1985, 100).

In the course of the 17th century, seats became more elaborate, and could be on one or both sides of the bowsprit, and on one, two or three levels. "Trunking" (a chute) could be fitted to the seat to "direct the discharge downward" (71). On the *Vasa*, there were two box-like structures on either side of the beakhead.

On HMS *Victory*, there were six "seats-of-ease" even though she had an 800-man crew (90). So it is not surprising to discover that the "head" was not the only option. A sailor could climb the lee fore or main chains (Little 2005, 87) and "let her rip," although that seems to me to be an even more precarious arrangement. Smith notes that the punishment for lying was having to "keep clean the beakhead, and chains" (Simmons, 71).

The term "roundhouse" may refer to a bow or a stern structure. Simmons dates use of a bow roundhouse as a privy to around 1700 (100). In Nelson's navy this private cubicle was used by midshipmen and warrant officers, and perhaps also by the men in sickbay (Adkins 2006, 43).

Ship architecture generally featured overhanging structures in the stern. For example, on the sides of the sterncastle of a warship there were balconies called quarter galleries. The sterncastle was officers' quarters, so the garderobes fitted into the quarter galleries were for their use. Of course, they didn't have to clean them. By the early 17th century, the quarter galleries were likely to be partially or fully enclosed structures. Large ships might have two levels of quarter galleries, with sanitary facilities on both, in which case soil pipes led down from the upper to the lower level (86).

It is likely that one quarter gallery would be reserved for the captain, and the other for the lieutenant or the officers on duty. Other officers would have to use chamber pots. These were made historically of metal, ceramic or wood (Museum of Historical Chamber Pots and Toilets).

Where there was no permanent overhanging structure, a temporary one might do. "Steep-tubs" (barrels or half-barrels) could be "slung over the upper bulwark or attached to the hull so that they had a relatively unhindered drop to the sea and, at the same time, were reasonably accessible." Typically, they were placed by the main chains (Simmons, 52ff). By the early 17th century, some English ships featured outboard side-shelves supporting the tubs (81).

"Pissdales were simple trough-like urinals placed at various locations forward and amidships, usually on the upper decks of ships." These most likely were provided only in larger ships (80) and possibly not until the mid–17th century (100).

"Necessary" tubs, into which the crew urinated, were provided on the lower decks and supposedly the urine was deemed usable for firefighting (Simmons, 53–4) or as a bleach (Adkins 2006, 44).

The human wastes that did not get dropped or washed into the sea eventually percolated down to the bilge.

The flush toilet. "The toilet in the residential quarter of the palace of Minos in Knossos ... is probably the earliest flush toilet in history." However, it was manually flushed; that is, someone had to pour the water into the piping (Yannopoulos 2017, 165). Sir John Harrington's water closet (1596) was more advanced; it had a pull mechanism that caused water to flow from a cistern, carrying the waste into a pot (which still had to be emptied by someone). Alexander Cumming (1775) added the S-trap, which sealed off sewer gases (Barksdale 2015).

The USS *Monitor* was equipped with a flush toilet, which functioned like a "mini-torpedo tube," with pressurized water being used to drive the waste out of the ship (the crew quarters were below the waterline). "One sailor turned the valves in the wrong sequence and was blown off the seat by a powerful jet of seawater" (Lienhard 1998).

The CSS *Alabama* had at least three personal flush toilets (Delgado 2019, 250; NHHC *Alabama*). These, presumably, functioned more like Harrington's. But even in World War II, a personal flush toilet was likely to be "strictly for ship's officers" (Squires 1992, 11).

On the HMS *Leander* in the 1930s, there was a communal trough-type head. The "captain of the head" would decide when to "pull a lever" and release a "flood of water," which could be disconcerting if you were still seated there (McKee 2002, 95).

Crewmen on one Korean War–era aircraft carrier had a head that consisted of a

V-shaped trough, with 12 seats, running athwartships. "One end was elevated higher than the other to permit sea water to constantly flow down hill to flush toilet tissue and feces down and out of the ship and into the ocean. A common trick was for someone at the high end of the trough to ball up a mass of toilet tissue, set it on fire…, and then send it flowing down the trough so that it would pass under the backsides of sailors on the downward end" (Wagner 2004, 88–9).

Toilet paper. While the Chinese invented toilet paper in the sixth century and were mass producing it by the 14th (Smyth 2012), in the West more primitive expedients were used. At least by the Napoleonic period, "on board ship, the officers commonly recycled paper, but the seamen would probably have made used of scrap fibrous material such as tow … or even oakum. In some ships a sponge rinsed in a bucket of sea-water might have been provided" (Adkins, 44).

Flush toilet from the wreck site of the CSS *Alabama*. "The porcelain toilet bowl features a popular 'Rhine' pattern with boating and country scenes. The bowl is encased in lead mounted on a brass base with flushing mechanism." (Naval History and Heritage Command. Catalog 91-217-AG. Artifact is on display at the History Museum of Mobile, Alabama, on loan from the National Museum of the U.S. Navy, Washington Navy Yard, Washington, D.C.)

Drinking Water Treatment

Microorganisms or other undesirable constituents can contaminate shipboard water in a variety of ways. First, they may be in the water that the ship takes aboard, if it comes from a contaminated source, or if contamination occurs during the watering process. Second, they may enter the water while the ship is en route.

In modern times, a particular concern is with inadvertent cross-contamination: "Space is often very limited on ships. Potable water systems are likely to be physically close to hazardous substances, such as sewage or waste streams, increasing the chance of cross-connections. Cold-water systems may be close to sources of heat, and this elevated temperature increases the risk of proliferation of *Legionella* spp. and the growth of other microbial life" (WHO 2011). Note that desalination by distillation will not separate fuel contaminants from seawater (Cassady 2000, 8).

While the existence of microorganisms was not known until the 17th century, and their role in disease until the 19th, people were aware in ancient times whether water looked clear or turbid, and tasted fair or foul. They learned to make water look and taste better by boiling, exposure to sunlight and filtration. Classical filtration methods included both straining through cloth and sand, and gravel or charcoal filtration. Pretreatment with alum to cause coagulation and settlement of particles was also known (Baker 1949, chapter 1).

Wooden casks could become fouled by bacteria and algae. The requirements for microbial growth are air, water and food. The water, obviously, is present. Air infiltrated through minute gaps between the staves and pores in the wood. The wood, itself, also served as a food source, decaying and in the process generating offensive decomposition products (Wilson 1879, 81).

There were a variety of countermeasures for fouling. In the early modern period these might have included charring and painting of the inside surface prior to use, and sulfurization. The latter practice is better known in its application to wine barrels; sulfur would be burned inside the barrels to generate sulfur dioxide, which kills bacteria.

Sulfurization was more effective than charring as a preventative. Ideally, sulfur was burned and then the barrel filled by a quarter, alternately, until the barrel was full. Then sulfuric acid was added (20 drops for every 10 liters water) (Beyer 1848).

In 1822, an experiment showed that water stored in a barrel pretreated with manganese dioxide was still good to drink after 33 months.

The water casks on ships could contain, not only microorganisms, but also inorganic matter, plant matter, insects and even (in the case of water collected from a river and presumably not distilled) fish (Brenckle 2018, 20). In the tropics, open water barrels served as breeding ground for the larvae of the *Aedes aegyptii* mosquito. On the 19th-century Boston-based sailing ship *Regulus*; there was an infestation of rats, which chewed holes in the water casks and then fell in (not exactly improving the taste) (Schultz 1999, 95).

Filtration. In the mid–18th century, James Lind alluded to the use of a "stone filter ... on ships to soften or clarify water." Porous stones would include limestone and sandstone. You would actually want a rock that is both porous (has void spaces) and permeable (the void spaces are connected so that water can pass through the rock, albeit slowly). Small pores filter out more particles but also reduce permeability. The porosity and permeability of rocks can vary even when they are of the same type, so there would be a preference for stones from particular localities, such as sandstone from Northamptonshire and limestone from Derbyshire.

Lind's preference was for filtration through sand. In 1762, he observed that a filter could be improvised "by placing a small cask inside a large one, both headless; then putting sand and gravel in the small cask and in the concentric space between the two casks. Water poured on the sand of the inner cask will pass down through the inner cask and up through the outer cask where it may be drawn off through a cock" (Baker 1949, 23). An 1800 French patent stated that a "marine filter" should be "suspended it to free it from the motion of the ship" (39).

James Peacock was a major innovator in the art of filtration. First, he found that filtering media (sand, gravel, broken glass, etc.) should be graded so the coarsest material was encountered first, and the finest last (67). Second, he cleaned the filter by reversing the flow (71). In 1801, one of Peacock's filters was tried out on the *Vengeance*, *Magnificent* and *Lancaster* (Evans 2014, 76). Peacock received a patent in 1791 and in 1815 Burney (145) claimed that Peacock's filter "has been fully proved to be highly useful on shipboard, particularly in long voyages." He rated its speed at 300 gallons a day.

Aeration. In 1755, Stephen Hales claimed to have "sweetened the stinking water in the well of some ships" by aeration. This was accomplished by using bellows to blow air upward through tiny holes in the bottom of a vessel containing the water (Baker 1949, 364–5). The opposite method, letting water fall through perforated sheets of tin and mix with air on the way down, was apparently devised by Osbridge and used in the British navy at some point (366).

There is a reference to a peculiar device, the "admiral's fiddle," for aerating water in Robinson's 1860 novel *Harry Evelyn* (177). If I understand it correctly, a cask, equipped with a plug pierced with 20 holes, was suspended in midair, with the plug facing downward. Strings of log-line, each terminated by a small grooved cone of wood, hung from the holes, and over the scuttlebutt. The water emptied from the suspended cask into the scuttlebutt "in a light misty shower, which the air purified and sweetened." The author noted that the operation was troublesome, especially "if the ship had any motion." I do not know whether this contraption was more than the product of Robinson's imagination, but he insisted that his novel was "founded on facts."

Gihon (1870, 92–93) suggested that the tank-fed scuttlebutt should, like the water tanks, "be of iron; it should be cleansed and waxed every month, and provided with a filtering diaphragm of sand and charcoal, which must be occasionally removed and renewed."

Distillation. In Robinson's *Harry Evelyn*, we are told that "the admiral had a great fancy for pure water; in his own case, his water was boiled and then passed through a charcoal percolator" (1860, 177).

While boiling is a standard method of disinfecting water, thermal desalination does not guarantee that the water is free of pathogens if the water is evaporated at a pressure low enough that the boiling point is under 80°C (WHO 2011).

Preservation and disinfection. In the mid–18th century, naval surgeon James Lind proposed to delay putrefaction of water at sea either by "fuming the casks with burning brimstone" [sulfur] or by "adding a little oil of vitriol" [sulfuric acid]. The burning of sulfur would generate sulfur dioxide, which is a broad-spectrum antimicrobial agent. As for sulfuric acid, it has been used in modern times as a wash or spray "applied to the surface of meat or poultry products to prevent the growth of spoilage microbes" (Agricultural Marketing Service 2012) and, in concentrated form, to clean cell culture containers (Wang 2018).

Beginning around 1800, efforts were made (on some ships) to deodorize the water with charcoal and powdered lime (Goethe 2012, 8). Various substances, even gunpowder, were also added as preservatives (Wagner 2006, 62). In the case of gunpowder, note that it contains charcoal, saltpeter (potassium nitrate), and sulfur.

Chlorination may use chlorine gas (which will form hypochlorite ions after reaction with water), or sodium or calcium hypochlorite. Hypochlorite was first used for disinfection in 1896, to fight a typhoid epidemic at an Austro-Hungarian naval base. The first continuous use in water treatment was in 1902 (White 2011, chapter 1). As of 2000, the aircraft carrier USS *Theodore Roosevelt* had an electrolytic disinfectant generator, which electrolyzes brine (a byproduct of desalination) to produce sodium hypochlorite (Cassady 2000, 8).

Ozone (O_3) is produced by the action of high electrical voltages or ultraviolet radiation on the normal oxygen molecule (O_2). It is highly unstable and hence must be generated on site for immediate use. It was first used experimentally for disinfection in 1886, and the first full-scale plant went into operation in 1906 (White 2011, chapter 15). A 1918 article describes an electric ozone sterilizer for purifying drinking water on ships (Groak 1918). A modern application is for reduction of bacteria, phytoplankton, zooplankton, invertebrates and fish in ballast water (Herwig 2006). Such water may be taken up in one part of the world and discharged in another, thus leading to transfer of invasive species.

Shipboard chlorination was practiced on at least an ad hoc basis (when contamination of the tanks was suspected) by 1929 (United States Public Health Service, 85). The U.S. Navy began equipping warships with chlorination systems shortly after World War II (White 1986, 92). Chlorination is currently the most common shipboard disinfectant. However, ultraviolet irradiation is more effective against *Cryptosporidium* (a protozoan) than normal doses of chlorine (WHO 2011).

Chloramine (the reaction product of chlorine and ammonia) has been used as a municipal drinking water disinfectant since 1929, and may be present in drinking water transferred to ships from a municipal source.

Bromine, a halogen like chlorine, is also a possible disinfectant. It has been used on naval vessels (Cassady, 8). These probably use a cartridge (patented in the 1960s) in which an anion exchange resin, impregnated with a solution of elemental bromine in aqueous sodium bromides, slowly releases the bromine (White 2011, chapter 16; NAVMED 2022, 6–21).

Drinking Water Handling

In 1848, a cholera outbreak occurred in London. "Case zero" was John Harnold, a crewman on the recently arrived steamer *Elbe*. The *Elbe* had come from Hamburg, where cholera was already raging. Harnold soon exhibited the symptoms of cholera and died. Then the man that took Harnold's room also died of cholera. The disease spread through the neighborhood and over the course of two years, there were 50,000 deaths.

In 1849, Dr. John Snow publicly hypothesized that cholera was a contagious disease, spread by drinking water contaminated by waste matter from someone already infected (Johnson 2006, 71). When cholera raised its fearsome head once more in 1854, Snow investigated, plotting the locations of the fatalities relative to public water sources. He found a statistical connection between those deaths and proximity to, or known use of, the water pump on Broad Street (153). This discovery inspired George Pinwell's 1866 cartoon, *Death's Dispensary*.

Dysentery (bloody flux) is also attributable to contamination of water (or food) by *Shigella* (bacteria) or *Entamoeba histolytica* (an amoeba). Five hundred men, including Sir Francis Drake, died of dysentery in the course of Drake's 1595–6 West Indies expedition. The bloody flux also assaulted Lancaster's 1600-3 East Indies venture, but "it was noted that it had broken out after taking in shore water" (Keevil 1957, 1:107–113). In 1770, Read "traced an epidemic of dysentery to the use of well water contaminated by adjacent latrines" (Woodward 1879, 613).

A third waterborne disease was typhoid fever. In 1847, in the Richmond Terrace neighborhood of Bristol, 13 of 34 houses experienced at least one case. "The only thing the 13 houses had in common was the use of a well; the 21 without fever had different water supplies." The local doctor William Budd concluded that the disease was waterborne, but his hypothesis and supporting evidence weren't published until 1859 (Moorhead 2002). Halsey Taylor was prompted to invent a drinking fountain in the early 1900s by his father's 1898 death from typhoid fever and later cases of dysentery among workers at the plant he supervised, both attributed to contaminated water (Halsey Taylor, 2023).

With modern drinking fountains, the principal health concerns are contamination of moist surfaces by noroviruses, rotaviruses, *Pseudomonas aeruginosa* and *Legionella*, and the presence of heavy metals (notably lead and copper) in the water.

The sailors lined up at the scuttlebutt were expected to use a common drinking cup. This might actually be chained to the barrel so it could not be removed (Chamier 1850, 74), which meant that germs on the lips, or in the saliva, of one sailor would be transferred to the cup, and from there to its next user. Bear in mind that the crew of an early 19th-century frigate in wartime might be 400 men (Adams 1999, 150).

Even after the germ theory of disease gained traction, some scoffed at the notion that this was unsafe, arguing that germs could not live on a metal cup, hence

George Pinwell, August 18, 1866. "Death's Dispensary," *Fun*, 10: 233. "Open to the Poor, Gratis, by Permission of the Parish" (University of Florida Digital Collections, https://ufdc.ufl.edu/UF00078627/00010/zoom/205).

the continued use of such common cups at public drinking fountains in the cities (Satran 2017).

Nonetheless, in 1902, MIT professor William Sedgwick warned against applying one's lips "to public drinking-cups which a few minutes before have been touched by the lips of strangers, possibly suffering from infectious diseases." And he noted that a cup-less "sanitary fountain has been devised" (Sedgwick 1902, 119–120). "'Ban the Cup' campaigns convinced almost every state to pass laws between 1909 and 1912 making common cups illegal" (Satran).

By the early 20th century, "the promiscuous use of unwashed drinking cups at ship's scuttle-butts [had] been universally recognized by Naval Medical Officers as unsanitary and dangerous." Some palliative measures were proposed, such as "steaming the cups" and "keeping them in an antiseptic solution when not actually in use," but these were considered to be only partially effectual (Gates 1907, 313).

In 1907, naval surgeon Manley F. Gates proposed "doing away entirely with cups." He designed a "sanitary scuttle-butt" and had it installed on the USS *Charleston*. Essentially, he fitted a fountain ("bubbling spring") attachment onto a more conventional scuttlebutt. The latter had a faucet "of an automatic closing type, preferably the 'rabbit ear' pattern," essentially a spring closure. The water traveled from this faucet through three elbows, downward, sideward and finally upward through a crosspiece. From the crosspiece, there was a final vertical tube leading to an "inner funnel." This tube was fitted concentrically within a larger tube leading to an "outer funnel." The water emerged from the inner funnel in a vertical jet, and the excess water overflowed into the gap between the two funnels and descended into the crosspiece, from which it could pass horizontally either to the next fountain (if there was more than one) or to the feedwater tank for the boilers (314–5).

Unlike the scuttlebutt mentioned previously that was pump-fed from water tanks below, the one on the USS *Charleston* was gravity-fed from a tank on an upper deck. The pressure, and thus the jet, was determined by the height of the tank above the fountain, as well as the inner diameter of a metal washer in the lower part of the tubing.

"When the faucet is opened a cone of water rises from one-half to three-fourths inches above the level of the inner funnel. In drinking, by slightly protruding the lips all risk of contact of the mouth with the funnel is avoided. If carelessly used only the outer, skin covered, surfaces of the lips touch the funnel" (316). Gates expected that the sailors would open the spring faucet before approaching the water jet, and thus the "first gush" of the latter would clean the rim of the funnel.

One flaw in Gates's design (and other early sanitary fountain designs) is that the jet is vertical. The general experience with public drinking fountains established that this encouraged people to put their lips on the spigot, which was just as unsanitary as using a common cup. Mouthguards were experimented with, but this just repeated the problem at a single remove: the users put their lips on the mouthguards. The solution was the slanting jet with a protected orifice (Satran 2017). In 1916, at the United States Naval Academy's Bancroft Hall, the navy was still using common drinking cups disinfected with formalin, and the surgeon general recommended

Manley Gates's experimental drinking fountain in use on the USS *Charleston*. Note the vertical spigots. Photo taken April 4, 1906, at New York Navy Yard (National Archives. Catalog 6880361).

that "the latest improvement of the drinking fountain be installed—that is, the inclined jet" (Braisted 1918, 111).

Wastewater Disposal, Reuse and Recycling

Typically, ships simply discharge wastewater when they are far enough from shore. In some instances, they treat wastewater on board so it may be discharged closer to shore. Or it is held for offloading to a land-based treatment plant.

Wastewater is classified as either blackwater (sewage from toilets) or graywater (everything else). The U.S. Environmental Protection Agency states that the latter includes "galley, bath, and shower water, as well as wastewater from lavatory sinks, laundry, and water fountains." Often, shipboard treatment facilities commingle the blackwater and graywater streams (EPA 2011). This has been criticized since the graywater potentially could be processed for non-potable shipboard use (Coursey 2013).

Shipboard reuse of even graywater is in its infancy. In 1976, Lent (1) proposed subjecting waste shower water to membrane filtration for subsequent reuse as laundry water. In 2010, Guildbaud conducted a feasibility study on recycling laundry graywater back to the inlet of the washing machines.

The Royal Caribbean cruise ship *Celebrity Flora* (launched 2019) has "the ability to reuse air conditioning condensation to provide water to the shipboard laundry facilities" (Zaltzman 2020). There is a similar apparatus on Disney cruise ships.

Scanship sells a water reuse unit to further "polish" water treatment system output so the water may be used "for wash down, laundry, sanitary flushing, machinery 'technical' water, fuel water emulsions, plant irrigation, and other non-potable and non-recreational water uses aboard ship."

I have not found any reference to the use of processed wastewater as shipboard drinking water. However, in more isolated environments—space stations, for example—there is recycling even of urine into potable water (Bobe 2016; Williamson 2020).

Effect of Ship Design on Health and Safety

In the 19th century, sailing and steamships coexisted, however uneasily. Notter (1908, 951) comments, "Sailing-ships appears to be more liable than other vessels to accidents by loss of men overboard and the general danger of the sea. On the other hand, accidents connected with machinery, scalding, &c., and the sickness resulting from continuous work in high temperatures, are characteristic of steamships."

The 19th century also witnessed the partial replacement of wood construction with iron or steel. This, too, had mixed effects. The "substitution … has rendered easier the cleansing and disinfection of vessels, but has made it more difficult to maintain a suitable temperature in cold weather and to prevent excessive heat in the tropics." Notter also noted "the disagreeable effects of condensation on metal surfaces from moisture-laden air between decks."

Chapter 6

Keeping Dry (and Afloat)

In Dumas's *The Count of Monte Cristo*, the sailor Penelon tells the story of a crisis at sea he had survived. His ship had pitched heavily for 12 hours, scudding under bare poles in a gale, and finally sprung a leak. Penelon continues, "'All hands to the pumps,' I shouted; but it was too late, and it seemed the more we pumped the more came in. 'Ah,' said I, after four hours' work, 'since we are sinking, let us sink; we can die but once.' Hearing this, Penelon's captain fetched a brace of pistols, and said, 'I will blow the brains out of the first man who leaves the pump.'"

Plainly, Dumas saw the dramatic potential of a ship being in danger of foundering. Inevitably, a ship takes on water. Waves crash over the bulwarks or surge through open gunports; rain falls on deck; enemy gunfire, icebergs or submerged rocks may pierce the hull; the hull itself leaks. The increased weight of the ship, attributable to the unwanted water, reduces the reserve (net) buoyancy. Ultimately, if the process is not arrested, the ship sinks.

The risk of foundering can be reduced by good ship design, but the time will come when the ship needs a good pump. Or more than one.

Floatability

In *On Floating Bodies*, Book I, Archimedes stated his Proposition 5: that "any solid lighter than a fluid will, if placed in the fluid, be so far immersed that the weight of the solid will be equal to the weight of the fluid displaced" (Heath 1953, 257). Thus, it will "not be completely submerged, but part of it will project above the surface" (Proposition 4, 256). If that solid is "forcibly immersed" in the fluid, it will "be driven upwards by a force equal to the difference between its weight and the weight of the fluid displaced" (Proposition 6, 257). If the solid were heavier than the displaced fluid, it would sink to the bottom, but if "weighed in the fluid, be lighter than its true weight by the weight of the fluid displaced" (Proposition 7, 258).

One of Archimedes's famous feats was a practical application of these propositions to determining the proportion of gold and silver in a crown. It was based on the knowledge that the ratio of weight to volume—what we would call density—was different for the two metals (260).

It is evident that by "lighter than a fluid," Archimedes meant lighter than the same volume as the solid, that is, less dense.

These laws of buoyancy were exploited by humans long before they were enumerated. The first boats made use of building materials—woods and grasses—that were less dense than water, and hence could float in their own right. *Homo erectus* is believed to have built rafts, possibly of bamboo, in order to spread across the Indonesian archipelago around 800,000 BCE (Vaucher 2023).

The oldest boat remains found by archaeologists is the Pesse canoe (ca. 8000 BCE), made from a hollowed-out trunk of a Scots Pine (Buchanan 2018). The

"Caulking bottom of ship with oakum and cotton to make water tight." Photo by Underwood & Underwood, circa 1900. The oakum can be seen hanging straight down. A caulking tool is held in the foreground worker's left hand. The original image was a stereograph but only the left-hand image is shown here (J. Paul Getty Museum, Los Angeles, object 84.XC.873.8736).

hollowing was an improvement over the simple raft, because it meant that some of the water was actually being displaced by air, rather than wood—and air is about one-thousandth the density of water.

A further innovation "was putting a waterproof skin over a rigid frame." The resulting hull was lighter and thus more portable than a same-sized hull made of solid wood. The skin could be leather, or an oiled cloth. Later advances replaced these skins with caulked planking, and permitted an increase in the size of ships.

The density of iron is about 7.85 times that of water. So the key to making a floatable iron vessel was to use thin plates surrounding a lot of air. While iron is heavy, its specific strength (ratio of strength to density) is quite favorable (and that of steel is better still).

Setting aside the controversial "Helton Tarn" boat, the first iron boat was a "small iron pleasure vessel built for the Foss at York" in 1777, builder unknown. It was carried by two men to the water (Barker 1987).

We know much more about John Wilkinson's *Trial*, a canal boat launched in 1787. It weighed about eight tons but drew (empty) just eight or nine inches. At the time, iron plates would have been made by hammering bar iron. Iron boilers and water tanks existed in the period.

The construction of rolling mills greatly facilitated the production of iron plate. "The first iron steamer was the *Aaron Manby*, built in London in 1821" (Fougner 1922, 1). The specific strength of iron made possible the building of ships longer than 300 feet, and it also helped ships resist "the forces developed by a propeller" (2).

"Steel ships were first registered [with Lloyd's] in 1878." By 1890, steel ships accounted for 95 percent of the tonnage built in British shipyards (4).

The first seagoing ferroconcrete ship was the *Namsenfjord*, launched in 1917 (7ff). Prompted by a steel shortage, about two dozen ferroconcrete ships were built during World War II (Bender 2011).

Load, Draft and Freeboard

Draft is the distance from the waterline down to the deepest part of the hull or keel. The deeper the draft, the more danger there is of the ship running aground in shallow waters. Freeboard is the distance from the waterline up to the lowest point at which water can enter the ship. (The variation in freeboard from bow to midships to stern is called "sheer," and freeboard is usually highest at the bow on modern ships.) The more heavily laden the ship, the greater the draft and the smaller the freeboard.

A shipwright. shipowner or captain might wish to know the relationship between the ship's lading (whether with cargo or armament) and its draft and freeboard. Knowing the density of seawater, one can calculate the change in displaced volume corresponding to a given added weight. Calculating the change in draft requires being able to determine the volume of a portion of a hull of arbitrary shape and size.

Here, too, Archimedes pioneered the way, showing how areas and volumes could be approximated as the sum of simple two- and three-dimensional shapes whose size could be exactly calculated (Heath, chapter VII). The difference in draft from launch empty to fully laden was estimated by Phineas Pett and Anthony Deane for British warships, and by Olaus Judichaer for the Danes, in the 17th century (Ferreiro 2007, 196–7).

In the modern method of numerical integration, the hull is divided into vertical or horizontal slices, and the volume is estimated as the total volume of these slices. The volume of each slice, in turn, is determined by similarly estimating the face area of the slice and multiplying by its thickness. Often, Simpson's rules (Thomas Simpson, 1710–1761) are applied.

Even in the 19th century, ships were lost as a consequence of overloading (leading to excessive draft). Samuel Plimsoll proposed marking the hull to show the load limit. If his mark, a circle bisected by a horizontal line, was submerged, the ship was overloaded. In 1873, he persuaded the British Parliament to survey ships going to sea, and "at least one in every ten was found to be so dangerously overloaded as to be in almost a sinking condition before leaving the dock." This led Parliament to mandate, over shipping industry opposition, the Plimsoll mark in the Merchant Shipping Act of 1876 (Ellicott 1896, 45–6).

The density of seawater varies with its salinity and temperature, and that will affect the draft for a given load. In 1892, Britain adopted a more elaborate mark adding separate load lines for fresh water (FW), Indian Ocean summer (IS), summer (S), winter (W), and winter North Atlantic (WNA). (Sailing ships only added the FW and WNA marks.) The modern Plimsoll mark includes one for tropical fresh water (TF).

If an Age of Sail warship had open gunports on a lower deck, that temporarily reduced its effective freeboard, and water could surge in through the open ports. Insufficient gunport freeboard (16 inches!) paved the way for the loss of the *Mary Rose* in 1545 (Sephton 2011). Reviewing the original 1664 plan of the *Warspite*, Charles II prudently insisted that the freeboard of the lower deck gunports (i.e., the distance from the waterline to the lower sill) be increased to 4.5 feet (Winfeld 2010, 54).

Heavy seas could reduce the disparity in fighting power between a frigate and a 74-gun two-decker, as a frigate built from the mid–18th century on didn't have guns on its enclosed decks (Allan 2023), and the freeboard of its main open deck was likely to be greater than that of the larger's ship lower deck gunports. Freeboard on Napoleonic Era British frigates was "often 7ft or more" (Gardiner 2011).

On the first-rate, three-decker *Queen Charlotte*, the "lower ports had only 4.5ft of freeboard and after the skirmish of 29 May 1794 her lower gundeck was full of water" (Gardiner 2011, 12). The USS *Independence*, one of the American navy's first

74s, "showed but 3 feet 10 inches of freeboard between waterline and the sill of her lower gundeck ports" when loaded for a six-month wartime cruise (Bearss 1984, 858).

Lloyd's at one time required that insured ships have three inches in freeboard for each foot of depth of hold (Muckle 2013, 46). Later, complex tables were introduced which considered ship type, length, depth of hold, the coefficient of fineness, the sheer, and enclosed superstructures. By way of example, an iron or steel sailing vessel with a length of 300 feet, depth of 30 feet, normal sheer, no enclosed superstructures and a coefficient of fineness of 0.68 would have a required freeboard in the summer North Atlantic of seven feet (Owen 1906, 21–48).

After World War II, it was recommended that freeboard for warships equal 1.1 times the square root of the length (feet) (Brown 2006, 214), probably taking into account the relationship of maximum speed to length and the resulting chance of taking green sea over the bow.

Closed Decks and Camber

A closed deck, first seen in some ships of ancient Egypt, increased seaworthiness. With an open deck, if a wave crashed over the side of the ship, the water would weigh down the ship until it could be baled (or pumped) out.

A closed deck, if watertight, would keep the water out of the interior. The water on deck would still weigh the ship down, but hopefully it would quickly roll off. This could be facilitated by camber—the deck being higher at the centerline than at the sides.

Thomas Harriot's *Notes on Shipbuilding* (1608) suggested that the camber should be about one-half inch for every foot of half-breadth, a 1:48 ratio (Lavery 1988, 16). On the mid–17th-century *Zeehaen*, the hull breadth was 22 feet but the camber on the upper and lower deck beams was ten inches, almost double the relative height (Hoving 2000, 127–8). Nicolaes Witsen taught that the camber of the lower deck beams should be one inch for ten feet of length (and the length was supposed to be four times the breadth) (Hoving 2012, 74, 250). On the Napoleonic-era *Bellerophon*, "the camber was six inches on the gun deck and an inch less for each deck above" (Pope 2013, 39).

Bulwarks

A further defense against wave action is the bulwark, essentially a wall on an open deck. The *Mayflower*, already an old ship when chartered in 1620, had solid bulwarks. Despite those bulwarks, it was a wet ship (Magoun 1928, 51).

Contemporary illustrations show that many early modern ships didn't have a solid bulwark, just a rail supported with stanchions. This might give a crewman something to hang on to but wouldn't keep out water. While sometimes canvas was hung over the rails, that would just keep out sea spray and not green water.

Even if a ship had a bulwark, if it were armed, it might either have a relatively low bulwark so the open deck guns could fire over it, or gunports cut into a higher bulwark. Normally, these open deck gunports lacked lids, so the effective height of the bulwark in terms of watertightness was the height of the lower sill.

Heck's *Iconographic Encyclopedia* (1852) said that bulwarks "are a continuation of the timbers lined with planks, and covered with a plank on the top. The bulwarks are usually from three to four feet high; the almost semicircular part surrounding the bow, the arch of the forecastle, is the highest…. The bulwarks in ships of war are differently arranged, and higher throughout, at the quarter-deck being from five to six feet high."

Holms (1918, I:345) states that the average height of the bulwark is 4.5 feet in sailing ships and 2.5 feet in steamers (which are not heeled over by the action of a beam wind on sails). By his time, ships had metal hulls and the bulwarks were thick plating.

Scuppers

The bulwarks that keep the smaller waves out also trap the water left behind by waves that crest the bulwark. Hence, ships are equipped with scuppers, deck-level openings through which the water can drain off. The scuppers are usually on the main open deck (the weather deck), which is above the waterline. However, they can be lower, and channels conduct the water from the higher deck to the one the scuppers are on.

That of course leads logically to the question, what keeps the water from entering by way of the scuppers? On the *Mary Rose*, a leather flap was nailed over the scupper hole. This acted as a one-way valve (McElvogue 2020, 21).

The scuppers must be dimensioned to take into account the volume of water that can be trapped by the bulwarks. Holms (1918, I:345) advises that the combined area of these freeing ports should not be smaller than 10 percent of the area of the bulwarks.

Transverse (Roll) Stability

A ship may heel over (roll to one side) as a result of wind or wave action, the centrifugal force experienced while making a turn, the reaction force from firing a broadside, or shifting a large weight (carriage guns, cargo, etc.) on board. Its design, weight distribution, and hull integrity then determine whether it recovers, remains at a list, or capsizes.

Archimedes recognized that the gravitational force (weight) acts downward through the center of mass (gravity) and the buoyant force upward through the center of buoyancy. The location of the center of mass depends on the shape and composition of the hull and superstructure, and how and where the ship is loaded (armament, provisions, fuel, ballast, etc.). The location of the center of buoyancy is determined by the shape of the submerged hull.

When the ship is upright, the center of mass and center of buoyancy lie on the same vertical line. But when the ship is heeled over, this is no longer true. A "wedge" of the hull, on the heeled side, moves below the waterline, and another "wedge," on

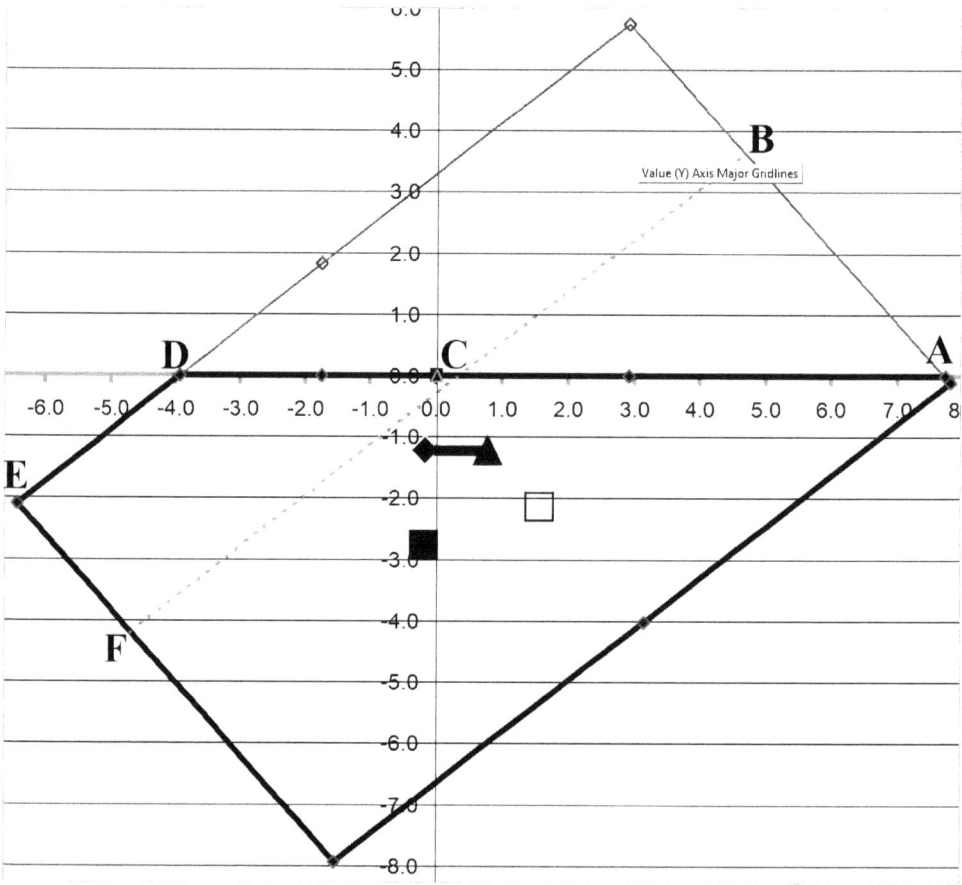

A ship with a rectangular cross-section is heeled 40o to port (left). The coordinates are in meters. The old waterline is shown by a dotted line, as a result of the heeling, the wedge ABC rises out of the water, and the "wedge" (quadrilateral) CDEF is newly submerged. (Note that this means that water has come over the deck edge E.) This shifts the center of buoyancy from its old position (open square) to its new position (filled square). This new position is on the "heeled" side of the new position of the center of gravity (filled triangle). The lever arm GZ is the thick line from the center of gravity to G (filled diamond). The drawing was made assuming height 25 feet, beam 40 feet, KG[KK1] 12 feet. GZ was 2.39 feet (0.97 meters) (calculations and drawing by the author).

the opposite side, moves above it. (The wedges may have different shapes and volumes.) Assuming that the weight distribution of the ship is unchanged, and there is no breach of any internal space, the buoyancy remains equal to the weight, and thus the total volume below the waterline remains the same.

If we examine the heeled ship in an external, fixed frame of reference, both the center of gravity and the center of buoyancy are shifted toward the heeled side. but usually not by the same amount.

Since the line of action of the buoyant force no longer passes through the center of mass, the force will tend to cause the ship to rotate. The rotational tendency is called a moment, and it equals the force times the distance of the line of force from the axis of rotation.

Now, remember that since the ship isn't sinking or rising, the gravitational and buoyant forces are necessarily of equal magnitude, and since they are acting on parallel lines but in opposite directions, they form what physicists call a "couple." For a couple, the total moment is independent of the choice of the "center" around which the moment is calculated.

If the center of buoyancy is on the "heeled side" of the center of gravity, then the moment is a restoring moment, that is, acting to rotate the ship back toward the upright position. But if it is on the other side, then the moment is an overturning moment, causing the ship to heel over further.

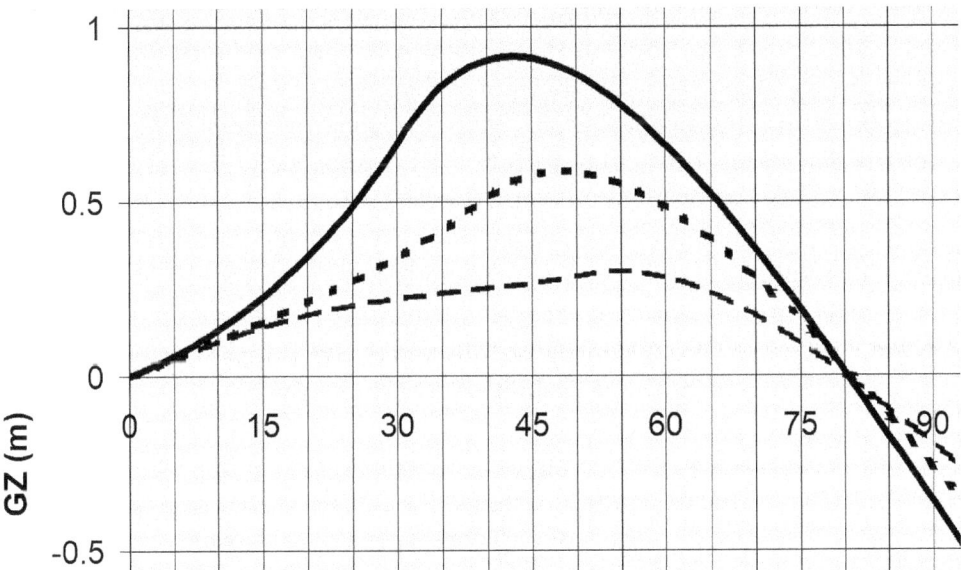

GZ curves (discussed below) for three ships with the same KG, the same initial submerged cross-section, draft and waterline beam, and the same total cross-sectional area (hence, reserve buoyancy). Note that they have the same initial slope (initial stability), but Ship 1 (flared above waterline, solid curve) has greater total stability than Ship 2 (wall-sided, dotted curve), and Ship 2 more than Ship 3 (tumblehomed, dashed curve) (drawing and calculations by the author).

A ship has weight stability if its center of mass is lower than its center of buoyancy. If so, then assuming these do not move (relative to the ship hull) when the ship heels over, they provide a restoring moment. The Swedish *Vasa* did not have weight stability, as will be seen in the next section.

A ship has form stability if its shape is such that when heeled over, the center of buoyancy shifts to a position from which it provides a restoring moment. Above the waterline, flared-out sides provide more form stability than vertical sides, which are better than inwardly curved sides. A cylinder has no form stability; there is no shift in the center of buoyancy.

Many premodern warships had "tumblehome." The bottom of the hull was more or less hemispherical but, a little above the waterline, the sides would angle straight inward as they rose. This had the advantage of increasing structural strength (to support the armament) and moving the armament closer to the centerline, which as we'll see reduces "stiffness," but it reduced the ultimate stability of the ship.

Initial Stability

Initial stability—the resistance to a very small change from the upright position—could be crudely evaluated by "sallying": shifting guns, or a group of sailors, back and forth between port and starboard. This was done for the *Vasa*; 30 men ran side to side three times. The result? The ship rocked back and forth like crazy. The outcome was not reported to the shipyard or the King of Sweden.

In the third *Pirates of the Caribbean* movie, Captain Barbossa deliberately overturns the *Black Pearl* by sallying with 16 crew members and 18 cannon (Mungan 2011, 268). But this is not a recommended nautical maneuver.

In 1697, Paul Hoste proposed an "inclining experiment," in which the change of inclination resulting from suspending a weight from a boom would be measured and used to determine the power to carry sail. Unfortunately, he got the math wrong. This was corrected by Pierre Bouguer in 1746, and the inclining experiment was used by shipwrights to estimate initial stability (Ferreiro 2007, 214).

In the 18th century, Pierre Bouguer and Leonhard Euler independently developed equivalent mathematical approaches to the calculation of a ship's initial stability. Bouguer's approach made use of the concept of the "metacenter."

You need to imagine the original line of buoyant force as rotating with the ship as it heels. The metacenter is the imaginary point at which that rotated line intersects the new, vertical line of buoyant force running through the new position of the center of buoyancy. Its position is fixed for small angles of heel.

The metacentric height GM is the height of the metacenter above the center of gravity. It is related to how the ship is loaded, and the intrinsic hull design. If KB is the height of the center of buoyancy above the keel, and KG is the height of the center of gravity above the keel, then GM= KB - KG + BM, the height of the metacenter above the center of buoyancy (also called the "metacentric radius").

The metacentric radius (BM) equals the second moment of area of the waterplane (the intersection of the water surface with the hull) around the longitudinal axis, divided by the volume of water displaced by the hull. (The term "second moment of area" means we sum the squared distances of each patch of the waterplane from its center.)

For an ellipse (admittedly a crude approximation of the platform shape of the hull), the second moment of the area around the long axis is proportional to the length along that axis, and the cube of the breadth. If we double the waterline beam of the ship, keeping length and draft constant, we increase the second moment of area eight-fold, and the displacement two-fold, for a net increase in metacentric radius of four-fold.

The Swedish crown had an unpleasant reminder that even kings are subject to the laws of physics. The pride of the Swedish navy, the *Vasa*, sank in 1628 on its maiden voyage, blown over by a gentle wind gust estimated as being just eight knots. Fairley (2007) says that according to modern calculations, four knots would have been enough to capsize it. Its maximum angle of heel was just 10 degrees.

The *Vasa* was top-heavy. It was the first Swedish warship with two enclosed gundecks. This was not part of the original plans, but rather a last-minute development in the Swedish-Danish arms race. There were also several upward revisions, during construction, of the number and weight of the cannon. All this meant that the ship was not only taller, but wider. Since the *Vasa*'s keel had already been laid, the width had to be added mostly in the upper part of the hull, which further raised the center of gravity. The keel was found to be a bit thin for supporting all the added weight, so additional braces were added in the hold. With space reduced, *Vasa* could only carry about 120 tons of ballast, and Fairley says it would have needed more than twice that amount to be stable. (But it was impossible to add more since the gunports were already only 3.5 or 4.5 feet above the waterline [Borgenstam 1984, 38]).

There were two problems with the *Vasa*. The first was that it had a high center of gravity, and hence a low initial stability. And the second was that it was sailing with open gunports close to the waterline, so when it heeled over, water flooded in. That caused it to sink deeper, which increased the water pressure at the gunport level, and thus the rate of water entry.

A very broad-beamed, twin-hulled ship (290 feet long, 60 feet wide), the steamer *Castalia*, was launched in 1874. She had a draft of only six feet, which let her freely enter "tide-controlled" ports. Nonetheless, because of her beam, she was resistant to rolling (Rogers 1936, 65–7).

Roll Period

The period of roll is inversely proportional to the square root of the metacentric height. A ship with a low metacentric height is "crank" or "tender"; it is slow to recover when heeled by the wind, and rolls slowly.

A ship may instead be "stiff," that is, have too high an initial stability. If there is a sudden gust, and it doesn't timely reduce sail, then since the ship doesn't heel much, the sails take the full force of the wind, and "the topsails are often carried away, or the sails torn to shreds" (Walton 1899, 215). Worse, if the ship heels and then rights itself too quickly, it could be dismasted (as happened repeatedly with the 1800 *Akbar*).

For a warship, you want a slow and easy roll, limited in angle, to make it easy to aim (Chappelle 1949, 24). Stiffness can be reduced by "winging" weights out to the ship's sides or raising the center of gravity (Walton, 168).

Girdling and Furring

As early as 1622, the English sometimes rebuilt ships by increasing their beam from several feet above to several feet below the waterline. Girdling simply added an extra layer of planking. Furring was more elaborate; the original planking was removed, a second layer of timbers added, and then planking added upon that (Johns 1922, 109; Wagstaffe 2010, 40). The terms were often used interchangeably.

The rebuilding was typically to correct an unexpected stability problem. For example, the 76-gun *Royal Katherine* (1664) "proved dangerously crank, having less than three feet of freeboard when her guns and stores were aboard. She had to be girdled before she could put to sea" (Hemingway 2002, 58). When lying hove-to after launch, the 100-gun *Royal Anne* (1704) "lay on her middle tier of guns in the water [and] had four feet of water on her lower gun-deck." Her captain opined, "she would not have righted had I not ordered her lower yards cut down ... she ought to have a good girdling" (123).

Mainwaring's *Seaman's Dictionary* complained that the shipwrights who needed to fur their ships should be punished for their miscalculations (Wagstaffe, 9).

The girdling would have two effects on stability. First, there would be a slight reduction in draft because the girdling was with wood of density less than that of seawater, and thus would boost buoyancy in salt water by 30 pounds per cubic foot of wood (Hemingway, 43). Second, the increased beam would increase the metacentric radius and therefore also the metacentric height.

However, it was also argued, at least in the case of the *Royal William*, that girdling would also "make her almost shot-proof between wind and water and consequently not in so much danger of being sunk" (42). (But if that were a real concern, rather than an afterthought, then girdling would have been part of the original building plan.)

The 96-gun *St. Michael* (1669) had its beam increased from 40 feet 8.5 inches to 41 feet 8 inches (48–9).

There were more extreme rebuilds: the *Defiance*, with a 32-foot beam in 1590, was increased to 37 feet in 1612 (Johns). Attwood (1908, 200) says that girdling substantially increases stability while adding "very little" to the draft.

Ballasting

The lower the center of gravity of the ship, the better its stability (although if too low, the ship may become too stiff, that is, roll frequently and violently). Stability is a particularly acute problem for warships, since guns are more effective if mounted high, and that raises the center of gravity. But all ships find it advantageous to carry ballast: heavy materials deep in the hold.

The disadvantages of carrying ballast is that it adds to the weight of the ship and takes up space, thus reducing the ship's ability to add armament or cargo. The merchantman-privateer *Angel Gabriel* (1619–1635), the subject of the ballad "Honour of Bristol," was 240 tons burden, and carried 10–15 tons ballast (Riess 1980, 75). The schooner *Flying Fish*, 150 tons burden, carried 13.25 tons pig iron and three tons of stone (Chapelle 1967, 167, 211). Some captains under-ballasted so that the guns would ride higher.

The earliest ballasts were gravel, coarse shingle, sand and rock (Chapelle 1949, 247). They had the advantage that they could be laid wherever desired. The construction contract for the *Nuestra Senora de Atocha* (1620) declared, "The lower hull and crutches from stem to stem must be filled with lime and sand and gravel of small pebbles between frame and frame" (Crisman 1999, 260).

Cast iron ballast came into use in the late 18th century. It was more efficient (higher density) but more costly (Dodds 2022, 23).

Lead ballast would have been even more effective. "In 1672, Phineas Pett, building the *Prince*..., suggested 300 tons of lead ballast.... As the cost of the lead—£5250—was as much as that of building a second rate, the proposal was not approved" (Johns 1922, 109).

In 1796, Samuel Bentham had pig iron ballast bolted *outside* the hull, beside the protruding keel. That moved the CG more than the same weight of internal ballast would have. This deep ballasting eventually evolved into the uncapsizable hull. (This has a deep fin with ballast attached at the end; the hull is uncapsizable if the external ballast moves the CG *below* the CB.) (Gougeon 1973, 39, 51; Chapelle 1949, 236.)

The crew's water and victuals may also be placed low in the hull, to augment the ballast, but of course they diminish over the course of the cruise. HMS *Andaman* (1797) carried 120 tons iron, 26 shingle, and 124 water (Gardiner 1989, 145).

Cargo can also serve as ballast, if dense enough, and has the advantage of earning revenue. The Portuguese found Chinese porcelain to be a useful ballast for their East Indiamen (Brigadier 2002, 54). Madeira is a fortified wine, and in the 18th century it was recognized that unlike other wines, its taste is improved if it spends, say, three months "cooking" as ballast for a ship traversing the tropics (*New Scientist* 2005, 114). The improved Madeira was called *vino da road* ("wine of the round trip").

It was sometimes necessary to shift ballast horizontally, to correct the pitch or roll of the ship. On small ships, the crew shifted a couple of tons of sand, in bags, to one side or another ("sandbagging") (Gougeon 1973, 55). They could also use human ballast; in the America's Cup race of 1895, there were 60 men on the *Defender* whose sole purpose was to move to a rail on cue (56). The ballast must be kept from shifting on its own, as a result of a heel; that would reduce the lateral stability.

Water Ballast and Trim Tanks

Merchant ships that carried heavy cargoes in one direction and light cargoes in the other had the problem that when they discharged the heavy cargo, they had to take on ballast, and then when they replaced the light cargo with a new heavy load, they had to dump the temporary ballast.

This was a particular problem for colliers (coal carriers). Their owners didn't like having to pay for the one-way ballast, and the port of origin didn't like the ballast dumps. A collier sailing from Newcastle to London with 250–400 tons of coal would have to pay one shilling a ton for the ballast in London, and another six pence a ton to put it on board. Then, back in Newcastle, it would pay one shilling a ton to the River Commissioners (a pollution charge?) and ten pence a ton for depositing it on the riverside (Holmes 1906, 2:162). Another estimate was related to a merchant steamer in the Mediterranean trade. Carrying 200 tons of ballast, if it had to load and unload it each voyage, that would cost 260 pounds (1877), the steamer itself costing 20,000 pounds and making four voyages a year (163).

In 1852, the SS *John Bows* was equipped with some kind of "temporary appliance" for carrying water ballast—water being free and environmentally acceptable. This experiment being deemed successful, the SS *Samuel Lacing* (609 register tons) was built in 1854, equipped with fixed iron water ballast tanks. The ship was double-bottomed and the tanks rested on the floor created by the top of the inner bottom (Holmes, Fig. 75). The next step, taken in building the SS *Rouen*, was to make the tanks an integral part of the ship structure: the top of the tank was the inner bottom.

Naturally, to fill and drain these water ballast tanks, pumps were necessary, but by the time they were introduced, pumps were steam-powered.

If the ship has not a single ballast tank, but rather separate tanks fore and aft, then by pumping water forward or backward, the trim of the ship may be adjusted.

Likewise, port and starboard tanks may be used to correct a list to one side (Rogers 1936, 29–30).

The principal disadvantage of water ballast is that water is less dense than iron (7.87-fold relative to fresh water) or lead (11.35-fold). Hence, it's less efficient (on a per volume basis) at lowering the center of gravity.

Also, it is very important to keep the tanks full; any partially full tank of liquid is subject to the "free surface effect"; the liquid sloshes in the direction of the tilt and moves the center of gravity in the "wrong" direction.

Stability at Large Angles of Heel

Unfortunately, a ship can have a respectable initial stability and yet be unable to recover if the perturbing force causes it to assume a large angle of heel.

In 1796, George Atwood presented a paper to the Royal Society of London in which he noted that "ships at sea are known to heel through angles of 10°, 20° or even 30°" (52) and argued that "the theory of stability, restrained to cases in which the angles of inclination, or heeling, are very small, cannot be relied for ascertaining the relative stability of ships in the practice of navigation" (119).

He recognized the significance of the righting lever GZ (57, 61, etc.). "Z" is the intersection of the horizontal line through the center of gravity with the vertical line through the center of buoyancy. The moment about the center of mass equals the product of the buoyant force (which equals the weight) and the distance GZ from the center of mass to the vertical line of buoyant force. If positive (Z on heeled side of G), the vessel will experience a restoring moment; if negative, it will heel over further. Thus, stability at any given angle of heel could be determined by "finding either by construction or calculation the length of the line GZ" (108).

Atwood further developed his theory of ship stability in a 1798 paper. He calculated GZ for a variety of simple shapes of constant cross-section, for which the value of GZ could be expressed analytically (i.e., as a formula).

Atwood also explained how one could calculate how much a wind of a given force would heel over a ship. The wind exerts a force that acts on the center of pressure of the sails (and superstructure). If the wind isn't coming from directly behind the ship, then a component of the wind force is transverse to the ship, so the wind creates a disturbing roll moment equal to the magnitude of that component times the height of the center of pressure above the center of gravity. The ship will heel over until its weight times GZ—the restoring moment—equals the disturbing moment (Atwood 1798, 217).

There are now a number of different measures of static stability: the "initial stability" (the slope of the curve at low heel, 5–10 degrees); the maximum GZ and the heel angle at which it occurs, and the angle of "vanishing stability" (GZ is zero). The initial GZ equals the metacentric height times the size of the angle of heel. Sailing ships typically had a maximum safe heel of 45–65°, depending on hull form (Chappelle 1967, 213).

The static stability characteristics of a ship may be represented by a "GZ curve," a plot of GZ against the angle of heel. The area under the positive range of the GZ curve is a measure of "dynamical stability," as it indicates the total work that an upsetting force must perform in order to heel the ship to that angle.

If we were to look, as textbooks do, at a "block" ship—one with a rectangular cross-section—we would see that its GZ curve increases up to and a little past the heel angle where the corner of the deck is immersed. Then it declines. The greater the freeboard, the greater that angle.

Previously, it was stated that the initial stability increases with breadth. But all else being equal, if you increase breadth, you also decrease the angle of deck corner immersion. And if, to keep deadweight constant, you also decrease the depth of hold (distance from the top of the highest enclosed deck to the bottom of the hull), that would decrease the freeboard and further decrease the critical angle.

The earliest example I could find of the actual plotting of a GZ curve was in Reed 1868 (Plate VI), when he explored the effect of changing the freeboard of the "Duncan" monitor. Reed observed that the reduction of the freeboard reduced the monitor's angle of maximum GZ, angle of vanishing stability, and dynamical stability (202–4).

The HMS *Captain*, a low freeboard turret ship designed by Captain Cowper Coles, capsized in gale force winds in 1870. Its GZ curve was subsequently compared to that of the HMS *Monarch*, a contemporaneous turret ship designed by his rival, Edward Reed (Attwood 1908, Fig. 184). The *Captain* actually had a slightly higher initial stability. However, its maximum GZ was at 21 degrees versus 40 for the *Monarch*, and its angle of vanishing stability was 54.5 degrees, as compared to the *Monarch*'s 70.

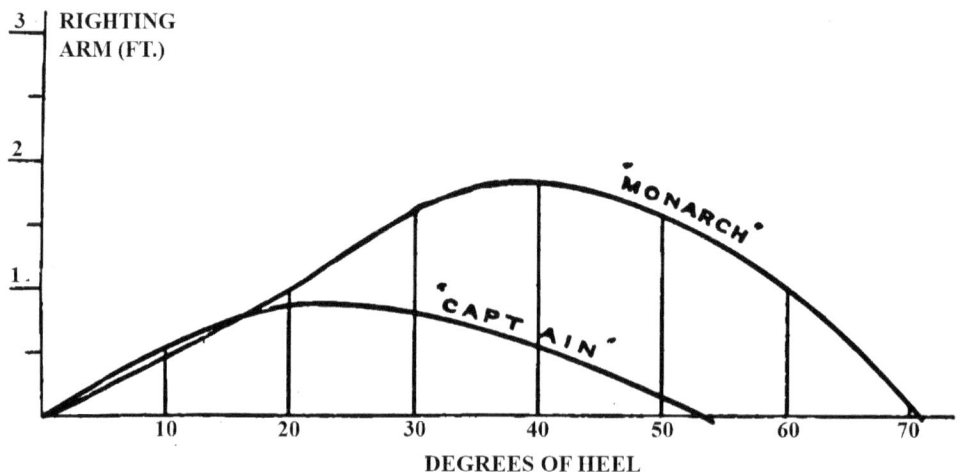

GZ curves for HMS *Captain* and HMS *Monarch*, per Edward Attwood, *War-Ships* (1908), 225.

International intact stability criteria, based on the GZ curve rather than freeboard, were first promulgated for passenger ships in 1948 (Francescutto 2011). The 2008 requirements for general cargo, tanker and passenger ships specify a minimum metacentric height (GM), righting lever (GZ) for heel of 0–30 degrees, heel angle corresponding to maximum GZ, and area under the GZ curve up to heel angles of 30 and 40 degrees.

Complications

Stability predictions are inherently more difficult to make for wooden ships because of the great variation in the specific gravity of wood, and thus in the location of the center of gravity (Reed 1885, 360).

If the ship is heeled over by the wind, the wind force on ship diminishes as the heel increases because the projected area of the sail in the plane perpendicular to the wind decreases.

Also, either the center of mass or the center of buoyancy will necessarily remain in the same positions relative to the hull. Objects on board can shift toward the heeled side, moving the center of mass in such a way as to reduce stability. One such object is liquid in a partially filled container; this is called the "free surface" or "loose water" effect.

The total volume of the enclosed (watertight) spaces that lie above the waterline when the ship is upright is called the reserve buoyancy. If they become submerged, they contribute to the buoyant force.

Once the ship heels to the deck immersion angle, letting water on deck, there is an increased risk that the sea will "break open some of the hatches and companionways," with dire results (Chappelle 1967, 213). If water enters a nominally closed space, through a hatch or port left open or as a result of battle damage, the weight increases as a result of the water taken on. In ship design, it is now commonplace to assess the "damaged stability" resulting from a compartment being flooded as a result of, say, a torpedo hit, running aground, or hitting an iceberg. The 1990 Safety of Life at Sea (SOLAS) convention required passenger ships to meet stability criteria despite the flooding of one compartment on a small ship, or two adjacent ones on a large one (Francescutto 2011, 24).

Bulkheads

These divide the ship into watertight compartments. Thus, if there is water leakage into one of them, the maximum loss of buoyancy is the volume of the affected compartment. Bulkheads may be transverse or longitudinal, and made of wood or iron.

Bulkheads were used on Chinese and Japanese junks for centuries before they entered European shipbuilding practice. There is some dispute as to whether these bulkheads were in fact intended to be watertight, as they had limber holes at the bottom, but I agree with Chinese scholars that these holes were intended to facilitate washing cabins (the boat could be trimmed by the stern, so the water would drain sternward and there be pumped out), and at sea the limber holes would be plugged (Cai 2010).

In 1787, Franklin proposed that the holds of packet ships be "divided into separate apartments, after the Chinese manner, and each of these apartments caulked tight so as to keep out water." In 1795, Bentham likewise advocated "partitions contributing to strength, and securing the ship against foundering, as practiced by the Chinese of the present day" (Wikipedia).

Watertight bulkheads remained uncommon in the West. Most 19th-century sailing ships had merely a collision bulkhead. Those that had more were mostly steamers converted into windships, and the most bulkheads on any unconverted ship was four. Even steamers weren't necessarily compartmented. In 1881–3, 120 British iron steamships were lost that "had a single compartment the filling of which would have caused the ship to founder" (Barnaby 1890, 445).

The 2009 SOLAS convention required calculating the probability that a ship would survive after collision damage to the ship's hull. This probability would be calculated for each damage scenario (single compartment flooding). The total probability ("attained subdivision index") had to exceed a "required minimum subdivision index" specified in the regulations. Under the 2020 convention, the requirement is dependent on the number of persons on board a passenger ship and the ship length of a dry cargo ship (Malinowska 2020; Francescutto 2011; International Maritime Organization 2023).

Caulking

Carvel planking has been used in shipbuilding since ancient times. The planks are fitted edge-to-edge and nailed to the frame. It was the dominant method of building wooden ships since the medieval period, but carvel hulls required more caulking than clinker-built (overlapped planking) boats to make them watertight (Denny 2008, 24). In premodern Europe, the seams were sealed by driving in oakum (old, untwisted hemp rope) with a mallet and then covering the seam with "hot melted pitch or resin" (Falconer 1830, 65, 254, 322). The pitch was usually pine resin, but the Spanish sometimes used asphalt from Trinidad's Pitch Lake (345).

Leaks

The rate of inflow from a hole below the waterline is usually proportional to (1) the size of the hole and (2) the square root of the depth of the hole below the

Chapter 6. Keeping Dry (and Afloat)

A caulker at work. He is hanging from a sling, over the ship's side, and he is nailing a plate, probably of lead, near the waterline. From J.G. Heck, *Iconographic Encyclopedia* (1851), Plates vol 2, part 6, plate 23, fig 1.

waterline. If the water rises inside the hold enough to cover it, the rate of flow is proportional to the square root of the distance between the water level inside the hull and the waterline outside (Oertling 1984, n4).

It follows that as the ship fills with water, the rate of water entry slows down (assuming the water doesn't create new holes). Thus, a point can be reached at which the rate of water removal (by bailing or pumping) equals the rate of water ingress—the ships stays afloat even though it is waterlogged. A leak in the bow was more dangerous than one in the stern because the forward movement increases water pressure at the bow.

In early modern naval warfare, while a cannonball could certainly create a large hole, it wasn't that easy for enemy fire to hit a ship below the waterline. The most common leak was the result of a planking seam that had lost its caulking. The leak could be located by listening for it with an ear trumpet (Oertling, 6).

Leaks could be plugged from the inside or outside. On the inside, one could use some sort of gelatinous mixture (like tallow and coals), pieces of raw beef, oatmeal bags, sheet lead, or canvas or leather backed with oakum. On the outside, one lowered a bag or net of oakum down over the leak, which then sucked in the oakum. Shot holes were usually closed by driving a canvas-covered wooden plug into the hole. One could also nail sheet lead over a hole, or "fother" it with a spare sail (7).

Bailing and Scooping

With 10-liter buckets, "one man can lift about 15 buckets per minute or 300 cubic feet an hour to a height of 3.3 feet" (Wood 1977, 29). Water may also be thrown by using a semi-enclosed shovel, with about the same rate of water transfer (41). A trick that increases the rate of scooping is to attach the shovel by a rope to a tripod so you get a "pendulum assist."

Ship Pumps

If the deck water doesn't escape by way of the scuppers, then it will eventually drain down to the bilge. Water entering by way of leaks or shot holes will do the same.

It was not very practical to use a bucket brigade to carry water all the way up to the weather deck in order to dump it out. Instead, pumps were used.

Pumps were under the supervision of the head caulker. "If the ship draws so much water that it cannot be controlled by the pumps, it is the duty of the chief caulker to give private notice of the fact to the commander" (Heck 1851, 3: 746).

The pump drew water up through a pump tube, whose length was dictated by the depth of the hull. The pump and pump tube were inside a compartment called the pump well.

The 1911 *Encyclopaedia Brittanica*'s "Pump" says that the simplest type of pump used to move a liquid is a plunger pump, characterized by a piston moving in a cylinder, and various valves. The plunger pump type is subdivided into suction pumps and force pumps. We'll discuss suction pumps in more detail shortly.

Force pumps have a solid piston (there is a valve inside the piston of a suction pump) and the outlet is below the piston (rather than above it as on the suction pump). Liquid rises on the upstroke, and is forced into the outlet by the downstroke (with the inlet closed by a check valve).

A more cogent, modern approach is to classify water pumps according to whether they work by volumetric displacement or by adding kinetic energy. The displacement pumps are classified as being reciprocating (piston, plunger, diaphragm, etc.) or rotary (gear, lobe, screw, vane or cam). Reciprocating pumps can be single- or double-acting, the latter pumping on both strokes. Kinetic pumps are classified as centrifugal (radial, axial, mixed flow) or regenerative. There were three basic types of premodern shipboard pumps: burr, common (suction) and chain pumps.

Burr pump. Described by Agricola in 1556, it had a vertical pole that moved up and down inside the pump tube. On the lower end, the pole was thickened (the burr) and to this was attached a leather cone ("shoe"). Strips of leather ran from the base of the cone to an anchor point above the base. On the downstroke, the cone closed and water entered the pump tube. On the upstroke, the cone opened and water was carried upward (Agricola 1912, 176; Ewbank 1876, 214; Oertling 1984, 24ff).

The burr pump was also equipped with a foot valve at the bottom of the pump tube. A foot valve is a one-way valve used to keep the pump primed (i.e., filled with liquid).

By the 17th century, the burr pump generally was "no longer in use on English ships, but could be found on Dutch and Flemish ships." However, it was occasionally seen, in modified form, as late as 1860 (Oertling, 29).

Oertling says that it was difficult to service; the entire pump tube had to be lifted off its base. On the other hand, according to Boteler (1634) and Manwayring (1644), it drew up "far more water and was less labor intensive than the common pump at that time" (Oertling, 30).

Common (suction) pump. First described in 1433 and used in mines and on ships in the 16th century, if not earlier. It features a fixed lower valve, an airtight piston that moves up and down in the box, and an upper valve inside the piston. On the upstroke, a vacuum is created between the piston and the lower valve, drawing water up through the latter. On the downstroke, the drawn water is forced up through the upper valve and ultimately to a spout (Oertling, 32ff).

The theoretical limit for raising water by suction is about 30 feet (assuming barometric pressure of air is 30 inches mercury). Because of friction, the practical limit is 28 feet, and this is measured from the surface of the water to the closing member of the upper valve. The pump was typically placed near the center of the pump tube to reduce the critical distance.

The lower valve was a lift check valve: it had a central hole closed by a vertically movable leather claque held down by a valve weight. Water pressure could lift the claque and weight, opening the valve, but a guide maintained its alignment. It was equipped with a staple so that it could be fished up with a hook at one end of a long pole, for repair.

The upper valve-cum-piston had a wood body, with a slot to receive a wooden shaft ("spear") at the upper end and a check valve covered with a leather gasket at the piston end. This could be a lift check valve, but one recovered from the *Machault*

(1757) had a hinged claque instead. Both lower and upper valves were usually made of elm or ash.

The spear was pivotably connected at its upper end, above the top of the pump tube, to one end of a lever ("brake"), whose fulcrum was provided by a "cheek" that curved away from the top of the pump tube.

Common pumps could be connected in parallel (two piston cylinders connected by a T to a pump tube communicating with the bilge, as in Dodgeson's 1799 pump) or in series (two pistons in a single cylinder, as in Taylor's 1780 pump) (Oertling, 64ff).

On the Taylor double piston pump, the shaft of the lower piston ran through the upper piston and was connected above it by a jog section to one side of a cog wheel, and the upper piston was connected at its periphery to a shaft that in turn was connected to the other side of the same cog wheel. Thus, when one piston went down, the other came up. The cog wheel was connected to a double action pump brake or drum. It would thus produce twice as much water as a single action common pump of the same bore and piston stroke (Oertling; Ewbank 1876, 226).

Suction pump from the *Vasa* (1628), operating between the upper gundeck and the hull. The pump is made of alder and still has the bark. Length is 8 meters; diameter 40 cm outside, 33 inside (CC BY 4.0, Vasamuseet, ID 29789).

Chain pump. Chain pumps have a curious history; while known in Roman Europe, the concept was lost, and was reintroduced via contact with the Tartars of eastern Europe in the 15th century. It was then used to drain mines. Whether as a result of technology transfer from the European mining industry, or from the Chinese in the 16th century, Raleigh reports the introduction of the chain pump in the late 16th century to the British navy (Oertling, 75), and it is also described by Manwayring (1644) and Boteler (1634). Dampier, in the late 17th century, had both a chain pump and a common hand pump. But in his day, the chain and wheel "were inaccurately and badly made," and there would be slippages and violent jerks that could rupture the chain (Ewbank 1876, 154–5).

The chain pump features an endless chain bearing circular discs ("burrs") through a vertical tube, open at the top and bottom, the latter being immersed. The chain runs around two sprocketed wheels, a drive wheel on top and a guide wheel on bottom. The chain enters the water, passes around the submerged guide wheel, and moves upward, the discs entrapping water when they enter the bottom of the tube. The water is carried by the discs up to the top of the tube, where it passes into a discharge channel, and the chain passes around the drive wheel and descends back to the bilge. Originally, the drive wheel was a solid wooden wheel with iron sprockets to engage the chain, and a crank attached to the shaft.

The chain pump was able to move more water and was easier to work than the common pump, but it required a large crew and the discs wore out quickly. It was used mostly on large warships (Oertling, 80).

One problem with the old chain pump was that the weight of the water pressing down on the discs would tend to cause the chain to slide back on the sprocket wheel. The links were not well united, and often broke. In addition, there was a lot of friction between the chain and the wheel, which, I imagine, increased the effort necessary to lift a given quantity of water.

In 1764–68 Cole and Bentinck tested a new chain pump design. It wasn't officially adopted until 1774, after further modifications were made (Oertling, 78). Cole, British Patent 911, issued December 16, 1768, just mentions the ease of repair and not how it was achieved.

Apparently, "every other link was formed of two plates of iron, whose ends lapped over those of a single one, and secured by a bolt at each end" (Ewbank, 155). The chain links were cast to the same size, and were therefore interchangeable, as were the link pins that connected the links (Oertling, 93).

The links were designed so that they could be undone and a worn link replaced easily. Ewbank's description of this is a bit difficult to follow, but Oertling (Fig. 25) has described the chain assembly from the HMS *Charon* (sunk 1781)' s chain pump. In essence, the link pin has a slot near one end, and an L-shaped cotter key is inserted into the slot. Thus, to unlink, just pull out the cotter key and then the link pin. In one experiment, the chain was deliberately broken and dropped in the well; it took just two and a half minutes to retrieve it, repair it, and resume pumping (Nicholson 1825, 268).

The burr ("saucer") was positioned every fourth single link and was composed of two plates of cast iron with leather in between. The leather plate was of the same diameter as the bore of the pump tube, and the flanking metal plates were slightly smaller to minimize friction (Cole and Bentinck, British Patent 982, issued January

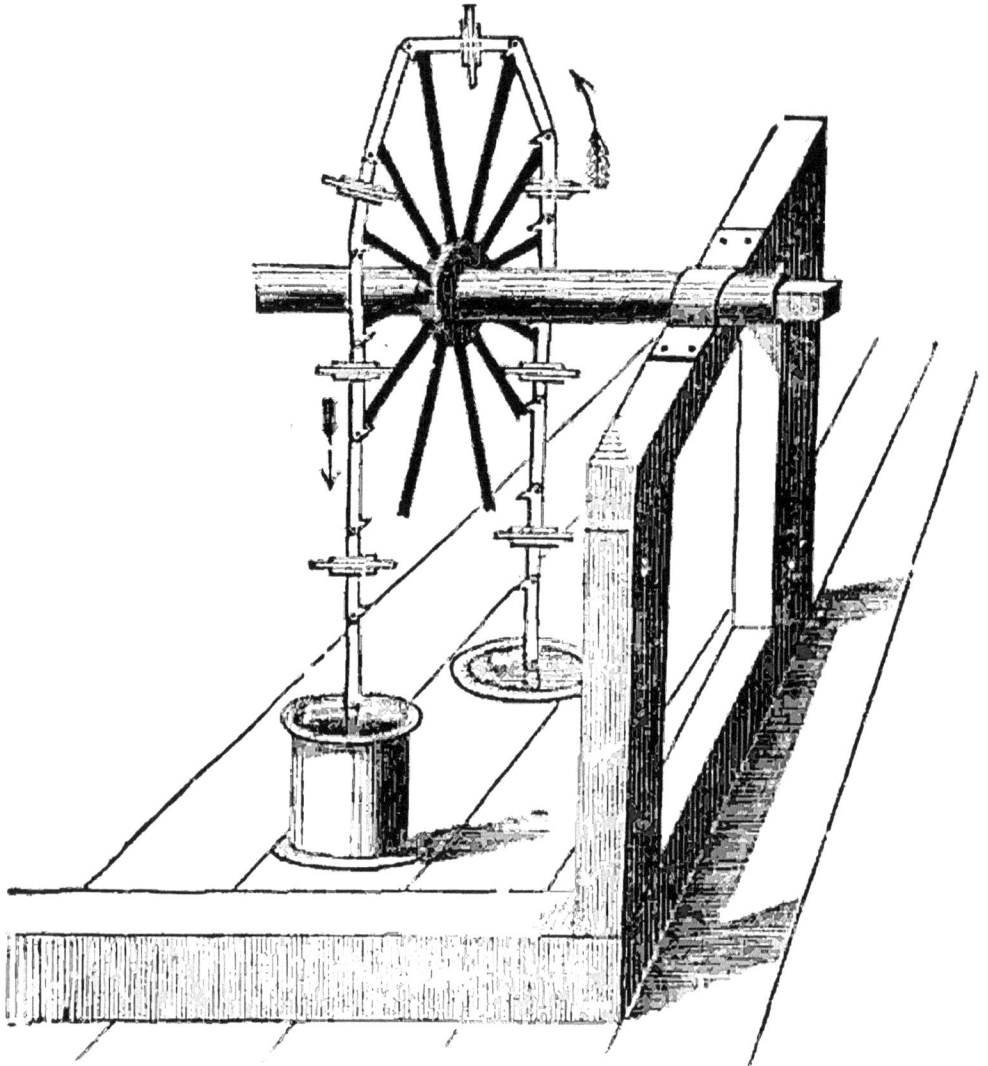

Ewbank's depiction (1842[KK2], Fig. 67) of the chain pump installed on the USS *Independence* in 1837. The chain passes through two copper cylinders, and one, through which it descends (note the arrow), is raised above the deck to prevent the water from returning to the wheel. It is cast iron and two feet in diameter. "When the pumps are used, long cranks are applied to each end of the shaft, so that fifteen or twenty men can be engaged at the same time in working them."

17, 1771). Ewbank says that even the leather doesn't actually have to touch the wall of the tube.

The drive wheel, instead of being a simple sprocket wheel, took the form of two metal (brass?) discs eight inches apart on a common axle, further united by peripheral (iron?) bolts parallel to the wheel axis—essentially a cage gear. (The 982 patent likened it to the "skeleton of a drum.") The links of the chain had hooks that engaged these bolts (teeth) (Nicholson 1825, 268) .

With four men at the crank, the Cole-Bentinck chain pump discharged one ton of water in 43.5 seconds, versus 83 (Oertling, 78; Ewbank, 155 says 55) seconds for the old design.

The cast iron links were replaced with brass ones in the early 19th century, and still later the lower wheel was replaced with a curved metal tube to reduce friction.

In the mid–19th century, British warships carried four chain pumps and three common pumps (Ewbank, 154).

Wood 1977 (70) says that the lower pipe end of a chain pump is "usually flared to facilitate entry of the discs into the pipe," but I haven't seen reference to this feature on ship pumps.

In a 1956 study, four men operating a chain pump with a four-inch pipe were able to achieve 40 cubic feet/hour discharge over a 20-foot lift, 72 cf/h over a 10-foot lift, and 110 cf/h over a five-foot lift (72).

Centrifugal pumps. These are probably the most important modern bilge pump type. These have a wheel with curved vanes ("impeller"), enclosed in a chamber. Water enters at the center of the chamber and spirals out under the influence of the rotating impeller.

Euler discussed its theory in 1754 and some sources say it was invented by Jordan (1680) or Papin (1689). There was a successful centrifugal pump design introduced in 1818 ("Massachusetts pump"), but it had straight vanes, and curved vanes proved much more efficient (Greene 1919, 43ff). The 1851 "Appold" centrifugal pump, with curved vanes, "raised continuously a volume of water equal to 1400 times its own capacity per minute." A further innovation was the "whirlpool" zone suggested by Professor Thomson, a free vortex space surrounding the wheel (EB 1911, "Hydraulics").

They are not self-priming, and thus must be sitting in water in order to pump. In theory the impeller could be rotated manually by a crank, drum or capstan. However, when they were introduced, steam power was already available.

The USS *Monitor* (whose freeboard was only 18 inches [Tucker 2013, 1,312]) had a steam-powered centrifugal pump capable of moving 23,000 gallons per minute, but it wasn't enough to save it from sinking in 1862; when it took on water, its coal became wet. This reduced effective steam power and hence its pumping ability (NHHC "Coal").

Initially, centrifugal pumps were driven by gearing from the main engine, but later these pumps were driven by auxiliary engines. If there was a long vertical shaft from the engine to the impeller below, they could be worked even if part of the hold was flooded. On the *Inflexible*, the pump engine was high enough so the pump could be worked even with 12 feet of water in the engine room (Smith 2013, 208).

In 1961, Charmonman improvised an axial flow pump by encasing the propeller of a Thai-style outboard motor in a cylinder (Wood 1977, 112).

Diaphragm pumps. These are sometimes used nowadays as backup bilge pumps. They have the advantage of being self-priming. Like a piston pump, they vary the volume inside the pump chamber. However, they accomplish this by moving a flexible diaphragm in the side of the chamber, rather than by moving a piston.

Pumps: motive power. Generally speaking, 17th-century shipboard pumps were human-powered, with sailors pulling down a lever, turning a crank, or pulling on a rope wrapped around a drum. It should be noted that "during short time periods (10–15 minutes), the legs can develop about 0.25 hp while the arms can only provide about 0.10 hp. Over a sustained period (say five hours), a grown man is capable of 0.06–008 hp" (Wood 1977, 122).

In the 19th century, Norwegian and Swedish ships were routinely equipped with wind-powered pumps (Leslie 1886, 52). However, they were less common in other merchants marine (even Dutch!). Wind pumps were used on ice barges in upstate New York, and one was rigged up on the *Henry Woolley* in 1871 after it sprang a leak (*West Coast Times* 1871).

An 1876 British writer estimated that the cost of the wind pump for a vessel of 800 tons would be about 40 pounds. He assumed that it would have sails six feet long, fixed to a revolving head mounted on a bipod mast (Wade 1876, 1028). The wind pump frees the crew from pumping duties, but it doesn't work in a calm and also takes up deck space.

The other major 19th-century source of motive power for a pump was the steam engine. Strictly speaking, devices that used steam or heated air to displace water were developed by Heron of Alexandria (first century AD), Giovanni Battista della Porta (1601), Jerónimo de Ayanz y Beaumont (1606), and Salomon de Caus (1615). However, here we speak of the use of steam as motive power (engine) for a drive wheel that drives a piston or chain pump. The steam pump, like the wind pump, was a labor-saving device, and unlike it was not dependent on the wind, but of course it needed fuel to operate, and steam engines were finicky enough so that the bilge would also be equipped with a hand pump.

Note that mechanical linkages can convert reciprocal motion to rotary motion, or vice versa.

Steam-powered pumps were replaced in turn by electrically powered pumps. Solar energy has been used to charge a rechargeable battery, which in turn powers a bilge pimp.

One interesting emergency expedient I found reference to was to use wave action to operate the pump. Captain Leslie of the *George and Susan* reported fixing a spar aloft, with one end over the spear of the pump and the other projecting over the stern. At each end he mounted a pulley and ran a rope over the pulleys from the spear of the pump to a counterweight (a 110-gallon cask holding 60–70 gallons water, i.e., half full) at the stern end of the rope. Supposedly, when a wave lifted the butt-end of the cask, the spear was depressed, and when the wave retired, the spear was raised (Nicholson 1825, 269).

It seems to me that for this to work, there would have to be a downstroke bias on the pump, that is, without an upward pull on the spear, gravity would be stronger than friction and the spear would descend. If so, the cask could be weighted to just balance the spear when the water was at a neutral height. When the wave lifted the weight, the rope would slacken and the forces on the spear would no longer be in balance, it would fall. When the wave dropped, the counterweight would drop, thanks to gravity, and through the rope exert a tensile force on the spear, pulling it back up.

I have also found U.S. patents (ex. Delaney, USP 3120212) dealing with wave-operated pumps—generally speaking, a float is connected to one end of a rocker arm, and the spear of a piston pump to the other—but I don't know whether any of these have been put into practice.

Pump tubes. While on ships, the bore of the pump tube was open on the bottom rather than plugged, the heel of the tube was seated in a hole cut in the mast step (the structure in which a mast is seated) or in the floor timbers. That would block the bore, so channels were cut through the wall of the tube at its heel to let water in. Bilge water being unpleasant from a sensory standpoint, a conduit ("dale") was used to guide it from the top of the pump tube to a scupper at the side of the ship, rather than just spilling it out on deck (Oertling 1984, 41). Or a hose could be connected to the pump outlet; hoses are discussed in chapter 9.

Debris from the bilge could be sucked up the bore and gum up a valve. The debris could be garbage, or cargo or ship stores that became wet and migrated into the bilge. The pumps of the *Sea Venture* (whose 1609 voyage inspired *The Tempest*) were clogged by biscuit fragments, and the HMS *Centaur* (1782) was lost when rising water caused its load of coal to infiltrate the pumps (46). To prevent this, lead, copper or tin sieves were installed at the lower end of the tube (43). Of course, the sieves would need to be cleaned from time to time.

Materials. In the 17th century, the tube was usually made of wood, most often elm but sometimes larch, beech or alder. A tree with a straight, knot- and branch-free trunk was found and cut. The center was then bored out with a hand auger. Alternatively, a tube could be constructed from the hollowed halves of a log, or with strapped and caulked planks (somewhat like making a barrel from staves).

It was of course possible to use metal instead, and lead tubes had been used in some urban water supply systems going back to Roman times. Agricola (1556) suggested that valves be made of iron, copper or brass, and lead ones may have been used on an early 16th-century wreck (48).

Metal tubes in the 17th century were made by taking a sheet whose width matched the desired circumference, rolling it so the edges met, and then soldering. Later, alternative processes were developed. One was to cast a section in a mold, draw the finished tube partway out, and then pour another section to join with the first (Oertling, 56–7).

Metal tubes and pump parts were more durable than wooden ones, so the objection to them was related to cost. Lead use became significant in the early 18th century, copper and bronze in the late 18th century, and iron in the 19th century.

Pump utility. Besides being used to pump out bilge water, pumps could be used to distribute seawater for washing and firefighting (see chapter 9).

Common pumps were of limited value for firefighting, as their pressure was limited to the head pressure (weight of the column of water).

Bilge Alarm

If the ship is taking on water rapidly, the progressive reduction in freeboard will be obvious and "all hands to pumps" might well be the command. But a small leak might go undetected until a sailor had reason to descend to the orlop deck, and found himself knee-deep. For example, in the case of the wreck of the *Protector* in the Bay of Bengal in 1838, while the ship was sailing under bare poles in a gale near the reefs off the mouth of the river Hooghly, a midshipman sent below for grog returned hurriedly to report that the hold was half full of water (*Pilot* 1839).

Once the presence of water was detected, the pump well could be "sounded" to determine how much water was in the hull, and this could be monitored periodically to determine whether the pump was holding its own with the leak. On one of Cook's voyages, at the change of shift, the new man inadvertently took the sounding at a different point, making it seem as though the leak had gained 16 or 18 inches in a short period of time, and inspiring the men at the pump to "redouble their vigor" (Lamb 2000, 84).

Modern ships are equipped with bilge water level detectors that trigger an alarm or even turn on pumps automatically.

The simplest sensor design is probably a float that activates a switch when it climbs to a set point. While in modern systems, this is an electric switch, it is also possible to use mechanical linkages to create a visible or audible signal. Mechanical bilge water alarms are described in Knight's 1884 *American Mechanical Dictionary* (1:281).

Henry Scott Tuke painted a picture entitled *All hands to the pumps!* (It is in the Tate Gallery, London.) We see five men frantically working the seesaw arms of a deck pump, with the discharge tube bringing water up from the bilge to the deck (and hopefully flowing out the scuppers). But was it working fast enough to save the ship?

The next chapter deals with what happens when the ship can't be saved, or a mariner falls overboard.

Chapter 7

Lest You Drown

On August 31, 1773, Samuel Johnson said, "No man will be a sailor who has contrivance enough to get himself into a jail; for being in a ship is being in a jail, with the chance of being drowned." In 1766, Wilkinson claimed that on average, 4,200 British seamen drowned annually. In 1866–1872, the average number of deaths by drowning annually among British merchant mariners was 2,641 (13.3 per 1,000), and among the Royal Navy, 140 (2.9 per 1,000) (Registrar-General 1875, xxi).

It is helpful to distinguish two situations. In one, the ship is intact, but one or more individuals have by accident fallen overboard. In the second, the ship itself is in distress; it may have run aground or even be sinking.

Man overboard. In 1620, John Howland fell overboard from the *Mayflower* during a storm, but "managed to grab hold of the topsail halyards, giving the crew enough time to rescue him with a boat-hook" (Johnson 2020). Most men overboard were less lucky. For example, Dana's *Two Years Before the Mast* reported that a young sailor was going aloft with a "strap and block, a coil of halyards, and a marline-spike about his neck" when he fell from the starboard futtock shrouds. "[N]ot knowing how to swim, and being heavily dressed, with all those things round his neck, he probably sank immediately." Nonetheless, his shipmates rowed about for nearly an hour trying to find him (37).

Abandon ship! If a ship were wrecked, the loss of life was likely to be much greater. From January 1814 to June 1816, 1,702 lives were lost as a result of the shipwreck of British transports, presumably by drowning (*Estimates* 1818). "The introduction of iron ships in the 1850s meant that (a) the ships sank faster and (b) there was little flotsam and jetsam to provide flotation. As a result, marine drowning statistics promptly increased" (Brooks 2001).

Nowadays, even with far more safety measures than in the 19th century, there is a recreational sailing-related fatality rate of 1.19 per million sailing days. "As many as 73% of sailing-related deaths are due to drowning as a result of falls overboard (39–44%) or capsizing the vessel (20–40%).... Leading contributing factors to fatal sailing accidents are high winds (12–27%), alcohol use (10–15%), and operator inexperience (8%)" (Nathanson 2019).

Some of the sections that follow are specific to either man overboard or shipwreck situations, and others are applicable to both.

Laws and Regulations

The International Convention for the Safety of Life at Sea (SOLAS) was adopted in 1914, in response to the loss of the *Titanic*. It was further revised in 1929, 1948, 1960 and 1974. The 1974 version introduced a "tacit acceptance" procedure (article VIII), which obviated the need to hold a new convention in order to adopt additional amendments.

Here, we are primarily concerned with chapter III, "Life-saving appliances and arrangements." Chapter III was amended in 1983, 1995 and 1996.

SOLAS chapter III is supplemented by the International Life-Saving Appliance (LSA) Code, adopted by the International Maritime Organization's Maritime Safety Committee in 1996 and in force since July 1, 1998.

Current U.S. Navy safety regulations are based on those of the Coast Guard, which in turn were based on those of the and the related LSA code (1998). Most of the Coast Guard regulations (Subchapter W, Part 199) ("CG199") were promulgated in May 1996, but some sections have been amended since then.

The Risk of Death from Immersion

Depending on the location and the season, water temperatures can vary widely, from over 90°F in the Persian Gulf and Red Sea, the 80s in most tropical waters, all the way down to just above freezing. Even in temperate latitudes, the water temperature is likely to be well below your body core temperature. Heat flows from where it is hot to where it is cold, and the greater the temperature difference, the faster the flow. Water transfers heat much better than air, and the more of your body surface area is exposed to the water, the greater the rate of heat transfer. The movement of water over your body helps to remove heat, too.

According to modern science (Brooks 2001; Bierens 2006), there are four stages at which death can occur as a result of cold water immersion.

Stage 1: cold shock (3–5 minutes after immersion). This can occur at water temperatures under 77°F. The shock of immersion in cold water can cause panic and loss of breath control. If you swallow water, that initiates the drowning process. You can drown by inhaling just 150 milliliters. The cold shock can also result in muscle spasms, numbness in the limbs, and loss of grip strength. These in turn can affect the ability to inflate a life jacket or swim to a buoy raft or boat. Finally, there are sudden changes in blood pressure, etc., and heart failure is a possibility.

Stage 2: swimming failure (3–30 minutes). The cold makes it progressively more difficult to swim, most likely because of the local cooling of arm tissue. Stroke length is reduced and the movements become uncoordinated. If you are trying to swim to safety, or you must make swimming movements just to keep your head above water, this is a problem.

Stage 3: hypothermia. As the heat gets sucked out of your body, you lose dexterity (and thus the ability to hold onto a floating object) and eventually consciousness. Without a personal flotation device (PFD) designed for an unconscious survivor, that means you drown. And further heat loss can kill you even if you don't drown.

After the Battle of Trafalgar (October 21, 1805), "sailors clung to flotsam and jetsam for up to 15 hours before rescue" (TransportCanada 2003). October water temperature off modern Tangier is 68–73°F, but water temperatures were probably colder then, thanks to the Little Ice Age.

The expected time to unconsciousness is 3–12 hours for water in the 70s, 2–7 in the 60s, 1–2 in the 50s, and 0.5–1 in the 40s (United States Power Squadron 2007). However, before losing consciousness, physical impairment and despair may lead a survivor to turn to face the waves and drown.

Stage 4: post-immersion collapse. This occurs during or after rescue as a result of the damage to the lungs by previously inhaled water, the movement of cold blood from the extremities to your core, or a sudden reduction in arterial blood pressure.

Safety Aloft

Prevention is better than cure. In the Age of Sail, the topmen were of course at the greatest risk. They climbed into the rigging with the aid of ratlines tied between the shrouds to form a ladder. In the 17th century, these were probably hitched at the ends (Anderson 1994, 129). In the mid–19th century, rat-boards (battens) were sometimes used in place of lines (Mondfeld 2005, 288). The ratlines themselves replaced the Jacob's ladder (a rope ladder with wooden steps) by the 16th century (Nance 2012), although the latter reappeared by the 19th century. If the wind was strong, you ascended on the weather side (Harland 1985, 93).

Once aloft, topmen had to move out on the yards in order to raise or shorten sail. Footropes ("horses") were "nonexistent until 1642 at the earliest, and probably not until 1680 on the topsail yards" (Harland 1985, 25; cf. Anderson 1994, 152). So that meant that the topmen of the early 17th century had to stand, sit, or lie on the yard itself. Lifts are lines used to support the weight of the yard, and it is possible they could be used as handholds.

A "lifeline" ("jackline") is a line (manila originally, later wire) strung along a part of the vessel for men to hold on to. Steele (1794) says they are "for the preservation of the seamen upon the yards." The first record of the proverb "one hand for yourself and one for the ship" is from 1799 (Manser 2007, 213).

It could be inconvenient to have to keep one hand on the lifeline, and so

Detail from William van de Velde the Younger (Dutch, 1633–1707), *An English Warship Firing a Salute*, 1673. Note the lack of footropes. (The Lee and Juliet Folger Fund. National Gallery of Art [Washington, D.C.]. Accession 2018.10.1.)

alternatively the sailor could be tied somehow to the yard or the lifeline. In the 19th-century Italian merchant marine, one end of a tether ("lasso") would be tied around the sailor's waist, and the other around the yard (Mercieca 2014). Brady's *The Kedge-Anchor* (1847) suggested wearing a belt to engage the lifeline.

Even in 1931, when sailors were no longer climbing rigging to adjust sails, in the U.S. Navy there were 91 sailors who experienced falls from elevations aboard ship, 17 who fell through an open cargo hatch, and 21 who fell overboard (Navy 1932, 179–180). Ship ladders are steep and ship decks are often wet. An unexpected ship motion may cause a sailor to lose balance. A sailor may take hold of a metal fitting and receive an electric shock as a result of static charges collecting on it, or a charge induced by radio or radar transmissions (NAVEDTRA 1979, 1–58).

OSHA refers to guardrails, safety nets, and personal fall arrest devices as "conventional fall protection systems." The fall arrest device "consists of an anchorage, connectors and a body harness" (OSHA 2015, 6–10).

In 1973, on an aircraft carrier, a sailor missed the first rung on a ladder and fell from the third deck to the sixth, "where a safety net caught him … his injuries were limited to cuts and bruises" (*Lifeline* 1973, 19). "Sailors working aloft must wear a standard Navy-approved safety harness with a safety line attached" (Harrelson 1992, 19-4).

The modern climber's safety harness, developed by a climber in the 1960s, transfers the load from the waist to the thighs (Alpinist 2008). Tall Ships America says that these are "good candidates" for work aloft, but it prefers an arborist harness or an industrial full body harness, as they reduce the risk of back injury or falling out of the harness if inverted (Boulware 2015).

Safety on Deck

On sailing ships, the first defense against being washed overboard was the bulwark, which is the extension of the ship's sides above the level of the deck. Period illustrations also show handrails, particularly on upper decks (Anderson 1994, plate 3).

Stormy weather could isolate the forward crew from the aft galley. To facilitate passage from fore to aft, a lifeline could be run from one end of the ship to the other. In the late 1930s, ships were finally built with "tunnels" (belowdeck passageways from bow to stern) (Boles 2017). If a modern ship lacked a tunnel, the sailors could wear chest harnesses and clip them to the lifeline. (Note that these harnesses are not suitable for work aloft.)

Boarding nets could in theory serve, literally, as an additional safety net, but in the event of a mass evacuation they could prove an impediment to escape, as they did on the *Mary Rose* (1545) (Foster 2019).

Man Overboard!

If you are unlucky enough to fall overboard, hopefully this is observed by someone else, who shouts, "man overboard!" If not, then perhaps the ship has posted a stern lookout who will see you floating in the ship's wake and sound the alarm. Either way, we hope that your comrade throws you a life buoy (see below) while this is still possible.

Assuming that there is no compelling distraction—a storm, an iceberg, enemy action, etc.—the ship should heave to (use the sails to halt forward motion) and if possible first come about. (For the detailed commands, see Totten [1862, 219ff] and Harland [1985, 289ff], and it's a slow process in a sailing ship.) In the meantime, the stern lookout and the men aloft will try to keep you in view. But it is unfortunately true that at best only your head is visible, and the view may be obscured by rain, waves and spray.

A boat may be sent to your rescue. But the two most dangerous moments for a boat's crew are when it is lowered into the water and when it is recovered. "The decision whether or not to risk the lives of a whole boat crew in the hope of saving one man, depended on the weather conditions" (289). And you will probably only be visible to the boat crew when both you and the boat are on the crest of a wave.

Yet another possibility is to have a crew member put on a life jacket, attach a safety line, and stand by to jump into the water to assist you (CG199).

If it is possible for your ship to communicate with other friendly ships or aircraft—either visually or by radio—they may be recruited to join the search for you.

Shipwreck

Shipwreck is an even greater test of the expertise of the officers and the discipline of the crew. When it occurs, the question is whether it is better to stay with the wreck (a wrecked ship is not necessarily sinking; it may be aground or dismasted and thus unable to proceed, but still able to preserve those on board from drowning) or to abandon ship. Sinking can take seconds, minutes, hours or days, depending on the exact circumstances.

The division of the ship into multiple compartments and decks, with limited exterior and interior openings, helps to keep the ship watertight and afloat, but it can be an impediment for sailors and passengers in the interior trying to escape a sinking vessel. The narrowness of the passages can cause bottlenecks, which is what happened on the *Mary Rose* (as evidenced by the positions of the skeletons in the wreck) (Foster 2019).

If the ship is to be abandoned, the best chance of survival is in ship's boats, lifeboats, or life rafts, as these keep the survivors mostly out of the water and thus reduce the risk of hypothermia. These may be guarded until the captain orders "abandon ship!" lest a panicked sailor or passenger decide to help themself to a boat before the captain has given up hope of weathering a storm or other calamity.

If you must jump into the water, your chances of avoiding drowning are best if you are wearing a personal flotation device (PFD). Even on a boat, a PFD may come in handy, as boats can founder, too.

Preparation for Disaster

Under CG199, all items of lifesaving equipment must be of a type approved by appropriate government authorities (199.40), and maintained and inspected periodically (199.45, 199.190). To be accepted as an able seaman in the U.S. Merchant Marine, you must pass an examination and have sufficient sea service (some of which may be on training ships). The examination includes demonstrating competence as a lifeboatman.

On board, there's a muster list that details the emergency signals, the duties of the crew during an emergency, the substitutes for key persons if they are disabled, and how the order to abandon ship will be given. Muster and survival craft embarkation and launching stations are specified (199.80). For passenger vessels, there must be procedures for locating and rescuing people trapped in their staterooms (199.217).

There must be trained people on board to muster and assist the untrained ones, and to operate the launching arrangements and survival craft. Training materials must be on board each vessel. Crewmen with emergency duties assigned by the muster list must be familiar with them before the voyage begins (199.180).

A safety briefing is given to passengers the first day. Emergency instructions must be posted in passenger cabins and at muster stations, and operating instructions in the survival craft (199.80, 199.90, 199.100, 199.110, 199.120, 199.220, 240). All crew members must participate in abandon-ship and fire drills at least monthly, and every three months, the abandon-ship drill must include an actual lifeboat launch and wearing immersion suits (199.180). Use of line-throwing appliances must also be drilled every three months. Man-overboard drills (launching rescue boats and maneuvering them in the water) are conducted every month (185.520). On passenger vessels, there are weekly abandon-ship drills for passengers and crew (199.250).

In the Age of Sail, sailors learned their jobs by aping the more experienced sailors, rather than in the classroom. There probably were no "man overboard" and "abandon-ship" drills. There was no dedicated safety equipment. There was overall a greater tolerance of risk.

Ships' Workboats

Sailing ships of a certain size carried boats. This was necessary because it was rare for ships to dock at a port; rather, they would moor (which means anchor fore and aft) at a convenient spot and transport men and goods between ship and port by means of the boats. The boats, of course, drew less water than their host ship. They might raise a sail, or be rowed.

Ship's boats could also be used to haul the host ship if it lost its masts or ran aground, or to carry the carpenter around to inspect the hull for damage. Those of a warship might be used to carry sailors and marines to an enemy ship, or a beach near an enemy fort or town.

An early 17th-century warship probably would only carry at most three boats: longboat, pinnace, and skiff (Oppenheim 1896, 339). A later source (Falconer's *Universal Dictionary of the Marine*, 1780), says merchant ships "seldom have more than two, viz., a longboat and yawl." (By then there were warships that carried five boats.)

Harland (1985, 282) reports that "until about 1800, all the boats were stowed on the booms amidship, sometimes nested one inside the other." However, that is not entirely true for earlier periods. Morison (1974, 344), writing of the early 16th century, says that the longboat was "usually towed." Oppenheim concluded that in 1625, the Cadiz fleet must have been towing its longboats astern, as they lost every one of them in crossing the Bay of Biscay. But Blomfield (1911b, 238) points to Captain Boteler's *Sea Dialogues* (1634), which asserts that the longboat was normally "hoysed into a shipp."

I suspect that a large ship with a relatively short boat might carry it on board, whereas a small ship might tow a relatively long boat behind. And since the length of ships increased over time, there would be a concomitant trend toward carrying all the boats on board. And that would reduce the risk of losing a boat in a high sea state.

Montaine's *The Seaman's Vade Mecum* (1761) gives lengths of 29–36 feet for longboats, 23–30 for pinnaces, and 15–22 for yawls. Davis (2012, 266) states that the length of the largest boat would be about 2.6 times the square root of the length of its ship.

"Before the introduction of davits in the 1790s [sic, see below], boats were hoisted using tackles attached to the yards.... The introduction of davits undoubtedly made this process faster, but their real value lay ... in recovering men overboard, where seconds represented the difference between life and death" (Willis 2008, 79).

Boats on davits were often stowed on the stern or the quarter (side near stern) of the ship, but that wasn't the only place a boat could be squirreled away. In the 19th century, a boat was laid upside-down on top of the paddle-box of a paddle steamer, as can be seen on a model of the steam frigate HMS *Tiger* (RMG).

Ship's boats can be used either to rescue a person who has fallen overboard or to escape a shipwreck. As an example of the latter, in 1629, when the *Batavia* struck an Australian reef, the ship's longboat and yawl were used to reach the nearby islands.

If a ship-boat wasn't available, it might be possible to use the wreckage to build a raft and reach shore that way, as happened with the *La Salle* in 1686. Even if one can't reach shore, you don't drown on a raft (although you may die for other reasons, as in the case of the *Medusa*, 1816). Even if you reach shore, you might not want to stay there. In 1609, the *Sea Venture* was deliberately driven ashore on Bermuda, and thus wrecked, and the castaways built pinnaces from local wood.

Survival Craft (Lifeboats and Life Rafts)

Unlike the traditional ship's boat, a lifeboat is one designed and earmarked for use in carrying off the ship and crew in the event of a shipwreck. According to the *Oxford English Dictionary*, "The first ship's lifeboat is generally considered to have been that designed by James Mather (1799–1873) of South Shields, Tyne and Wear in 1826, for use on his father's ship the *Mary*." And in 1831, the *Times* printed a letter suggesting that "all vessels employed in steam navigation, for conveyance of passengers, &c., shall henceforth, by act of Parliament, be furnished with two life-boats."

Lifeboats (and rafts) may be classified according to their structure (rigid vs. inflatable) or means of propulsion (oars, sails, engine). Nowadays, lifeboats are typically rigid and motorized, and life rafts inflatable and rowed. Normally, life rafts can be launched faster than lifeboats.

An early objection (1870) to providing enough lifeboats for all on board was that "they would encumber the decks, and rather add to the damage than detract from it" (Watson 2015, 84). Of course, shipowners were probably more worried about the effect on the bottom line. Lifeboats used only in an emergency occupied deck space that could carry cargo or passengers. (Davits did help somewhat, as they had a small "footprint" and they held the lifeboats over the side of the ship where they didn't consume deck space.)

In some cases, the ship was merely aground and not sinking or breaking up, the weather was tolerable, and a safe haven was close by. In that case, the lifeboats merely served to ferry people to the shore or rescue ships, and could make multiple trips, so their limited capacity was just an inconvenience.

But a shortage of lifeboat capacity meant that if the ship was sinking fast, some of those on board would end up in the water, not the boats. Sometimes the crew controlled who got to go into the lifeboats, and other times it was every person for themself. On the *Titanic*, class (and hence berth deck) made a difference—61 percent of first class, 42 percent of second class, and 24 percent of third class (and crew) survived. Gender, too: 20 percent of male passengers and 75 percent of female passengers survived.

On the other hand, on the *William Brown* (1841), 31 passengers were left on the foundering ship. Another 32 made it onto the longboat, but when that began to leak, and it started raining, the nine crew members in the longboat decided to toss out 14 of the 17 male passengers, and two of the female passengers. In *United States v. Holmes* (1842), one of the longboat crew was subsequently tried for homicide. The court declared that the sailors had the obligation to make their safety secondary to the safety of the passengers and hence, provided there were enough sailors left to manage the boat, "the supernumerary sailors have no right, for their safety, to sacrifice the passengers."

Given time and materials, rafts could be improvised. But Rear Admiral Ryder cautioned in 1871 that "no one ever supposed that a raft or rafts for a large ship's company ... could be properly and securely put together, stored and provisioned in less than a few hours," and once constructed it would not be ready to shove off, even if there had been "rehearsals," in "under a quarter of an hour." He furthermore complained that in his day, "very few spare spares and yards are now supplied, and some ... are ... made of iron. Casks also are much diminished in number" (Ryder 1871, 637).

The 1912 *Titanic* disaster prompted the adoption of the 1914 SOLAS. The chapter on lifesaving appliances was applicable to passenger ships on international voyages. It did not apply to cargo ships or warships.

The 1914 SOLAS provides that "at no moment of its voyage may a ship have on board a total number of persons greater than that for whom accommodation is provided in the lifeboats and the pontoon life-rafts on board" (Art. 40, but see regulation XLII). The boats and rafts had to be stowed so they could be launched "in the shortest possible time," and so a large number of people could be embarked on them "even under unfavorable conditions of list and trim." In particular, the davits (or equivalent) had to be capable of lowering boats with their full complement of persons and equipment, despite the ship listing 15 degrees (Art. 49).

Several different types of lifeboats and life rafts ("pontoon boats") were contemplated. Class 1 boats had entirely rigid sides, while class 2 were partially collapsible and thus easier to stow. The boats were required to have internal buoyancy (provided by watertight air-cases, and these were preferably at the sides of the boat or raft). Those of class 1A relied solely on internal buoyancy (with one cubic feet buoyancy for ten cubic feet capacity), while classes 1B and 2A could rely in part on external buoyancy (cork or any other equally efficient material, but not "rushes, cork shavings, loose granulated cork or any other loose granulated substance, or by any means dependent upon inflation of air").

Rafts (classes 1C, 2B, 2C) had to be fitted with one-way valves for quickly clearing the deck of water. They also had to be reversible, fitted with bulwarks to provide at least six inches freeboard when fully loaded, equipped with air-cases or equivalent buoyancy (three cubic feet per person), and capable of being "thrown from the vessel's deck."

The 1914 SOLAS regulations assessed passenger capacity based on either the cubic volume (one person per 9–10 cubic feet) or deck surface area (one person per 3.35–3.5 square feet), whichever yielded the higher (!) passenger capacity. Cubic volume was determined by "exact measurement" (at three sections, then applying Stirling's rule, or any alternative method of the same accuracy, rather than by a coefficient formula such as 0.6 × length × breadth × depth. There was also a penalty to the volume calculation if the boat were too deep).

The registered length of the ship determined the minimum number of sets of davits, of open boats of the first class, and the minimum capacity of lifeboats in cubic feet.

All of the lifeboat types set forth in the 1914 SOLAS were of the open type. The totally enclosed lifeboat was invented by Ane Schat and it was "initially not allowed since rules and regulations prescribed lifeboats to be open, presumably to make rowing possible and to make it easier for swimmers to get on board, but ignoring" the risk of hypothermia (Gunsteren 2013, 13).

The 1974 SOLAS required lifeboats to be partially or totally enclosed. Obviously, this was to provide more protection from the elements. The totally enclosed lifeboats were required to be self-righting.

Typically, late 20th-century U.S. Navy vessels were equipped with inflatable life rafts rather than rigid-hulled lifeboats. (The larger ships did carry auxiliary craft that could serve in a pinch.) The kind used in the Vietnam War era were inflated by carbon dioxide and had a drop-down floor (NAVSHIPS 1967).

Mike Sparks of Life Support International complains that "the inflatable life rafts are in barrels that pop open and the raft unfolds itself into the water. Whether you can get to the raft before burning to death from flaming aviation or ship fuel, drowning or exposure is not guaranteed." He proposes that "sailors should dress for NAVAL COMBAT starting with a fire-resistant Nomex CWU-23 flight suit." Over that he would have them wear reversible, floatable body armor, with one side "blaze orange for rescue and the other Navy blue for combat camouflage." And finally, he would have them wear a "small LRU-18 pilot's inflatable lifeboat instead of a chest life preserver" (Sparks 2005).

There are a few survival craft requirements of the CG199 regulations that are worth singling out:

- Life rafts must be arranged so they can be dropped into the water from the deck on which they are stowed [199.03(b)(7)].
- Each life raft must have a capacity of six persons or more [199.201(a)(5) and 199.261(a)(4)].
- For passenger vessels, with some exceptions, the total survival craft capacity must cover at least 125 percent of the total people on board, and the lifeboat capacity on each side must be at least 37.5 percent [199.201(b)]. There must also be at least one lifeboat (or rescue boat) for every six life rafts (nine for a short international voyage), for marshalling purposes [199.203]. It must be possible to launch enough survival craft to disembark all people on board "within a period of 30 minutes from the time the abandon-ship signal is given" [199.245].
- Lifeboats must be open boats with rigid sides having internal buoyancy (typically from polystyrene or polyurethane) only. They may be made of steel, aluminum, fibrous glass reinforced plastic or other approved material (no mention of wood!). One cannot carry more than 150 people and usually those carrying more than 100 are motor-propelled. For lifeboats of normal proportions, the cubic capacity is determined as $0.64 \times length \times beam \times depth$. The number of cubic feet required per person depends on the length and ranges 10–14 [46 CFR 160.035].
- For cargo vessels, lifeboats must be totally enclosed. For cargo vessels at least 85 meters in length, there must be enough lifeboats on each side to carry off everyone on board. Additionally, either there are enough life rafts for everyone on board and they are stowed in a position providing for easy side-to-side transfer at a single open deck level, or each side has enough lifeboats for everyone on board. Shorter cargo vessels are not required to carry lifeboats, but must carry enough life rafts on each side for everyone on board [199.261].
- It must be possible to launch enough survival craft to disembark all people on board within a period of 30 minutes for passenger vessels and 10 minutes for cargo vessels, from the time the abandon-ship signal is given [199.245, 199.280].

- Also, survival craft must be stowed as near the water surface as is safe and practicable (although at least two meters above), and sufficiently ready for use so that two crew members can complete preparations for embarkation and launching in less than five minutes [199.130(a)].
- Lifeboats must be stowed as far forward of the propeller as practicable and protected from heavy seas [199.130(b)].
- Life rafts are stowed not more than 18 meters above the waterline, unless a different maximum is specified on the life raft container. They are stowed so they can drop into the water from the same deck, that is, either outboard of the rail or bulwark, or there is an opening allowing them to be pushed directly overboard. Manual release from the securing arrangement must be possible, and the launching arrangement must ensure that the released and inflated life rafts are not dragged under the sinking vessel [199.130(c)].
- There must be enough survival craft capacity so that if the largest one on either side is lost or unserviceable, everyone on board can still be disembarked on that side [199.201].

Survival Craft Equipment

Per 1914 SOLAS, both boats and rafts were required to be equipped with oars and related items, one kilogram of food and one quart of fresh water, in watertight receptacles, for each person on board (seems low to me, but this was written at a time when rescuers could be summoned by radio), a vessel containing five liters of vegetable or animal oil (for quieting the surrounding water), a watertight box of matches, a number of self-igniting "red lights," a sea anchor, a towrope, and a lifeline becketed around the outside. Boats were also required to have a compass, two hatchets, an oil lamp, and, if they didn't have a motor, a mast and sail.

The CG199 regulations have additional requirements, depending on whether the voyage is an international voyage or a short international voyage, and the craft is a lifeboat, rigid life raft or rescue boat. The items not mentioned by the 1914 SOLAS include a bailer, bilge pump, boat hook, bucket, can opener, dipper, drinking cup, fire extinguisher, first aid kit, fishing kit, flashlight, hatchet, heaving line, jackknife, knife, boarding ladder, signaling mirror, manually operated pump, radar reflector, rainwater collection device, repair kit (if an inflatable), sea anchor, searchlight, seasickness kit, smoke signal, sponge, thermal protective aids, tool kit, and whistle. One unit (2,390 calories) of food and three liters of water (1.5 for a life raft) are required per person [199.175].

Launching and Loading Boats

Launching appliances transfer the boat from the stowed position to the water. The most common sort of launching appliance uses tackles, falls, and a pair of davits. The davit is a crane that suspends the tackle over the water.

Davits were first used to hold whaleboats on ships engaged in the Greenland whaling trade, possibly from the early 1600s, although they weren't migrated to warships until much later (Harland 1985, 284). On the HMS *Victory*, the davits were simple baulks (wooden beams), hinged at the deck end and held by lines (topping lifts) at the outboard, tackle-holding end. The latter were steadied by fore and aft stays, and a jackstay between the two outboard ends. "Lifelines" were suspended from the jackstay; the boat crew could grab these to transfer their weight temporarily from the boat to the davits (285).

The radial davit was a metal bar in the shape of a hanging post with a rounded corner, with the horizontal limb holding the tackle. The fore davit was pivoted around its vertical axis, by means of turning-out gear, to bring the bow of the lifeboat outboard between the davits, and then the aft davit was pivoted to move out the stern. It typically took six men to work a radial davit, and it was intended for relatively light boats.

The quadrantal davit (Welin 1909) is pivoted near the foot, with the pivot held above the deck by a sturdy, heavy frame. The foot acts as a counterbalance as the davit is inclined to bring the tackle end of the arm outboard. In addition, the pivot bar itself moves outward on a worm drive built into the frame. This davit was designed for use on warships in which the lifeboats were stowed on superstructures some distance inboard.

Gravity-type davits have a gooseneck profile, and are spanned together. The curve part of the arm parallels the curve of the boat side, cradling it. The base of each davit has two rollers riding on a track. When the davits are released, they roll down the tracks, which carry them first obliquely and then directly downward. At the end of the track, the lead rollers act as hinges, and the hook end pivots outward. Gravity davits can be operated by a single crewman. (See generally U.S. Bureau Naval Personnel 1958, 52ff; Army 1958, 227ff.)

The "falls" (lines) running over the tackle were originally manila rope but were replaced with wire. They were stowed on reels, or flaked (loosely coiled). While they can be eased or drawn by men holding the fall, the lowering or hoisting motion is smoother and safer if the falls are winched.

With independent falls, lowering a boat could be a five-man job—two on each fall, and a fifth to watch the descent (Schat 1929, 452). It was also easy for one fall to be paid out faster than the other, inclining the boat, so common drives were developed. Also, with hand-operated falls, there was the chance that the operator could lose his grip, and so some sort of automatic brake is desirable.

When a lifeboat is launched down the side, there is the danger of it running afoul of protuberances on the hull, or of being stove in against the side of a rolling ship. "Schat Skates" (late 1920s) may be used to overcome these problems, even when the ship is listing sharply.

The skate is "a curved iron, with a narrow edge or runner, which follows the line of the lifeboat from gunwale to keel. The inner part of the skate has a wood lining

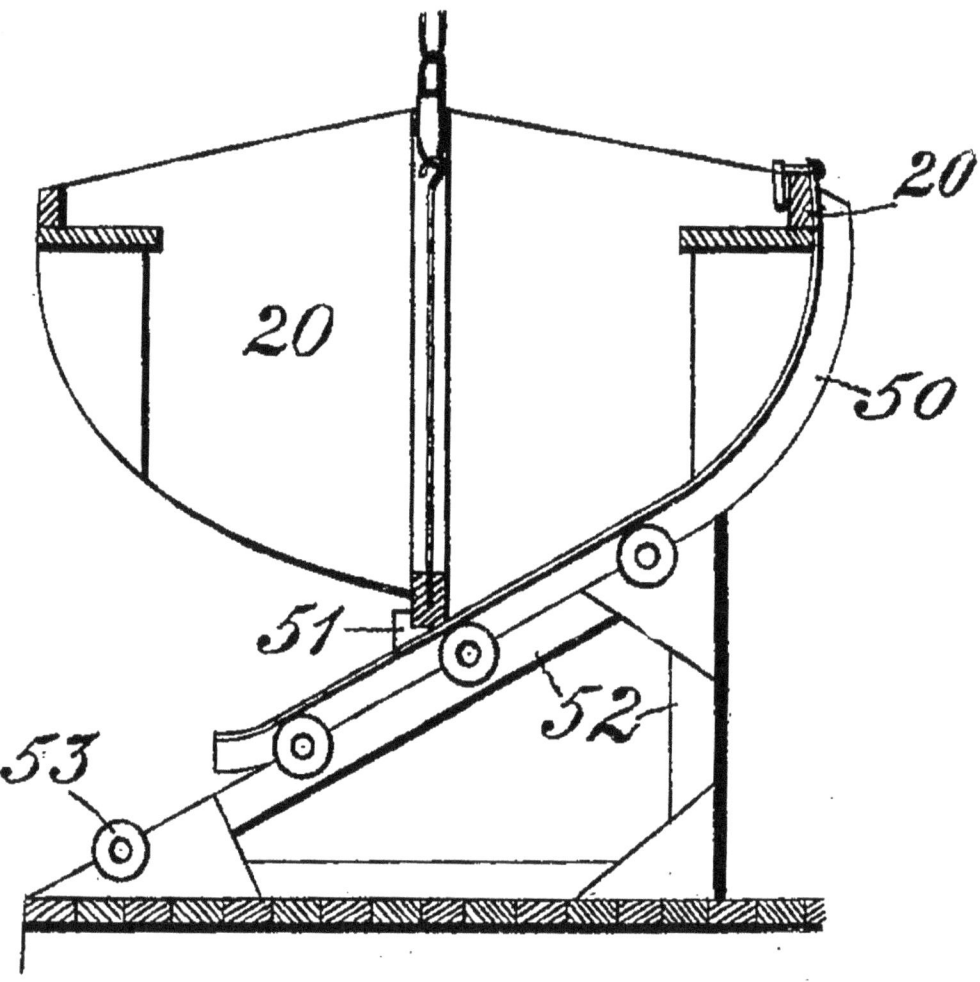

Schat received over 200 patents, and his "skates" took several different forms. This is Fig. 1 from his US Patent 1,747,795, Means for Handling Lifeboats, filed June 26, 1925, but issued Feb. 18, 1930. It is a side elevation; there are two parallel "skids" (skates) 49, 50, "each consisting of a piece of T-iron bent to fit with its flange flat against the inboard side of the boat." When the boat is launched, the skids slide down the inclined ways 52 and over the rollers 53, "but when the boat is over the side, the sharp edges of the skids serve to guide the boat over the hull of the vessel." Schat noted, "only the relative sharp edge … of the T-irons can contact with the hull," and thus "the frictional resistance and the canting moment are reduced to a minimum."

which fits snugly against the strakes … of the boat…. The upper part of the skate clamps over the gunwale and the lower part hooks around the keel" (Richards 1942, "Introduction"). A photograph showing Schat skates in action, on a lifeboat being lowered from a ship at a 45-degree list, is available online (Septer 2021). If skates weren't available, sailors were advised to "use the strongback and oars as skates, to get the boat over rims, ports, gangway fittings, and other projections, and past the promenade deck on the high side" (Richards, "5. Lowering").

Coast Guard regulations require a launching appliance when the embarkation station is on a deck more than 4.5 meters above the waterline [46 CFR 180.150]. Launching appliances had to be "arranged that the fully equipped survival craft or rescue boat it serves can be safely launched against unfavorable conditions of trim of up 10 degrees and list of up to 20 degrees" either with just its operating crew or with its full complement. They must also be capable of recovering the lifeboat with its crew. They must not depend " on any means other than gravity or stored mechanical power which is independent of the ship's power supplies to launch." They are designed to require minimum maintenance, and that on readily accessible elements. There are structural strength requirements with a high safety factor.

There are basically two choices for getting people onto the survival craft, load-launch or launch-load. If the latter, then you have to get the people down from the ship to the boat, probably under rough sea conditions. The simplest means is an embarkation ladder, and modern ones use manila side ropes and hardwood steps. It is essentially a traditional Jacob's ladder. Ideally, there would be spreaders to keep the ladder from twisting and trapping the user against the ship's side.

"Marine evacuation systems" are a lot fancier, e.g., an inflatable slide and floating platform.

Prepositioned Floating Shelters

During World War II, the Luftwaffe dropped *Rettungsboje* (rescue buoys) in fixed locations within the English Channel for use by downed flyers. The improved version (1940) had a signal mast (with a light, flags, and radio antenna), a cabin equipped with storage batteries, medical supplies, clothing, fresh water, food, a stove, signaling equipment, double-deck beds (for four occupants), and a "tubular" lifeboat for transferring the airmen to a rescue vessel (Patoway 2020). The British deployed a similar contrivance, the "Air-Sea Rescue Float," along bombing routes (Gross 2013).

Personal Flotation Devices (PFDs)

There are basically two types of PFDs, those that are worn or carried by the seafarer before immersion (life jackets, vests and belts), and those tossed to one already in the water (life buoys).

The first PFDs were probably the oars used by rowers on ancient vessels. Floating pieces of wreckage, such as spars and hatch gratings, might also be used (Cleborne 1876, 1,033).

While it is possible to float in water without a PFD, the PFD provides extra buoyancy to help keep your head above water. Modern Type I (offshore) adult life jackets are required to provide at least 22 pounds buoyancy, and type II (near shore), 15.5.

PFDs may be inherently buoyant (i.e., containing cork, balsa wood, kapok, or plastic foam), inflatable (with air or other gas), or a hybrid of the two.

PFD Flotation Materials

Inflated skins. A gypsum wall panel relief found at Nimrud, and dated to about 860 BC, shows the Assyrian army of Ashurnasirapal II crossing a river, some using inflated skins (British Museum). Other ancient soldiers resorted to similar expedients for water travel, and the inflated skins could be the goatskins used normally to carry water, or oxhides used as tents. Sometimes several skins were used together to provide additional buoyancy to a wooden raft, as can be seen on Trajan's column (105 AD), depicting Romans crossing the Danube (Vaucher 2018).

Cork. Air-dried cork has a density of about 15 percent that of water (Pereira 2011, 187), but there is considerable variability. In the 1940s, it was required that block cork used in life preservers weigh not more than 12 pounds per cubic foot (46 CFR 37.6–3[a]), whereas the density of pure water at 60°F is about 62.4 pounds per cubic foot (and seawater is slightly denser). It floats; indeed, its density is lower than the woods used in ship construction because of its numerous air pockets. While it is hydrophobic, it is capable of absorbing water, especially when warm. If it absorbs too much water it will no longer float. Hence it might be waxed to retard the water intake.

In 1538, Wynmann wrote a Latin treatise, *Colymbetes*, on how to teach swimming. The reason for mentioning it here is that Wynmann said that if a student wanted to practice without a teacher, they could go into water no deeper than mid-chest and use a flotation device, which could be "crafted from reeds, cork or two inflated leather bladders" (McManamon 2016; Tabaczek-Bejster 2019, 263). However, Wynmann said nothing about the possible utility of such devices for saving seamen from drowning.

In 1758, John Wilkinson proposed to the Royal Society that seamen be furnished with cork jackets, which he had first made in 1757. He first took this proposal to the general public in *The Seaman's Preservation* (1759) (citations here are to his fourth edition, 1766). He considered cork jackets to be superior to the simple cork floats used by those learning to swim, since there had been fatal accidents when the water had wrested these away from the learner (Wilkinson 1766, 20).

Wilkinson conducted experiments to determine the buoyancy of cork in fresh and salt water, and "the specific quantity of cork necessary to sustain a man in the water." He found that it required a little more than 12 ounces of cork to keep above water a man that was five feet two inches and 104 pounds (Wilkinson 1765). The same year, he also received an English patent on cork jackets or waistcoats. His concept was that cork would be cut into pieces and sewn in. He recommended that the pieces be somewhat rounded for greater comfort, and noted that the "greater their number is the smaller they will of course severally be, and of consequence the more complyable to the motions of the body of the wearer." Several versions of the jacket are depicted in his book. One proposed feature was tapes on the hip of the jacket that could be tied around the thighs so that the jacket would not ride up against the armpit because the jacket was more buoyant than its user.

Wilkinson mentions two objections made to providing cork jackets. First, that seamen would be encouraged to abandon ship in time of danger sooner than they should. Second, that the cork jacket would make it easier for impressed men to desert, at least when the ship was near land.

A sailor reported to Wilkinson that in 1760, the frigate he was on deliberately ran aground on a sandy beach to avoid a worse calamity, and he "leapt off the weather fore-chains with my cork-jacket on." The sea was so rough that he was rendered insensible, but the cork jacket kept him buoyed up, and the sea and wind drove him onto the shore. He noted that 18 of the crew drowned, including the best swimmers, these being unable to cope with the mountainous seas experienced. The jacket in question weighed three pounds, of which the cork contributed one pound. Wilkinson commented that a jacket for a middle-sized man, made of old canvas, would cost no more than five shillings. In 1763, the Royal Society recommended to the Admiralty the adoption of cork jackets.

In 1817 there was a comparison between the "common cork jacket" and William Mallison's "Seaman's Friend." The Admiralty concluded that "any trifling superiority of buoyancy of the Seaman's Friend clearly resulted from the greater weight of cork being contained in it," and declined to consider it further (Estimates 1818).

However, it does appear that Mallison's device featured a crotch strap that was not part of the standard jacket (Brooks 2001). Its importance is illustrated by the case of the yacht *Ouzo* (2006). Not wearing the straps, the "jackets slipped up the torso, thereby altering their floating angle to near vertical." In two cases, the heads then slipped through the collar, and they ended up face down. "The third remained vertical, since he had fitted his lifejacket more tightly, but when he lost consciousness through hypothermia, his neck muscles could no longer support the weight of his head" (Gelder 2013, 8).

A simpler PFD was touted by Bosquet in 1800. This "Marine Girdle" was a long canvas bag stuffed with cork shavings. It could be wrapped about the sailor's upper body and the ends fastened together with straps and buckles. Bosquet thought every person going to sea should have one. But it seems inferior to the cork jacket, as it is more cumbersome to secure.

In 1854 Royal National Lifeboat Institution Inspector Captain John Ward proposed a vest-style jacket that "consisted of small blocks of cork sewn onto canvas, which allowed sufficient freedom for rowing or swimming" (Pickthall 2016, 66). It had a buoyancy of 25 pounds (RNLI).

Cork had its problems. Its bulk and rigidity made it uncomfortable to wear and thus discouraged it being worn as a matter of course. Also, when you jumped into the water, it could recoil against you. In the *Titanic* sinking (1912), it "broke the jaws of some people and rendered others unconscious" (Pickthall). The U.S. Army (Army 1950, 10) later warned against wearing a cork jacket when jumping from a vessel, as it could injure the neck or head.

Balsa wood was also used, particularly after World War I. It is even less dense than cork (approx. 16 percent water) and was native to the New World (unlike cork, which was from an Old World tree). That had advantages to America in times of war. Otherwise, it shared cork's disadvantages.

Bamboo. A stem of bamboo has a low effective density (50 percent that of water [Kumar 2015]) because it is composed of sealed chambers filled with air. The Chinese were aware of the buoyancy of whole bamboo at an early time. Indeed, Needham believes that the "earliest vessels of east Asia were rafts of bamboo" (1971, 695). In 1656 Nieuhoff "sketched a 'floating village,' i.e., a great bamboo raft, on the Yellow River" (390d).

In 1730, a ship on the Batavia-China run encountered a typhoon, and it was feared that the ship would founder. A British passenger observed that the Chinese merchants came on deck in a "bamboo habit." This consisted of eight bamboo stems,

Captain of Tenby Lifeboat (1853–6), albumen silver print by Mr. Gonne (sepia tone removed by author). The lifejacket is most likely stuffed with cork. The Tenby Lifeboat Station was established in 1852 in Tenby, Pembrokeshire, Wales. In 1852 the station commander was Robert Parrott and in 1856 it was Richard Jesse, but I don't know when the changeover occurred (print from Gilman Collection, gift of the Howard Gilman Foundation, 2005. The Met [New York]. Accession 2005.100.382 [44]).

four horizontal (two in front and two in back), and four vertical (two on each side). They put this over their heads and tied it to their body (Wilkinson 1766, vi). Cleborne (1876, 1,035) provides a depiction. An even simpler bamboo life preserver used on 19th-century Chinese houseboats was made of just four stems, crossed to form a "#" (Forbes 1902).

Kapok. The seed hair fibers obtained from the fruit of the kapok tree (native to the neotropics and west Africa) weigh only one-eighth as much as cotton. Also, the fiber contains a water-repellent oil (*Popular Mechanics* 1915). They were first approved by the U.S. Coast Guard as a stuffing for life jackets in 1902 (Stein 2017, 652). I believe that the kapok was originally stuffed inside canvas, but later nylon was used (Army 1950).

It was softer and more flexible than cork or balsa. Unfortunately, kapok had some problems. First, "there is a great difference in buoyancy in the kapok grown in different parts of the world, though it all looks pretty much alike" (*Popular Mechanics*). Second, it was more flammable than cork. Third, if compressed for long periods (during storage, or when sat upon), or if it came into contact with fuel oils, it lost buoyancy (Brooks 2008, 9B-2).

Coast Guard regulations specify that kapok should be of a quality "equal to that grown in Java," processed without undue pulverizing, and provide a fresh water buoyancy of at least 48 pounds per cubic foot [49 CFR 164.003].

Gas-inflated. Early air-inflated life preservers were unreliable and "became in a short time life preservers in name only," and in 1855, American inspectors were instructed not to accept new ones of this kind (USMM 1944, 91).

The big advantage of an inflatable PFD is that when uninflated, it is very comfortable. The modern inflatable life jacket was invented by Peter Markus (1928) and nicknamed the "Mae West." It was "made of rubberized fabric and inflated by carbon dioxide cartridges or by mouth." The gas filled tubes in the front of the jacket and around the back of the neck (to keep the head upright) (Army 1950).

The use of confined air for buoyancy of course was an old idea. The basic problem was making sure that the device did not collapse or leak under water pressure. In 1783, MacPherson reported that he had constructed an "air jacket" (presumably using leather or oilcloth since rubber wasn't yet available) and found that "when first made, and in perfect order, [it] was well adapted for enabling a man to float and swim; but it was a bulky affair, not easily made, still less easily kept in order, [and] expensive" (MacPherson 1783, 10).

Another issue was whether the victim would be able to inflate it if it wasn't self-inflating. Nowadays inflatables are used mostly by aircrews and not on marine vessels.

Fibrous glass. During World War II, it became difficult to obtain kapok. A research program led to the development of silicone-coated fibrous glass as a substitute (Webster 1954; King 1954). A fiber diameter under 0.00015 inches is preferred for life jackets, and this may be harder to achieve. (The smaller the fiber diameter, the smaller the capillary passages and the greater the pressure required to force water into them [Blefeld 1953].) The silicone renders the surface non-wettable.

Closed cell foam. Closed cell foam was first used in PFDs in the 1970s (Lim 1994, 289). The foam is usually polyethylene, polyurethane, or polyvinylchloride. Thanks to the trapped air pockets, densities run 2–3 pounds per cubic foot, versus 62 for fresh water.

The Hammock as a PFD

In 1870, Royal Navy Rear Admiral Ryder proposed the use of the naval hammock as a life preserver, in the situation of shipwreck with help near at hand, so it was "only necessary to float a ship's company for a very limited period (say for thirty minutes or an hour)." In response, Commander Bridge of HMS *Caledonia* conducted an experiment, and reported that a new, "well lashed up hammock, containing only a bed and a blanket, supported ... for a considerable time four men" (640). Captain Wilmshurst (HMS *Valiant*) observed that the weight of the hammock, bed and blanket, when dry, was 24.5 pounds, and it displaced 2.16 cubic feet (138.24 pounds) of water. Thus, its initial buoyancy was 113.74 pounds. (This would decline as the hammock absorbed or captured water.) Also, "the ordinary hammock floating horizontally will support six pounds of iron for nine minutes." However, if the bed-cover were oiled, the hammock floated the same weight for 2.5 hours (Ryder 1871, 641).

It is worth interjecting here that the hammock did not need to support the full weight of a man, but only compensate for the excess of the man's weight over that of the volume of water he displaces, i.e., for his "negative buoyancy." The lean density of the human body is 1.096 +/- 0.007 grams per cubic centimeter (Heymsfield 1989) and thus very close to that of surface seawater (about 1.03) (Open University 1997, 5). The buoyancy depends on lung volume and fullness and it was found that all of 98 male subjects could float in seawater with their lungs full of air, and 69 percent would float with the lungs at "functional residual capacity" (Donoghue 1977), although increasing buoyancy would bring more of the body out of the water.

The normal bed used in the hammock was stuffed with horsehair. Ryder proposed that it instead by stuffed with cork shavings (for maximum buoyancy) or coconut fiber (less buoyant than cork, but more comfortable) (641).

The editors, in a preface (635–36) to Ryder's article, commented that recent experiments had shown that a cork mattress weighing 7.5 pounds could provide more than 37 pounds of buoyancy, and asserted that this would "support an average sized man with his head and entire shoulders above the water ... after twenty-four hours immersion, its buoyancy is not greatly diminished." They noted that cork mattresses had "already been adopted to a considerable extent in the Russian Imperial Navy." They also suggested replacing the coarse hempen material of the hammock itself with a closely woven cotton.

Hammock mattresses, they added, "being stowed on the upper deck would always be immediately at hand." They advised bending the hammock, "as lashed up and stowed in the netting, and bringing the two ends together, thus forming a species of life-buoy ... which the person using it would pass over his head and under his arms," and then lash the ends together.

Forbes (1871, 796) proposed that hammocks be supplied with a mattress of vulcanized rubber and this mattress be equipped with valves so it could be inflated. He subsequently decided that it would be "too expensive" except for admirals and commodores, and it was questionable whether they ought to be saved. "Instead, he proposed furnishing the men with an inflated 'air pillow,' to be kept in the hammock."

PFD Legislation and Regulation

In the Steamboat Act (1852), the United States Congress required every passenger steamer to be provided with "a good life-preserver, made of suitable material, or float well adapted to the purpose, for each and every passenger, which ... shall always be kept in convenient and accessible places ... in readiness for the use of such passenger." The first British life jacket legislation was in 1888.

Article 51 of the 1914 SOLAS required that ships carry life jackets, or equivalent appliances capable of being fitted on the body, for every person on board. The regulations (Art. XLV) required that the life jacket be capable of supporting at least 31 pounds of iron in fresh water for 24 hours. It also required that they be placed where they were "readily accessible" (there was much dispute as to whether they should be in the living quarters or up on the main deck) and "known to the persons concerned."

In the 20th century, there were numerous instances in which rescuers found sailors who were wearing life jackets but nonetheless drowned, floating face down in the water. It was not until 1940 that it was demonstrated that an unconscious human wearing a standard life jacket would sink, and the life jackets were redesigned to provide self-righting and head support (Brooks 2008; Macintosh 1957).

SOLAS 1974 required that passenger ships must carry a 5 percent surplus of jackets. The life jacket had to be of a highly visible color and equipped with a whistle. Finally, it had to be constructed to eliminate as far as possible all risk of being put on incorrectly, and capable of lifting the face of an exhausted or unconscious person out of the water and of turning the body in the water from any position to a safe floating position (inclined backward).

Why inclined backward? Remember the assumption is that the person is unconscious. In supine (face up) horizontal, there is risk of death from choking on the tongue. In prone (face down) horizontal, the life jacket must be very bulky to have enough buoyancy to keep the mouth and nostrils clear of the water. In vertical, there is less resistance to vertical oscillations (TransportCanada 2003).

To achieve the desired position as quickly as possible, the life jacket was designed so its center of buoyancy was in front of the chest (to turn from prone) and as far from the center of gravity of the wearer as possible (to maximize the turning moment). Buoyancy is also needed to support the head. The jacket also should shield the face from spray.

By 1996, the Coast Guard was phasing out cork and balsa wood life preservers. Buoyancy could still be provided by kapok, "fibrous glass," or "unicellular plastic foam" pad inserts. (Note that this does not include inflatables.) The inserts are inside a pad formed by heat-sealing two pieces of vinyl film together. The kapok and fibrous glass preservers are a vest type, with each side being a "cotton drill" envelope having two front pockets and one back pocket to receive the pads [49 CFR 180.71, 160.002, .005, .055]. They must provide at least 22 pounds' buoyancy in fresh water for 48 hours [49 CFR 160.001]. The required amount of buoyant material was 24 oz (kapok) or 46 oz (fibrous glass). The foam preservers are a single piece of vinyl-coated foam with a neck hole and body slit, or three sections in a cloth envelope.

CG199 requires that life jackets bear retro-reflective material and lights [199.70]. For passenger vessels, an extra 5 percent life jackets are required [199.212].

Clothing, Swimsuits and Survival Suits

Ordinary clothing has the advantage of potentially trapping air and thereby reducing heat loss. However, it does hinder movement in the water. Simple swimsuits reduce drag when swimming, but probably provide less thermal insulation than ordinary clothing.

Nowadays, a passenger told to abandon ship might be handed a "thermal protective aid." Typically, it is a suit made of aluminized polyethylene (Mylar), which is waterproof and reduces heat loss.

Wet suits and dry suits provide protection from hypothermia, but can be difficult to put on quickly. The original wet suit (1952) was made of closed cell foamed neoprene, typically containing bubbles of nitrogen gas. Later versions sandwiched this between layers of tougher material, such as nylon and spandex (Lycra). Both thermal protection and some buoyancy are provided by the gas pockets within the foamed neoprene (Westwick 2013, 94–97).

In a wet suit, water does seep in at the neck, wrists and ankles, and possibly also at the seams and zippers, but is trapped between the suit and the body and warmed. Ideally, flushing (replacement of the warm water with cold) as a result of movement is minimized.

A dry suit features additional sealing features in order to exclude water. The first dry suit was made in the late 1800s by sandwiching a layer of vulcanized rubber between layers of twill canvas, and was used as a diving suit (Pressly 2020). Dry suits are waterproof but not necessarily insulated.

An immersion suit is a dry suit intended for use in involuntary immersion situations. Merriman (1872) developed a suit composed of "pair of rubber pants and shirt cinched tight at the waist with a steel band and strap. Within the suit were five air pockets the wearer could inflate by mouth through hoses." It was used by the "Fearless Frogman" (Paul Boyton) when he crossed the English Channel in 1875 (Quinn 2021, 74). In 1930, it was possible to buy a "safety suit"—a pair of "coveralls" modified to include "locking levers," "life-preserving pads" (flotation), and weighted soles (*Popular Mechanics* 1930).

An anti-exposure suit is "a protective suit designed for use by rescue boat crews and marine evacuation system parties" [46 CFR 125.160]; it provides extended thermal protection as well as inherent buoyancy.

Wearing a survival suit (with clothing underneath) can double or triple survival time in cold water (GBCAA 2005, 3). CG199 required cargo vessels operating outside tropical waters to carry immersion suits and to conduct donning drills [199.03]. Immersion suits and thermal protective aids were to be carried by passenger vessels beginning in 2003 [199.10]. The rescue boat crew and marine evacuation system member crew for vessels operating outside tropical waters must carry immersion suits or anti-exposure suits [199.70].

Under CG199, each passenger vessel must carry at least three immersion suits for each lifeboat on the vessel, and thermal protective aids for each person not provided with an immersion suit [199.214]. Cargo vessels must carry an immersion suit for each person on board, and extra ones for remote watch stations [199.273].

Wearing a dry suit (including a survival suit) for long periods out of the water may result in hyperthermia (overheating) if air temperatures are high.

PFDs: Life Buoys

Back in 1538, Wynmann taught that a good swimmer could equip himself with a flotation device, such as a plank of wood, and swim it out to a person in distress (McManamon 2016).

Of course, it could take time to swim out to the victim, and sea conditions might render swimming too dangerous, so better if the PFD can be tossed to (but not *at*) the person. In early times, these were improvised PFDs. Lucian mentions throwing cork markers for anchor cables to men who had fallen overboard (Pitassi 2012), and in the 1850s, a sailor fell off the Russian brig-rigged steamer *Possocchob* and was tossed a hatch grating (Mercieca 2014).

Eventually it was recognized that there was an advantage to having a PFD designed for this usage. These life preservers (ring buoys, Kisbee rings) are usually of a donut or horseshoe shape, and have a connecting line so the swimmer can be pulled to safety.

An early example was Abraham Bosquet's "ring" (1800). It had a "wedding band" shape, and was to be hung up by the mast. It contained cork shavings and a long, light hauling line was attached to it. An inflated bladder or ball of cork was to be attached to the line (Bosquet 1818, Fig. 2, 56–7). The donut-shaped version is usually attributed to Lieutenant Thomas Kisbee (1792–1877) (Compton 2014, 144).

"Rather than trying to climb up on the buoy, the man pushed his head up through the center, or capsized it so it came back over his head, and he then supported himself on his elbows" (Harland 1985, 289).

A variety of improvements were made on the basic concept during the 19th century. For example, making it easier to grab hold (by putting a rope around the circumference of the buoy, with possibly hanging grab lines attached to it), or bringing the user up out of the water (by providing a frame to sit on). "In the German Navy, cans filled with oil-soaked oakum were attached to the buoy, to help smooth the surrounding sea."

Cook's life buoy (1818) "consisted of two casks with a pillar, between which was a floating light or port-fire" (Cleland 1879, 1040).

The "Franklin lifebuoy" featured two stanchions, several feet tall, holding water-ignited calcium phosphide flares. These would burn for 20 minutes or so, a great advantage for a night rescue, but it should be noted that the reaction with water produces phosphine, a poisonous gas.

SOLAS 1974 required that a life buoy be inherently buoyant rather than buoyant as a result of inflation, and provide at least 32 pounds' buoyancy for 24 hours. They must be capable of being rapidly cast loose and of a highly visible color. At least half had to have self-igniting lights (these are batteries activated by contact with saltwater) and at least two of these had to have self-activating smoke signals. The required number of life buoys was based on the length of the ship. (The idea being, I believe, to minimize the distance a sailor on deck would have to run in order to reach a life buoy.)

CG199 required that "each lifebuoy must be capable of being rapidly cast loose" and "readily available on each side of the vessel," with at least one "near the stern." Lifelines attached to buoys were required to be buoyant, at least 8mm diameter and 30 meters long, non-kinking, and strong [199.70]. Even the shortest vessels must carry at least eight life buoys [199.211, 199.271].

Distress Signals and Radio Beacons

Distress signals. The point of a distress signal is to attract helpful attention, and in a pinch, anything out of the ordinary will do.

An early practice—I am not sure how early—was to hoist a flag upside-down, assuming that it looked sufficiently different in an inverted position. Another was to tie a knot in the ensign, making what was called a "waft" ("weft"/"wheft"/"whift"). Mainwaring (1644) refers to it as "a common signe of some extremetie."

If ships are traveling in a group, they hopefully have a prearranged accident signal. Admiral Howe (1776) directed his ships to fly "white over red flag in the mizzen topmast shrouds" when "in want of immediate assistance" (Prothero 2010).

International distress signals in the 20th century included firing a gun at one-minute intervals, continuous sounding of a foghorn, raising a square flag having above or below it a ball or anything resembling a ball, flames on the vessel (e.g., a burning oil barrel), slowly and repeatedly raising and lowering arms outstretched to each side, orange-colored smoke signals, red light flares, and slowly and repeatedly raising and lowering arms outstretched to each side (NIMA 2003, 136).

Radio beacons. During the 1920s, after the sinking of the *Titanic*, some lifeboats were equipped with spark-gap transmitters. The antenna wires could be held up by kites or balloons (I. Walker; Museum of Technology 2007). In World War II, German aircraft were equipped with vacuum tube-based emergency radio transmitters which were powered by a hand-crank and automatically generated an SOS. This technology was copied by the Allies (Meulstee).

Rescue Boats

Edwin Louis Cole wrote, "You don't drown by falling in the water; you drown by staying there."

A rescue boat is one "designed to rescue persons in distress and to marshal survival craft" [46 CFR 199.30]. These could be ship's boats used to rescue a "man overboard" from the same or a different ship, or shore-based boats that put out to sea to rescue castaways.

The 1911 *Encyclopaedia Brittanica*'s "Life-Boats" credits the first design to Lionel Lukin (1785); his "insubmergible boat" was a Norway yawl fitted with cork gunwales and bow and stern air-filled compartments ("end-boxes"). It saw only limited use. After the *Adventure* was wrecked on the River Tyne and all lives lost despite its being just 300 yards offshore, a design competition was held. The winning Wouldhave-Greathead design, 30 feet long and 10 feet wide, had almost 800 pounds (7 cwt) cork padding. It was double-ended and had a curved keel. It could carry 20 people, including a crew of 12. By 1803, Greathead had built over 30 rescue boats.

By the late 19th century, improvements had been made in self-righting and self-emptying capability, including relieving valves for automatic discharge of water off deck, side "air-cases," centerboards, drop-keels, water ballast tanks, and an iron keel ballast.

A "transporting carriage" was used to transport large rescue boats to the water's

Life-Boat on Its Transporting Carriage, from Richard Lewis, *History of the Life-Boat, and Its Work*, opposite 96 (Macmillan 1874). Lewis says, "On the withdrawal of a forelock pin, the fore and main bodies can be detached from each other."

edge. From 1888, they were equipped with "Tipping's sand-plates," composed of an "endless plateway or jointed wheeled tire" for easier movement over soft sand.

According to CG199, both passenger and cargo vessels must carry at least one rescue boat (one on each side if the passenger vessel is over 500 tons). A lifeboat is accepted as a rescue boat if it meets the requirements for such [199.202, 199.262]. Those requirements are that the boat be square-sterned, 11–14 feet long, readily maneuverable, of open construction, suitable for use by three persons, and possessing internal buoyancy imparted by unicellular plastic foam (the amount per a formula). There are also drop survival and freeboard requirements [49 CFR 160.056].

Line-Throwing Devices

These were mostly used by rescuers on shore to throw a line to a stranded or wrecked ship. Both rockets and mortars were employed.

The use of a mortar to fire a rope-bearing shell and thereby deliver a rope from the shore to a vessel in distress was first proposed by John Bell in 1791, and he carried out an experiment "in which a deep-sea line was carried to a distance of 400 yards" (Lyle 1878, 111). Captain George Manby independently developed a similar plan and this was used to rescue the crew of the brig *Nancy* in 1809 (112).

The immediate difficulty Manby had to overcome was that the rope would part as a result of stress, or burning by the combustion gases, and he solved this problem by use of "stout strips of hide, plaited extremely close at the eye" (119). Also, to prevent fouling, the rope had to be "laid in alternate tiers…, no part of it overlaying" (114). While Manby recommended a pear-shaped shot, it appears that a spherical one was more commonly used (SSVLB 2015).

Manby's mortar was of a conventional design, but it was fired at a low angle of elevation, contrary to normal mortar use. The Lyle gun was an evolution from Manby's. David Lyle reduced the caliber and weight of the projectiles and used an improved line. He also gave the projectile a long shank to protect the line. And he used a bronze cannon because of bronze's ductility, corrosion resistance, and high strength-to-weight ratio (Lyle 1878, 23). (For pictures of the cannon and its projectile, see USCG/Farley.)

Lyle tested both rifled and smoothbore shot, and concluded that the latter were better (22). These were elongated cast iron projectiles with a wrought iron bolt "screwed into the base of the projectile to serve as a point of attachment for the shot-line" (58). There were several guns, with bores ranging from two to three inches. These were short-barreled (for the gun with a 2.5 inch bore, the bore length was 20 inches). The service charge for the 2.5 inch gun was four to six ounces, and the charge used depended on the weight of the line (34). The shotline was preferably a waterproofed braided linen, not hemp (26).

The extreme range achieved was 695 yards, but Lyle pointed out that accuracy was more important than range. The lighter the line, the greater the range and the less the accuracy (because of wind effects). He considered a range of 400 yards to be sufficient (28).

Robert Parrott proposed, and patented, inserting the projectile so the eyebolt stuck out. When fired, "the check which is produced as the shot feels the draw of the line causes the shot to be turned over end for end." This eased the strain on the line while still achieving long range (130; quoting Parrott USP 75,742, 1876).

Line-carrying rockets were proposed as early as 1807, and came into use in 1832. Rockets accelerated more gently than mortar-fired projectiles and therefore didn't break the line. However, their range was initially inferior to that of the mortar.

In the 1860s, Colonel Edward Boxer's two-stage rocket was introduced. "Its two-stage design gave it extra range and gentler acceleration, which reduced the chances of the rope breaking.... The Boxer rocket was still in use at the beginning of World War II" (Van Riper 2007).

Typically, a "beach cart" was used to transport the line-thrower and other equipment to the water's edge, often across soft sand.

Regardless of the type of device used to deploy the shot line, the subsequent procedure was much the same. The shot was fired so it overshot the wreck, and then hauled back until it was caught on the wreck or by a sailor on the wreck. The shot line was used to haul out a tail block (pulley) to the wreck, where it was secured. An endless "whip line" was run around the tail block and, on the beach cart, the whip block. The whip line could then be used to transport an even heavier line, the hawser, and its traveling block.

Once the hawser was in place, either a breeches buoy (invented by Kisbee) or a life-car could be hauled over. "The breeches buoy is a cork life buoy to which is

Above: Illustration of line-throwing rocket from page 173 of George J. Hagar, "The United States Life-Saving Service: Its Origin, Progress, and Present Condition," *Frank Leslie's Popular Monthly* 5: 165–178 (1878). Note that the line is being pulled out of the "faking box," where it was laid around "pegs or partitions [so] that it runs out without hindrance." Also note the box with two spare rockets. Hagar notes that the rockets are particularly useful at night, as they light up the wreck.

Chapter 7. Lest You Drown   169

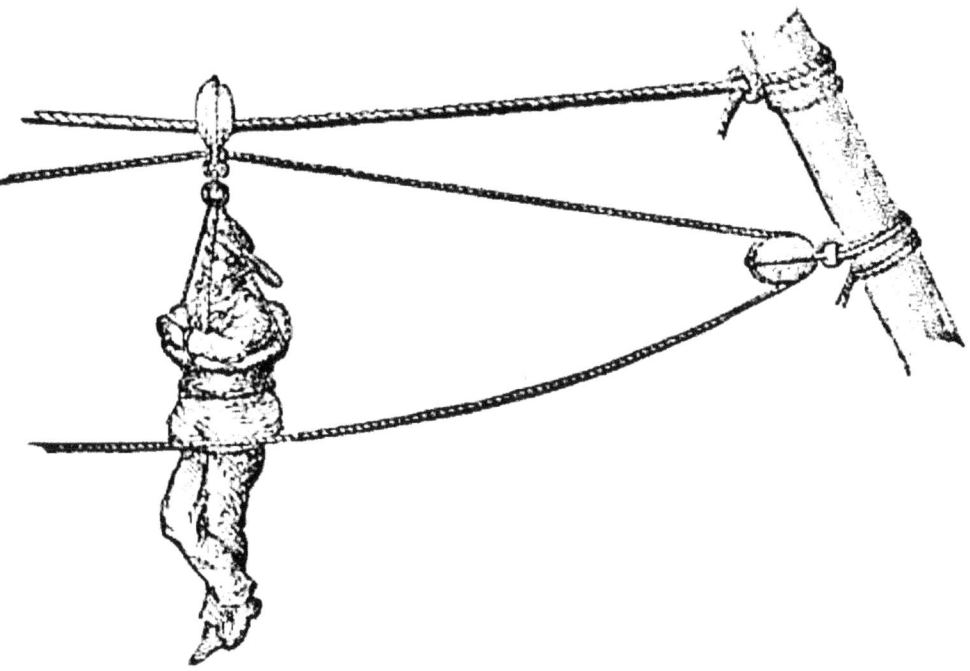

While less artistic than the Homer painting, this illustration from *Chambers Encyclopedia* (1875) shows the "breeches" part of the breeches buoy much better.

attached a pair of short canvas breeches.... The life-car ... is a boat of corrugated iron with a convex iron cover, having a hatch in the top for the admission of passengers ... and a few perforations to admit air, with raised edges to exclude water" (EB 1911).

In the 20th century, some modified smoothbore shoulder guns (M1903, M16) were used to fire a rodlike projectile with a line attached, the line being drawn from a canister attached below the front end of the stock (Firearm Blog 2018).

CG199 requires that vessels carry a line-throwing appliance [199.170]. Those of the shoulder gun type fire eight-ounce projectiles with an eye at the muzzle end. The service lines are of nylon, at least 600 feet long [49 CFR 160.031]. There is also the impulse-projected rocket type [160.040].

Opposite: The Life Line (1884), oil on canvas painting by Winslow Homer (American, 1836–1910), showing use of the breeches buoy. The person secured by the breeches is holding another individual who is apparently unconscious. According to the Met, which hosted it in an exhibition, the painting was "inspired by an event Homer had witnessed in Atlantic City, New Jersey, during the summer of 1883." (The George W. Elkins Collection, 1924. Philadelphia Museum of Art. Accession E1924-4-15. Image available at https://philamuseum.org/collection/object/102970.)

Water Survival Lessons

Floating

Human buoyancy depends on the salinity of the water, and the build (musculature and body fat) and the amount of air in the lungs of the person. "Tests on Service personnel in Great Britain indicate that about 10% are negatively buoyant in fresh water and about 2% in salt water" (TransportCanada 2003).

Floating requires being able to stay calm (as opposed to flailing about in a panic). It is unlikely that someone who has never practiced it before will be able to do it in a crisis situation.

When you float horizontally on your back, in "starfish" position, it is likely that only your nose and mouth, not your whole head, will be above water. According to the U.S. Navy's training on survival floating, "floating on your back ... works only in calm water. If you are in a rough ocean or lake, water can come over the top of your face and enter your mouth and nostrils, causing you to aspirate" (Morel 2022). The navy suggests a face down float. When you need to take a breath, you lift your chin and make compensatory movements with the rest of your body.

There is also the "vertical back float," where your arms are extended out to the sides and only your head and upper chest are above water.

If you are wearing a PFD, you don't need to use your arms and legs to keep your head above water, and the HELP (Heat Escape Lessening Position) is recommended. Essentially, you bring your knees in close to your chest and cross your arms tightly across the chest (Harris 1985).

Treading Water

Treading water keeps you in a vertical position with your head above water by virtue of leg (and possibly also arm) movements. It of course uses more energy than merely floating and your body core temperature will cool down about 35 percent faster than if you are just floating (United States Power Squadron 2007).

Swimming

In Ovid's *Heroides*, Hero tells Leander, "What you are eager for, Leander, to swim, the sailors fear: Always for them it follows the wreck of their ships" (Orme 1983, 6).

Would learning to swim help save sailors' lives? It depends. Perhaps no one is looking for you, and you need to swim to reach shore or another ship. But swimming ability doesn't help if you are unconscious or too injured to take advantage of it, and is of limited value if the water is cold. And good swimmers might foolishly decline to wear a life jacket. Still, there are definitely instances in which those who swam survived and those who didn't perished.

Attitudes toward swimming varied with time and place. In the Augustan imperial navy, basic training included "being taught to swim" (Pitassi 2012). And Vegetius, in the fifth-century treatise *De re Militari*, Book III, said that in summer, if the sea or river was near a Roman camp, the legionaries "should all be made to swim." Swimming ability was even considered a civic virtue, with the Romans looking down on those who could not swim.

During the Middle Ages, the notion developed in Europe that swimming was dangerous to one's body and soul. Swimming ability became rare even among mariners. The "son of Christopher Columbus reported that his father had jumped from a burning ship during a sea battle, swimming several miles to shore, while most of his companions, unable to swim, either died on the boat or in the water" (Goldsmith 2014). This was in 1476, in Atlantic waters off the Portuguese coast, so his companions were mostly Portuguese sailors. Summer sea temperatures in that region are presently mid–60s.

In the 1530s, some German schools and universities instituted a total ban on swimming, apparently to reduce the incidence of student drownings. (I suspect drunkenness was often a factor; it still is the major risk factor in recreational boating drownings [Maxim 2015, 25].) Cambridge imposed a similar ban in 1571, with "two public whippings, a fine of ten shillings and a day in the stocks for a first offence and expulsion for the second" (Chaline 2018).

Nonetheless, there were some advocates of swimming lessons, most notably Wynmann (1538) in Ingolstadt, and Everard Digby, author of "The Art of Swimming" (1587) and "A Short introduction for to learne to swimme" (1595) in England.

It is possible that swimming was more common among Europeans who grew up by the sea—especially warm waters such as those of the Mediterranean and Caribbean—and who had occupations that took them regularly onto the water.

In the early 19th century, a mariner who could swim was still the exception rather than the rule. The crew of the sloop *Childers* presented a sword to the shipmaster, George Wilson, "as a mark of their esteem for his jumping overboard at sea, and saving, at the risk of his own life, one of their shipmates from a watery grave, who had fallen from the fore-yard arm, and was in the act of sinking." And the incident was considered worthy of mention in the December 1810 issue of the *Naval Chronicle* (Age of Sail 2008).

In early 1918, 4,000 navy recruits arrived in San Diego and the navy discovered that 40 percent of them couldn't swim. The navy pleaded for conversion of a Balboa Park lily pond into a training pool, and the local newspaper editorialized, "A sailor who cannot swim is a sailor whose life is constantly in extra danger." Forty sailors were taught at a time, taking "instruction for thirty minutes each day until they could swim unassisted for five minutes" (Crawford 2013).

The modern U.S. Navy requires, as part of basic training, that every sailor qualify as a Third-Class Swimmer. In essence, this is about being able to survive without a PFD in open water long enough to be rescued after falling overboard. The specific testing includes a deep water jump from at least five feet above the water, a fifty-yard uninterrupted swim using four different strokes, a five-minute prone float, and shirt and trouser inflation.

One reason that accidents occur—whether those affecting a single sailor, or an entire ship—is because of our limited ability to see in the dark. This is addressed in the next chapter.

Chapter 8

Seeing and Being Seen in the Dark

The enclosed decks of ships are manmade caves, and were it not for openings in the hull and decks, or artificial lights, they would be as dark as night, even in the daytime. At night, of course, sailors would be dependent on moonlight or even starlight if they didn't have artificial lights.

Adequate lighting is essential to safety. In a Hungarian mine, the accident rate was 60 percent lower in a lighted section than in one where only cap lamps were used, and increasing the lighting level from 20 lux to 250 lux decreased the accident rate by 42 percent (Lewis 1986, 3). Mine studies have also shown that productivity increases if lighting is improved (4).

While there is an obvious need for lighting below deck, at night, the lighting above may also be deficient. The 1915 British Home Office Illumination Report advised that "a man was coming along the ship's deck to start work at about six o'clock" on a "very dark night, rainy and a high gale," and "he did not notice an opening, and fell about forty feet…. The lighting in this case was admittedly bad" (Electricity 1918, 302).

The lux is a measure of illuminance, the "luminous flux" falling on a unit surface area. Direct sunlight provides 32,000–100,000 lux; an overcast sky, 100–1,000 lux; the full moon, 0.002 lux. The ILO's 2006 Maritime Labor Convention requires (ABS 2014, 39, 41) that both interior and, at night, exterior walkways, passageways, stairways and access ways receive 100 lux. General lighting in cabins must be 150 lux; the mess room and cafeteria, 300 lux; the medical laboratory, treatment/examination room and dispensary, 500 lux. Also, it must be possible to reduce the illumination in cabins (by turning lights off and closing doors, curtains and shutters) to less than 30 lux to facilitate sleeping.

Natural Lighting

Openings provided both natural light as well as ventilation. The problem was how to let light and air in and keep water (and projectiles) out. The openings could be vertical (on the sides of the ship hull or superstructure) or horizontal (on the deck or the top of a cabin).

Gunports date back to the late 15th century. There was variation in the size and spacing of gunports, in part due to the variation in the caliber and crew requirements of the guns. On the British capital ship *Sovereign of the Seas* (1637), the ports were 38 inches square, with spacings of 9.5–11 feet (Sephton 2011, 69). Both circular and square gunports were employed (sometimes on the same ship, as in the case of the *Vasa* [1628]); in general, the square gunports were on the fully enclosed decks, and the round ones on the poop and forecastle.

The first gunport lids were merely boards placed over the ports from the inside and secured in some way. In the early 16th century, hinged lids were introduced. They were made of two pieces of wood, with the outer piece matching the curvature of the hull and the inner one the opening in the port frame (Mondfeld 2005, 176).

Based on Georges Fournier's *Hydrographie* (1642), there were regional variations in lid design, with side-mounted lids on Spanish ships, top-mounted ones on French, British and Dutch ones, and removable lids on ships from certain other countries.

Further adjustability was provided by replacing the single lid with two half-lids, joined by a hook and an eyebolt. In the late 18th century, the British added ventilation scuttles, holes with sliding metal covers. Around 1809, the British experimented with inserting small glass windows ("illuminators") into

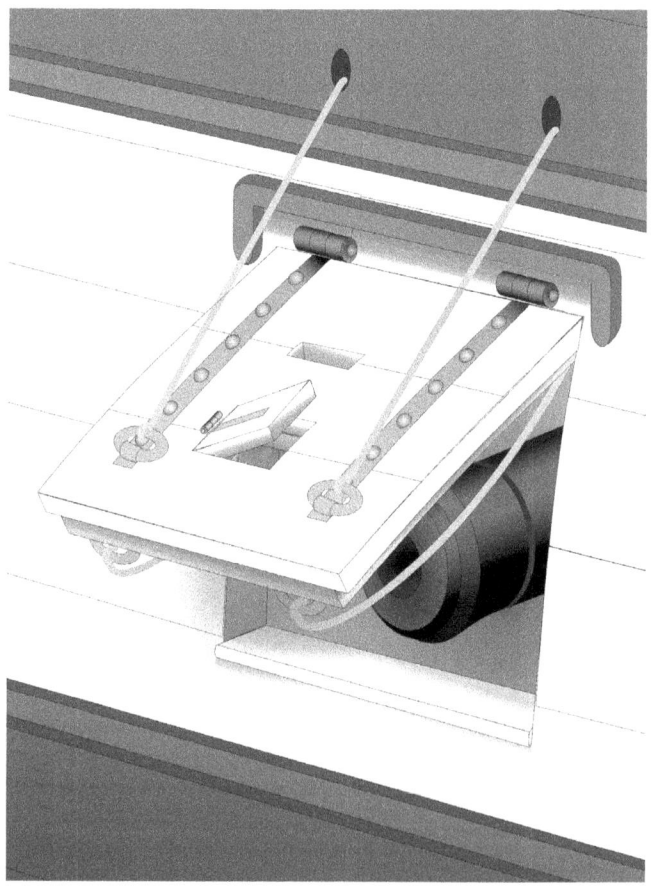

Artist's depiction of a top-hinged gunport lid with ventilation scuttle, cropped and converted to black and white by author (original artwork by Wikimedia Commons user "Rama." Creative Commons Attribution-Share Alike 2.0 France license, https://upload.wikimedia.org/wikipedia/commons/2/2a/Gunport.svg).

gunport lids (Blake 2005, 52), but these proved unacceptably vulnerable to enemy fire (Quinn 1999, 83).

Nonetheless, the U.S. Navy's 1864 "Allowances" stated that "the upper half-ports on the main decks of ships-of-the-line, frigates and spar deck sloops will be hinged, and each one will be fitted with a long patent light, or two bull's eyes" (USN 1865, 80).

During the 18th century, gunport lids on the gunports of the great cabin and the wardroom were replaced by sliding sash windows (Quinn, 84)—these can be seen on the HMS *Victory*—not unlike those found in a modern home. (However, few of us contemplate running cannon out through our windows, even if we don't like solicitations.) After the Napoleonic Wars, the guns were removed from officer's quarters, and the sash windows were replaced with "fixed lights," essentially windows that couldn't open (Id.).

Stern lights were windows composed of small diamond-shaped panes of glass or muscovy mica set in lead or tin-plate, which in turn was tacked and sealed into a wood frame. They provided illumination for the "great cabin" and perhaps also the wardroom on a warship. The stern lights might have a gutter and drain structure at the bottom to carry away spray, and be equipped with wooden covers ("dead lights") to keep out waves. Prior to 1690, mica was cheaper, so the use of glass was a prestige signifier (e.g., it was used on Henry VIII's *Katherine Pleasaunce*, 1519) (Quinn, 85ff).

The development of cast glass brought the price of large panes down (making possible the large mirrors at Versailles), but shipbuilders used smaller panes because of the stresses placed on the panes as the result of the flexing of the hull in response to wave action. If you doubled the length and width of a pane, you would want to double its thickness too so it wouldn't break. That would mean using eight times as much glass, but the cost would probably be more like 9–10 times as much because of the higher discard rate. The cost of stern lights was high enough so that a ship might have a combination of stern lights and mock lights (hull sections painted to look like stern lights during the day).

Portals (port-lights) were introduced in the 19th century, and were either fixed windows or a pair of hinged doors, one glazed and the other solid. The first type were common on fishing schooners and the latter on the outboard cabins of passenger and merchant ships (Quinn, 87–88).

Oliver Lang's tube scuttle (1823) is probably best classified as a type of portal. A copper tube, four to six inches in diameter, with thick glass on the outboard end, fitted into a conical opening in the side of the ship. A long "iron leg" was attached to the inboard end of the tube, and was supported inboard by a brace on a shelf. The leg could be turned to screw the tube into the opening, so only light was admitted, or unscrewed, so air could flow in around the tube. It was installed on the orlop deck of the 84-gun *Thunderer* (*Nautical Magazine* 1835, 167) as well as other vessels (NMNC 1841, 103).

Gratings over hatches provided light from above, but the structure had to be strong enough so the crew could walk over it. In the 17th century, the gratings were made out of wood. Notched slats fitted laterally into the hatch frame, and cross battens fitted longitudinally into the notches. The hatches had a raised border so that water on deck wouldn't run into the hold.

There was an 18th-century dispute as to proper spacing; the Dutch and English favored four-inch square spaces to maximize lighting and ventilation, and the French preferred 2.5-inch

Patent model of a ship side light, supporting Enoch Hidden's 1853 patent (US 9811). His model features captive screws, a lead sealing ring, and a pivoting window (Smithsonian, National Museum of American History. Catalog 308552).

spaces so the gratings were easier to walk on. The French appear to have won this debate since in the 19th century the British reduced the spacing to three inches.

Skylights, while known since ancient times, did not appear on ship decks until the early 18th century. Usually, these were not simple glazed windows but rather square or rectangular box structures, perhaps three to six feet in their longest dimension, and six to 12 inches high. Either the top or the sides of the box would be a glazed window. In some instances the window was hinged so that it could be opened to let in air, too. A *light well* was a cross between a grating and a skylight. It was a wooden box with windowed sides and a grating on top.

Lenses were first installed in decks in the early 19th century (Quinn, 95ff). (I prefer to reserve the term for optical elements with a curved side for more efficiently collecting or distributing light, but flat lenses were known.)

The first one was the Pellatt "Illuminator" (1807), known also as the "Patent-Light" or "Bull's Eye Light." Dana, in *Two Years Before the Mast* (1841), mentions that the forecastle (crew quarters) of the merchant vessel *Alert* is "tolerably well lighted by bull's eyes" in the daytime. The Pellatt lenses were used not only as deck-lights, but also as portals and in gunports.

In its original form, it was a lens with a hemispheric top and a flat bottom, five

A self-wiping porthole on the bridge of the USS *Brooklyn* (CL-40), photographed January 18, 1938 (Naval History and Heritage Command. Photo catalog NH 56628).

or six inches in diameter and one or two inches thick (centerline), placed in a wood frame (later, a brass or copper collar). Since the convex side was on top (and protruded above the deck), this was a collector lens. The frame could be hinged for ventilation purposes (95).

By 1818, it was sometimes installed in an inverted configuration, i.e., flat side flush with the deck, and convex side down, arguably making it a "distributor" lens (96).

However, this didn't seem to be considered a big advantage because "double flat" lenses (I'd call them flush skylights) were sometimes installed (e.g., on the Confederate submarine *Hunley*) (99). A problem with a flush flat glass surface was that it was very slippery when wet, and some ships had textured or roughened deck lenses to improve traction.

Prisms. There was less leakage if a deck fixture fit into a single plank. A square or rectangular lens of plank width transmitted more light than a circular lens of the same width. And rather than give them a curved surface, they could be faceted. The faceted side could face up or down, more often the latter (100–2). On the whaler *Charles W Morgan* (now a museum ship at Mystic Seaport), the deck prism was a hexagonal pyramid (Macky 2023).

Fuel-Burning Artificial Lighting (and Accessories)

Prior to the discovery of electricity, artificial lighting was provided by burning some kind of combustible fuel. This of course meant that lighting came at a price: not just the cost of the fuel, but the risk of an uncontrolled fire. (For fire prevention and control, see chapter 9.)

There is some archaeological evidence of shipboard lighting devices from the early 17th century. They include candlesticks or holders on the armed merchant ship *Sea Venture* (1609), the treasure galleon *Atocha* (1622), the East Indiaman *Campen* (1627), the warship *Wasa* (1628), the East Indiaman *Batavia* (1629), and the galleon *La Concepcion* (1641); iron oil lamps on the *Mayflower* (1620); and brass gimbaled oil lamps with three wicks on the East Indiaman *Witte Leeuw* (1613) and the *Batavia* (Quinn, 61).

Candles provide the fuel in a solid form. Tallow candles were made from animal fat; the process removed excess protein. Mutton drippings were the preferred source. Wax candles were made from beeswax or, later, from certain plant sources (bayberries, coconut palms, West African palms, and the Lisoea tree). Beeswax candles had a higher melting point, burned 15–20 percent brighter, and produced light for as much as twice as long as tallow candles of the same size. Unfortunately, in the 17th century they cost three to four times as much (Quinn, 30–33).

The *Vergulde Draeck*, a 260-ton, 28-gun "jacht" constructed in 1653 (Bander 2014, 218), carried 80 pounds of tallow candles, 80 pounds of wax candles, and 80 of a wax-tallow blend. It is estimated that it would have used 64–128 candles per month (figuring four to eight candles to the pound) (Quinn, 66).

By the mid–18th century, there was an elite alternative: the spermaceti candle (Irwin 2012). Spermaceti is a waxy substance (primarily cetyl palmitate) found in the head of a whale, in an organ of uncertain function. Spermaceti candles cost almost twice as much as beeswax candles, but burned 15 percent brighter. It was found to be advantageous to add a small amount of beeswax to the spermaceti to inhibit crystallization (Quinn, 33).

The now-obsolete luminous intensity unit candlepower was defined in 1860 as "the light produced by a pure spermaceti candle weighing 1/6 pound (76 grams) and burning at a rate of 120 grains per hour (7.8 grams per hour)" (Brown 2022). The modern unit is the candela.

In the 19th century, new manufacturing methods reduced costs. Water-cooled molds (1801) hardened the tallow faster, pistons (1823) could eject the finished candle faster than could be done manually, other machinery provided continuous wicking (1834), and a device could taper the candle base to eliminate the need for manual shaving (1861) (Quinn, 33).

New candle materials were also developed. In 1823, a method of producing stearin (stearic acid) was discovered by Chevreul and Gay-Lussac. (This stearin-making process is described in Faraday's classic *The Chemical History of a Candle*.) They patented a composite candle, a mixture of stearin and tallow. This was inexpensive but burned like beeswax.

Finally, in 1857, the first paraffin candles appeared. These were 20 percent brighter than spermaceti candles and cheaper to boot. Paraffin is a mixture of saturated hydrocarbons (alkanes) of 20 to 40 carbon atoms. They can be derived from petroleum, coal or oil shale. Paraffin has the disadvantage of a low melting point (115–154°F), but this was improved by adding stearic acid (melting point 157°F).

Candle holders. Since ancient times there have been two types of candle holder, the socket holder and the spike holder. Both were in use in the early 17th century (Quinn 1999, 34–5). Some 16th-century socket holders had a spring inside the socket, to push the candle up as it was consumed (Quinn, 36).

Oil lamps. These provide the fuel in liquid form. Sixteen oil lamps were found on the *Uluburun* shipwreck in Turkish waters, dated to 1316–18 BCE. We know that six were used by the crew, rather than were mere cargo, because they had blackened rims. Oil lamps have also been found on ancient Greek and Roman shipwrecks (Quinn, 13).

The oil in oil lamps was usually vegetable (olive, linseed, radish, castor bean sesame seed, or nut) oil; olive oil burned bright and was nearly smokeless. Fish oil has been known as a fuel since ancient times but created "a poor light, large amounts of smoke, and a disagreeable smell" (Quinn, 20).

In the early 17th century, whale oil was considered the best fuel; a single wick in such oil "burned with the brightness of two tallow candles and could last 12–15 hours without trimming." However, it was very expensive.

Oil lamps would have had slanted wicks. The upright wick was introduced in the 1770s (Quinn, 23).

The Argand lamp was invented in 1780. It used a hollow, upright wick mounted inside a cylindrical glass chimney. The oil was supplied by a reservoir mounted above the burner since it couldn't travel far up the wick. The improved air flow increased brightness, and the more complete combustion meant that one didn't have to trim the wick as often. The brightness was equivalent to six to eight candles (Dempster 2002, 30; EB 1911, "Lighting," 651–2). In 1879, the wardroom on the sloop-of-war USS *Kearsarge* was lit by "a swinging argand lamp of two burners" (Wales 1881, 113).

In 1846, Abraham Gesner distilled kerosene (a mixture of alkanes of 6–12 carbons) from bituminous coal, oil shale, and petroleum, and in 1850 he formed the North American Kerosene Gas Light Company. Kerosene "burned cleaner and brighter than whale oil, and didn't have a pungent odour." Over time, its cost dropped below that of whale oil. Kerosene displaced whale oil as the preferred lamp oil (Butts 2019).

In one test, a "hurricane style" kerosene lamp (22mm-wide wick) when clean provided 82 lumens (but 52 after 10 hours' soot accumulation), and the maximum output in a particular direction was 9–10 candelas. A simple kerosene lamp provided only one-tenth the light (Mills 2003). (The lumen is a unit of luminous [visible light] flux, whereas the candela is a unit of luminous intensity—the light emitted in a particular direction. A light source that radiates one candela in all directions has a luminous flux of 12.57 lumens.)

A modern source states that "as a general rule, oil lamps will burn about ½ an ounce of lamp oil per hour"; the oil was not specifically identified but based on a link, paraffin oil was contemplated (Bishop 2013). Consistently, another one reports half a gallon (64 ounces) consumed in 154 hours (Alpharubicon 2004). For olive oil, I find 2 ounces in five hours (Modernsurvivalblog). And for kerosene, "A kerosene lamp producing 37 lumens for 4 hours per day will consume about 3 litres of kerosene per month" (Rao 2011, 36 n30).

A gimbaled lamp was pivotably attached by pins to the two ends of a U-shaped piece, the center of which was attached to a wall mount (Quinn 1999, 68, Fig. 20). While the device as described would pivot only on one axis (defined by the ends of the U) a second gimbal could be used to allow rotation around a second axis. Also, elements could be added to limit how far the lamp would swing.

Wicks are fibrous, porous elements that deliver fuel to the flame of a candle or oil lamp. In either case, liquid fuel is drawn up into the wick by capillary action (in the case of a candle, the heat of the flame liquefies the candle material). "Too much fuel and the flame will flare and soot, too little fuel and it will sputter out" (NCA 2020).

Depending on the region and period, wicks were made of flax, wool, hemp, oakum, mullein, linen, castor plant, reed, papyrus and even asbestos (the last is attested to by Plutarch) (Dilek 2003, 449; Quinn 1999, 18). In the 17th century, cotton was preferred.

The wick is treated with a flame retardant solution (salt, borax) so it isn't immediately consumed by the flame. Curiously, the borax also raises the flame temperature.

Wicks require monitoring. If the wick in an oil lamp burns down to the fuel line, the flame is extinguished. Hence, a tender needs to periodically pull up the free end. For a flame of maximum brightness, you occasionally need to clip off the charred end.

A flat-ribbon cotton wick was introduced in 1773. It produced a large and brighter flame but at the cost of more rapid fuel consumption. The plaited cotton wick—several woven fibers with one tauter than the others—was introduced in 1825. It burned more evenly and curled over to remain in the outer mantle of the flame, causing it to be incinerated rather than charred (making it self-trimming) (Quinn, 34).

A wick could be stiffened with a metal wire; this also helps conduct heat down into the candle. Lead is now banned in the United States, but copper, zinc or tin can be used.

Lanterns are housings, made of metal, wood, ceramic or leather, that protect the flame of a candle or lamp from the weather and also reduce the risk of fire. There were ventilation holes on top. Light was allowed to escape through an opening, or a window of thin horn, mica or glass, and there might be means to shutter the opening or window. Mica windows had the advantages that they "did not yellow with age like horn, and were less expensive than glass." Unfortunately, one could not usually find pieces larger than six inches and thus to make a larger window, pieces had to be soldered together. In later times, they were superseded by sheet glass made by casting (late 17th century) or drawing (19th century) (Quinn, 37ff).

A variation on the glass window was the lens, shaped so as to concentrate the light. One could use the bulb formed from hand-blown crown glass, resulting in the "bull's eye lantern." This lantern could have "bull's eyes" on

Ship Lantern, **painting by Samuel Philpot, circa 1939 (Index of American Design. National Gallery of Art [U.S.], Accession 1943.8.12321. Some background cropped out by author).**

three faces, illuminating the front and flanks but not shining back into the carrier's eyes (42).

A "dark lantern" had a hinged or sliding shutter so the light could be completely hid until one wanted illumination (35). The first use of the term recorded by *Oxford English Dictionary* is from 1650. However, I strongly suspect that the "absconce" used in medieval monasteries was a dark lantern. The lantern may have some sort of hanger so it may be hand-carried or hung overhead from a hook.

When the yacht *Vergulde Draeck* sank in 1656, there were over a hundred lanterns on board. Twelve wood-framed "horn lanterns" hung in the mess, and there were eight brass-framed ones for the guns. The steward's chest contained two dark lanterns and two brass powder-lanterns (Quinn 1999, 63).

Mirrors may also be placed behind a light source so the light radiated in that direction is not wasted. The powder-room lantern used by the Dutch East India Company in the 1740s had mirrors (Quinn, 81). A tin reflector was found in an early American colonial lantern (Hayward 1962, 70). A modern lantern might have a stainless steel or aluminum reflector.

Gas-lamps were invented in 1792, with gas distilled from coal. In the 19th century, certain cities piped coal gas for use in street illumination. There was only sporadic use of gas lighting on large passenger ships in the 19th century; Quinn 1999 (76) attributes this to concerns over volatility—if the gas escaped and mixed with air below decks, it could eventually build up to levels constituting an explosive mixture. There is also the problem of storing or generating the gas on shipboard prior to use.

Gas mantles were first used in conjunction with gas lights, but in theory could be used in conjunction with another heat source such as an oil lamp or an incandescent electric light. The mantle is a cotton bag impregnated with magnesium nitrate (Clamond basket, 1881) or a mixture of thorium and cerium nitrates (Welsbach mantle) (EB 1911, "Lighting"). When heated, the fibers burn away but the nitrates are converted to refractory oxides that glow brightly with little infrared emission. In essence, they are converting heat energy to light energy.

Lighthouses have used high-pressure oil lamps together with a Welsbach mantle, and some thought them preferable to electric arc lights (Corbin 1917, 83). A 1944 article on Coast Guard navigational buoys and lighthouses remarked, "a Welsbach mantle, heated to incandescence by a gas flame, is a common source of light. Its rays are greatly magnified [*sic*, concentrated] by lenses for visibility from far out at sea" (Maisel, 109). The gas used was most likely acetylene from a pressurized tank (191), see below.

Alternative Chemical-Based Artificial Lighting

Acetylene lighting. Carbide lamps rely on the reaction of calcium carbide with water, producing acetylene. The latter is burned, producing light. Calcium carbide itself is made in an electric arc furnace by reaction of lime (calcium oxide) and coke (carbon).

A ship may either carry acetylene in gas tanks, or it may generate it on board.

The latter seems to have been the preferred approach. A 1906 article urged that "an acetylene installation costs about ⅕ of that of electric light" (Rawson, 53).

It appears that shipboard acetylene lighting began in British vessels around 1898. In 1906, John Alexander Smith reported to the British Acetylene Association that "there are now between 90 and 100 boats using acetylene with the greatest satisfaction." He proudly reported that acetylene lighting was used in cargo vessels on the Liverpool-Glasgow run. The big advantage seen for acetylene lighting over electrical lighting is that it was still available when the engine was stopped (*Acetylene Journal* 1906, 434, 470). Bear in mind that this was stated at a time when batteries were very crude.

In 1921, the Marine Department of the British Board of Trade issued guidance for ship inspectors with respect to acetylene generators. It noted that the calcium carbide was usually delivered to ships in commercial drums weighing 50 or 100 kilograms, and if so, the drums should be lowered into the engine room by block and tackle. It also called for the calcium carbide to be stored in a hermetically sealable container, with a posted warning that "no naked light is allowed near the generator or the carbide container." Moreover, "in handling carbide and charging the generator, care must be taken to avoid the creation of sparks" (*Marine Engineering* 1921, 587).

In March 1922, a carbide-related disaster befell the steam trawler *Betty Johnson*. One of the firemen, going down a ladder, slipped and dropped the drum of carbide he was carrying into the engine crankcase. It was crushed "with the next revolution of the engine. With water in the bilge and a lamp burning in the [engine] room, the acetylene generated by the contact of the carbide with the bilge water was immediately ignited and an explosion followed." The Board of Inquiry held that this was an avoidable accident, because the carbide should have brought down below by a "proper" means and before the engine was started (Johnson 2012).

In response to this incident, use of compressed gas cylinders, rather than gas generators, was urged: A steel cylinder "of only a few cubic feet water capacity can hold 200 cubic feet acetylene." Moreover, "the gas is stored in a porous substance with which the whole cylinder is tightly packed," allegedly rendering it "completely non-explosive." Another advantage enumerated for the "dissolved acetylene" was that it avoided the need to dispose of "sludge" (calcium hydroxide) (Harrower 1922, 234).

Limelights, invented by an ordnance survey officer in the 1820s, were first used theatrically in 1836. They relied on the reaction of oxygen and hydrogen gases with quicklime (calcium oxide). That reaction is potentially explosive, and the safest format is one in which the two gas jets meet at an angle where the lime cylinder is located (Woodbury 1896, 310).

Limelights were used by the Union navy during its bombardment of Charleston in September 1863, and to spot blockade runners in early 1865 (Nortum 2016). Drummond used the limelight (supposedly equivalent to "about 265 flames of an ordinary Argand lamp used with the best Sperm Whale oil") in conjunction with a 21-inch parabolic reflector for geodetic purposes; the combination produced about 92,000 candlepower. While he urged its use in lighthouses, the American Lighthouse Board reported in 1868: "The Lime light required much labor, there was danger associated with the production of the gases used, it required expensive apparatus, and the liability of the lime to become deranged far outweighed any advantages in the way of superior illumination, which could be derived from it" (Tag 2023).

Some sort of chemical-based searchlight was still available for military use in the early 20th century, but its useful range was something like one-eighth that of the 36-inch electric searchlight (Ordnance 1920, 37).

Electric Artificial Lighting

To implement electric lighting, one needs either batteries (which store electrical energy in chemical form) or a generator or dynamo (which converts mechanical energy to electrical energy). A dynamo provides the electricity as direct current, and the generator as alternating current.

The first shipboard installation of electric lighting was on the 1880 SS *Columbia* (Quinn 1999, 76). It used an Edison type Z dynamo measuring "58½ in × 24 in × 39 in" (NMAH).

The first dynamo was built in 1832, but major industrial use (e.g., in carbon arc furnaces) didn't come until after improved designs were patented in 1866–7.

In *Columbia*'s dynamo, two vertical coils support a circuit box. At the base there is a drum-type armature, which would have been turned by a steam engine. The vertical coils are electromagnets that create a fixed magnetic field, hence they are called the "stator." The armature bears coiled copper windings, and the rotation of the armature ("rotor") moves the windings through the stator's magnetic field. When a conductor is moved through a magnetic field, the latter induces the movement of electrons through the conductor, thus creating a current. At the end of the armature is a commutator, a rotary switch. It periodically engages conductive "brushes." If there was a continuous electric contact, the electrons would flow first one way and then the other (alternating current). The commutator completes the circuit only part of the time, so there is no reversal of current. This is called "direct" current.

An electromagnet also works by induction. The movement of a current through a wire creates a magnetic field; the curvature of the coiled wire concentrates the magnetic field in the hole in the center of the coil. The current passes through the stator coils, thus generating the fixed magnetic field, as well as through the lighting to which the circuit box is connected.

This leads to the question: how is the stator's magnetic field generated initially? There are a couple of possibilities. The device may have a residual magnetic field, so the rotation of the armature produces a small amount of current, enough to energize the stator coils. Or a battery or magneto may be used to provide the start-up current. (A magneto is an electric generator in which the magnetic field is provided by a permanent magnet rather than an electromagnet.)

The first American warship with electric lights (installed 1883) was the steam sloop USS *Trenton* (Bauer 1991, 71). The Edison Company for Isolated Lighting "submitted a bid of $5,500 to install one L dynamo & one Armington-Sims [steam] engine complete to supply light via insulated wiring to 104 16-candle power lamps, 130 10-candle power lamps, and four 32-candle power lamps. The ensuing contract also included 238 key sockets, six extra brushes, one automatic regulator and one dynamo foundation." The system was only run at night. The installation was paid for by the Bureau of Navigation, which commented in 1886, "[t]his method of lighting ships of war, owing to the small amount of heat given off, the absence of disagreeable odors, and the more perfect illumination, adds much to the health and comfort of the officers and men, tends to make them contented and happy during their long absences from home and friends, promotes discipline and prevents crime" (NHHC/Trenton 2019).

Lighting could be powered either by the main engine ("integrated electric") or by an auxiliary engine ("segregated electric"). In the modern U.S. Navy, propulsion and ship service power are provided almost always by separate prime movers (Amy 2002, 331).

Carbon arcs were the first practical electric lights, commercialized in the 1870s. A high voltage is applied to carbon electrodes, causing the ionization of the air between them. They were long used in searchlights, which are discussed below. Here we are concerned with their use in shipboard lighting. Arc lamps, run by a dynamo, were installed in HMS *Inflexible* in 1881 (De Steiguer 1900, 257).

Incandescent light bulbs. The basic principle of the incandescent light bulb is that you run enough electricity through a filament to heat it to incandescence. The heating is the result of electrical resistance (which is maximized by giving the filament a long length and a small cross-section). All objects at a temperature above absolute zero (°K) emit radiation. As the temperature increases, the peak intensity wavelength shifts toward the short end (according to Wien's displacement law, it is at 2898 microns divided by the temperature°K); visible light has a wavelength of about four to seven microns.

The first filaments were made of platinum, but that material proved unsuitable. Experiments were made with a carbon filament in a vacuum as early as 1838 (Shavinina 2003, 161), but the poor quality of the vacuum limited the lifetime of such filaments for several decades.

"By October 1879, Edison's team had produced a light bulb with a carbonized filament of uncoated cotton thread that could last for 14.5 hours. They continued to experiment with the filament until settling on one made from bamboo that gave Edison's lamps a lifetime of up to 1,200 hours" (DOE 2013). A later filament based on carbonized viscose was even better.

As for luminous efficacy (brightness per unit power), Edison's first bulb produced 1.4 lumens/watt, and the best carbon filament bulbs yielded 3–4 lumens/watt (4–5 for metallized carbon). Carbon filament bulbs are still available for niche "vintage" lighting applications.

The bulb may be made of glass or quartz, and the bulb evacuated or filled with an inert gas to protect the filament from oxidation (combustion).

In 1900, De Steiguer said that the HMS *Inflexible* was also equipped with "glow lamps" in 1881, and that they proved more practical, at least for warships, than carbon arc lamps. The term "glow lamp" would nowadays be interpreted as referring to a fluorescent lamp, that is, one that glows as a result of ionization of a gas. However, it is evident from Maier (1886, 269) that the term "glow lamps" was once applied to incandescent lamps, and this author is confident that is what De Steiguer contemplated. In 1907, the "working lamp for general lighting throughout the ship" in the United States Navy was the 16-candlepower incandescent (Walling, 4).

Tungsten has the highest melting point (about 3000°K) of all the elements, which means that it is capable of glowing white-hot (as opposed to the yellowish color of the carbon filament). The problem was that it couldn't be produced readily into filament form. The difficulty was overcome, in stages, by William David Coolidge. He found that tungsten powder could be mixed with mercury and other soft metals to form an amalgam, which could be squirted out. However, it became brittle with age. Later he developed a process for making a ductile tungsten filament. Essentially, a combination of heating and "swaging" (hammering) of pure tungsten changes its crystal structure to one that could be drawn (Ramirez 2020).

With tungsten filaments, it is advantageous to use an inert gas (argon, nitrogen, krypton) rather than a vacuum inside the bulb, as it retards evaporation. Adding a halogen (iodine, bromine) to the bulb gas sets up a chemical reaction that redeposits the evaporated tungsten back onto the filament. Bulb duration depended on various subtle issues, including how good the vacuum was, the use of getters like argon gas, and the filament structure (tight coiling desirable) and material (ductile tungsten was best).

Squirted tungsten filaments (1907) produced eight lumens/watt, and drawn tungsten (1910), 10. A typical modern tungsten filament bulb produces 16 lumens per watt (Andrews 2005), so a 40-watt bulb should be 640 lumens (Andrews).

Tungsten filament incandescents have a higher luminous efficacy than the old carbon-based ones, although their luminous efficiency (visible light output as percentage of total power) is still low (about 2 percent).

In 1907 a naval manual remarked that the tungsten filament offered an energy savings of 71 percent over the carbon one, but "the type has not as yet been extensively introduced, and it will probably be many years before the carbon filament lamp is importantly supplanted" (Walling 1907, 8).

Gas discharge lamps send electricity through a contained gas, ionizing it. Some ions are excited by the electrons and, when they fall back to a rest state, fluoresce. If the fluorescence is in the ultraviolet, then one needs a phosphor in the lamp envelope to convert the ultraviolet light to visible light.

The first gas discharge lamp was made in 1705 by Francis Hauksbee; he placed mercury in a partially evacuated glass globe and excited it with static electricity, producing enough (blue) light to read by.

Gas discharge lamps may be filled with carbon dioxide, nitrogen, a noble gas (helium, neon, argon, krypton, xenon), or a vaporized metal (mercury, sodium). They may operate at greater or less than atmospheric pressure (Lazaridis 2011).

In the "fluorescent light," the ultraviolet fluorescence of mercury vapor excites a phosphor coating, which in turn emits visible light. Fluorescent lights weren't commercialized until the 1930s (Whelan 2013). In 1943, the U.S. Navy's *Manual of Navy Hygiene* suggested that shipboard fluorescents would be beneficial because they are more than twice as efficient and produce cooler light than incandescents (i.e., more like daylight), with less glare (63). In 1948, a Bureau of Ships manual noted that "fluorescent lamps are used aboard some naval vessels in applications specifically approved by the Bureau of Ships." A 1950 article declared, "for the first time fluorescent lighting has been used extensively aboard ship," referring to passenger ships. It appears that for them, the past drawbacks of fluorescent lighting had been their "initial seasick green" emission and slow start-up (Stromsted).

Light-emitting diodes (LEDs). A semiconductor material may be doped with an electron acceptor, turning it into a p-type (positive) material, or with an electron donor, rendering it an "n-type" (negative) material. Placing these two materials together forms a p-n junction. Under appropriate conditions, the p-n junction acts as a diode, meaning that electricity flows across it in only one direction (electrons will flow from "n" to "p") (Harris 2021).

When that happens, the electrons entering the "p" material interact with the electron acceptors ("holes"). This briefly excites the electrons and when they fall back to the ground state, they release energy in the form of light photons. This is called electroluminescence, and the color of the light is determined by the energy difference between the excited and ground states.

Unfortunately, in early light-emitting diodes, that energy difference was small, corresponding to infrared light. The first visible spectrum (red) LED was demonstrated in 1962. Blue LEDs became available a decade later. White lighting was achieved either by combining the three primary color (red, green, blue) LEDs or by using a monochromatic LED to excite three different phosphors providing those primary colors (Aggarwal 2018, 105).

Early LEDs had a low power output and were suitable only for indicator lighting, but gradually power outputs were improved.

LEDs consume less energy, and emit less heat, than fluorescent (let alone incandescent) lights of the same light output. It has been reported that almost 100 percent of lights on new cruise ships are of the LED type (Aarnio 2022, 178).

Interior Paint Color

Paint in general is used to inhibit the corrosion of metal and the deterioration of wood. However, in the interior, it can also affect illumination, as there will be indirect, diffuse reflection of natural or artificial light off the interior ceiling, sides and deck of the ship.

Wilson (1870, 21) said, "the ship, below deck, should be painted white." White paint would maximize indirect lighting, but at the risk of causing glare and eye strain. Gatewood (1909, 20) suggested that the sides be "a light buff or what might be called a white stone yellow."

The interior finish specified by the United States Navy in 1924 was primarily "inside white" for both top and sides. A glossy white was to be used in a few compartments, notably the armory, broadside director towers, conning towers, fire-control tower, radio room, voice-tube relay station, pantries, dental office, surgeon's examining room, prophylactic room, treatment room, and various bathrooms. Green was specified for the central stations, chart house when used as a steering station, distribution room, dynamo room, interior communication room, pilot house, and switchboard room, "to prevent glare." Light gray was to be used for the blacksmith shop, and a "bituminous composition" for the battery charging station, coal bunkers (primer only), chain lockers, sand lockers, and part of the sides of evaporator and pump rooms. Photographic darkrooms were painted, understandably, a flat black. Finally, a glossy green was in order for the admiral's quarters, emergency cabins, officers' dining rooms, isolation wards, officers' quarters, sick bay, officers' staterooms, and wardroom country (Navy 1925, section 2538).

Red lead (lead oxide and linseed oil) was used as a primer. The "inside white" was a mixture of white lead, zinc oxide, a small amount of ultramarine blue, linseed oil and petroleum sprits. The glossy white was achieved by coating the inside white with first a flat white (zinc oxide- and ultramarine blue-based) and then a white enamel (zinc oxide, titanium oxide, and Damar varnish). The green paint contained white lead, zinc oxide and chrome green, and the enamel used in the glossy green was similar but also included Damar varnish (Navy 1925, section 2571).

When we say that the darker colors reflect less light than the lighter ones, we are also implying that they absorb more light energy, and this is reradiated as heat, although it is less of a problem in the ship interior, which does not receive much sunlight to begin with.

The illumination of cabins, corridors and the hold on ships is not very different from that of buildings on land. But now let's consider lighting concerns more specific to seafaring.

Compass Lighting

At night, the compass must be illuminated in such a way that the helmsman can read it, preferably without destroying their night vision. The compass was set in a non-ferrous housing (binnacle) with a hole or window through which to view the compass. In the 17th century, an oil lamp would be positioned inside the binnacle so as to fully illuminate the compass card without blinding the helmsman. At nightfall, the binnacle lamp would be lit, and a crewman would be responsible for making sure that it remained lit. Ideally, the oil chosen for the binnacle lamp would be of the best quality and spare binnacle lamps would be carried (Quinn 1999, 77ff).

In the 19th century, the British navy improved on the traditional binnacle lamp, first by using an upright-wick lamp positioned to illuminate the compass from above, and then by interposing a condenser lens between the lamp and the compass so as to concentrate the light. By the end of that century, compasses were lit by electric lights (79).

Powder-Room Lighting

A particularly ticklish issue was how to light the room in which gunpowder was stored on an early modern warship. In the 17th century, the solution was usually to just permit a single candle in a horn-lantern (and put the powder room far away from where fire was normally used [79ff]).

In the British navy, beginning shortly after 1702, rather than bringing the lantern into the powder room at all, it was usually placed in an adjacent "light room," and the light from the lantern would shine through a glass window into the powder room (in the bow). Sometimes the light room was above the powder room and the window in the floor (thus providing overhead lighting). Other times it was a triangular room that protruded into the powder room and had windows on the two entrant sides. In this case, it might be built around the foremast. The light room and the powder room had to be accessed from separate hatches in the orlop deck. In 1805, copper wire guards were added to the magazine light window. In 18th-century French and Dutch practice, a light well with a lantern for nighttime use was used to illuminate both the passage to the magazine on one side and (through a thick window reinforced by a brass grating) the magazine on the other (Lavery 1987, 146–9; Quinn 1999, 80–1).

The alternative to the separate light room (or well) was to use a specialized powder-room lantern. In the British and Dutch examples, the light source was positioned over a lead container filled with water.

The Color of Instrument, Bridge and Deck Lighting

The ability to see in low light levels (scotopic vision) is conferred by the rods of the retina, but the rods are inactivated (their photopigment, rhodopsin, is bleached out) by exposure to bright light. The intensity, wavelength and duration of the bright light exposure all affect the time to recover "night" (scotopic) vision ("dark adaptation"). After 5–10 minutes in the dark, the rods are more sensitive than the cones. After about 40 minutes, you reach maximum light sensitivity (Kallionatis 2007).

We are probably all familiar with submarine movies in which we see the crew garishly lit by red lights. The red (>620 nm) light doesn't readily bleach the dyes in the rods of our retinas and therefore doesn't destroy dark adaptation. Hence, we may see electric compass lights equipped with red glass filters.

While red lighting helps preserve dark adaptation, if the light in question could be visible to an enemy ship, there's a case for using blue lights instead. The atmosphere scatters blue light more than red light; hence, at a distance, the luminance of a red light will be higher than that of a blue light of equal intensity. On the other hand, the dark-adapted eyes of enemy lookouts would be more sensitive to blue light than red light (Pearce 1979).

Another problem with the red lighting was eyestrain, and in 1981 the American submarine command ordered that sonar room lighting be converted to blue. However, there was the problem that the blue light was deleterious to dark adaptation. Ultimately, in 1991, the submarine force switched over to low-level white lighting (achieved by placing neutral density filters over the lights) (Elliott 1992).

Running Lights

Stern lanterns. When ships were traveling in formation at night, there needed to be a way for the helmsman on one ship to see the ship in front of him (rear-end collisions and meandering off both being frowned upon). Hence, sailing ships carried stern lanterns (Laughton 2012, 159). This practice was not limited to warships as, if there was a threat of piracy or enemy cruisers, European trading ships often sailed with escorts.

In Edward III's navy, the number of stern lanterns indicated the status of the commander; three or more for the King, two for the admiral, and one for the vice admiral when off his normal station (Traill 1902, 252–3). On 16th-century Venetian galleys, those commanded by a squadron commander had a single stern lantern, and the flagship of the Capitano Generale da Mar or the Provveditore Generale da Mar had three. Indeed, the flagship was sometimes referred to as a *lanterna* (Motture 2011).

The 68-gun warship *La Couronne* (1626) had three lanterns above the taffrail; the center one was 12 feet high and 24 feet in circumference, illuminated by 12 pounds of candles (Sephton 2011). On the *Sovereign of the Seas* (1637), there were five lanterns on the stern (Sephton, 57, 61), two apiece on the port and starboard quarter galleries, and the fifth and largest on the aft end of the poop above the taffrail. It was six or seven feet high, and 4–4.5 feet wide. In 1661, Samuel Pepys, then clerk of the Naval Board, gave a tour of the *Sovereign* to his patron's wife, Lady Sandwich, the Lady Jemimah, and their seven companions and servants, and persuaded this tour group to join him inside the stern lantern (Dill 2006, 12)—plainly the 17th-century equivalent of squeezing into a phone booth.

While a single stern lantern reveals the position of the ship, it says nothing about its heading. But if you were looking at the stern of *Sovereign*, you would see three lights in circumflex (^) arrangement,

Octagonal ship lantern believed to have come from a 17th-century Dutch warship. Height 355 cm × diameter 121 cm (Rijksmuseum. NG-MC-1052). "Each of the eight side bars is ornamented with a human figure, some are of women in classical dress, some are of Dutch merchants" (van der Vliet 2016).

whereas broadside you would see a rotated "L." Nonetheless, this does not seem to have initiated a general trend toward use of multiple lights to show orientation.

In the early 18th century, all British first-, second- and third-rates carried three lights, and this privilege was extended to fourth-rates in 1722. In 1804 it was decided that only a flagship would carry two lights, and all others just one (Willis 2008,

56). However, I believe that the flagship's second light was a top-lantern (see next section).

At least some early lanterns had panes of green-tinted mica, but these were displaced by glass, which rendered the light easier to see. Hexagonal and octagonal designs were the most common, but the lantern on the *Merhonour* (1622) was seven-sided (Howard 2002, 114). It cost over 11 pounds, not even counting the glass plate, but almost half of that was attributable to gilding (Laughton 2012, 142).

Top-lantern. When William, Duke of Normandy, sailed across the English Channel, he "had a lantern placed at the top of his ship's mast, so that the other ships could see it and hold their course behind him." (Musset 2005, 196). On the 1564 Legazpi Pacific expedition, a ship in need of assistance at night would place a lantern in the main mast and fire a shot, and if it were an emergency, it also hung a lantern in the foremast and fired two more shots (Licuanan 1991, 64). In 1595, Drake ordered his fleet that if they had to unexpectedly make sail on a night that it had previously shortened sail, it would show "a single lantern with a light at the bow, and another at the fore-top" (Maynarde 1849, 64).

Later, it became customary that a British navy flagship leading a squadron would display a lantern at the aft edge of a masthead: the main top (full admiral), foretop (vice admiral), or mizzen top (rear admiral) (Lavery 1987, 255). It was supported on each side by iron braces (Falconer 1815, 569).

In 1762, Admiral Howe ordered that a ship tacking at night was to hoist a light and keep it visible until the maneuver was completed (Willis 2008, 56).

Lightships of course also displayed lanterns on high, but early lightships suspended small lanterns from a yardarm or dedicated crossarm. Robert Stevenson proposed a lantern that surrounded the mast of the vessel, and could be lowered to the deck to be trimmed and then raised back (Stevenson 2014, 39). Presumably, the vertical traversal of that lantern would be limited by the yardarm above. It is conceivable that the lantern had a dedicated mast, i.e., one that did not ever carry sail.

In 1838, the U.S. Congress enacted legislation providing that between sunset and sunrise every steamboat must carry one or more signal lights that could be seen by other boats navigating the same water. A three-light system was privately adopted by Liverpool steam packets. In 1847, a different system—red on the port bow, green on the starboard bow, and a bright white light on the foremast head—was adopted for mail steamers on the west coast of England. Finally, in 1848, a similar system was applied to all British steam vessels between sunset and sunrise (Grosvenor 1921).

By the 1870s, it was proposed that the masthead light be electric (Trowbridge 1874, 723). This was met with numerous objections—the ships met would be blinded by the light; the carrying ship's side lights would be rendered inconspicuous by comparison; the ship would be mistaken for a lightship, etc. (Thomson 1879, 190).

The *Titanic* carried a single electric masthead light on her foremast, 145 feet above the water. It was 32 candlepower, and its Fresnel lens concentrated the light into a horizontal arc with a vertical amplification factor of 25. It thus would have been as bright, at a distance of 17 miles, as a first magnitude star (Halperin 2007).

There is an obvious downside in wartime to the use of any lights on shipboard, let alone lights intended to reveal one's presence to other vessels. Drake ordered, "you shall keep no light in any of the ships, but only the light in the binnacle, and this with the greatest care that it be not seen, excepting the admiral's ship" (Maynarde 1849, 64). Even today, there are waters where small boat captains don't switch on their mast lights (Liss 2011, 62).

On the other hand, in 1800, Thomas Cochrane in the brig-sloop *Speedy* was able to evade a frigate at night by placing a lantern on a barrel and letting it float away (Harvey 2008, 141).

Lighting the Waters: Star Shells

Sometimes it is desirable to illuminate the surrounding waters at night, in order to spot navigational hazards or enemy craft.

The star shell ("light ball") is fired by a mortar (high-trajectory gun) and contains a small explosive charge and a time fuse. The charge in turn ignites the illuminating composition. Early compositions included mixtures of sulfur, saltpeter (potassium nitrate), and realgar (arsenic tetrasulfide), orpiment (arsenic trisulfide) or antimony.

Appier's *La Pyrotechnie* (1630) gives a formula for "fire balls … so white that one can scarcely look at them without being dazzled," that comprises saltpeter, orpiment, gum arabic and, strangely enough, ground glass and brandy (Skylighter 2018). In its original form it was not very useful at sea as the "stars" would fall into the water and be extinguished within a few seconds. Even in land warfare, the enemy could be expected to throw water or sand over it.

Boxer proposed modifying this shell to be composed of two hemispheres, one containing the illuminant ("stars"), and the other a calico parachute connected to the first by ropes or chains. The explosion of the charge not only ignites the illuminant, it separates the hemispheres, but only insofar as the connector permits. The parachute slows the descent of the illuminant (Griffiths 1859, 83, 91). Boxer was probably unaware that there had been experimentation during the time of Louis XIV with rockets equipped with parachute flares (Faber 1919, 181). For that matter, Congreve had a rocket light ball with a parachute (Sterling 2008, 401).

I have documented use of magnesium flares in photography of the Comstock Lode mine (1868) and the Great Pyramid (1865). I wasn't able to determine when magnesium, aluminum, or magnalium ribbons were first used in star shells, but the first reference I found was from just before World War I (U.S. Army, 2–11). The parachutes were also miniaturized, so that six or eight parachute-illuminant combinations could be fitted inside a single shell.

Lighting the Waters: Searchlights

Searchlights are essentially a military development of the spotlight—that is, they combine a highly luminous source, a light concentration system, and a pivotable and tiltable mount. The biggest reason for equipping naval warships, especially capital ships, with searchlights was the introduction of the motor torpedo boat, which could launch a night attack either stealthily or at high speed. Searchlights were proposed as a solution by Wilde in 1873, and experiments showed that they were an effective means of detection. This led the Royal Navy to equip "the *Minotaur*, the *Alexandra* and the *Temeraire* … with Wilde's apparatus" (Van der Kooij 2015, 75).

Small searchlights could be advantageous for nighttime civilian use, too: spotting navigational hazards, rescuing men from the water or a disabled craft, and signaling. Decades later, Wilde argued that if the *Titanic* had been equipped with searchlights, it might have spotted the iceberg that sank it.

There is a strong kinship between ship searchlights and lighthouse lights. Of

The left image shows a Sperry 36-inch searchlight. "Standard Navy type as used on warships. The searchlight has a 150 ampere high intensity arc, producing a beam of 400,000,000 candle power" (Naval History and Heritage Command, Photo catalog NH 115165). The 36-inch was the largest of the U.S. Navy's three standard searchlights in World War II and its light source was then a carbon arc. The right image is of an unidentified Navy crewman operating a battleship searchlight (Naval History and Heritage Command, Photo catalog NH 124097. This image was cropped).

Chapter 8. Seeing and Being Seen in the Dark

Unknown artist's depiction of the steamship CW *Morse* "using her searchlight while operating at night on the Hudson River, New York." This appeared on a circa 1903 postcard. Courtesy of Alfred Cellier, 1977 (Naval History and Heritage Command, Photo catalog NH 85930).

course, the latter can be much larger and heavier. Lighthouse searchlights are covered to some degree in this chapter because they make a substantial contribution to improving life at sea.

Light Sources

Electric searchlights, with light generated by a carbon arc, were used at the siege of Paris (1870-1). In a carbon arc, a strong electric current is made to flow across a short air gap between two carbon electrodes. The proof of concept was made by Davy in the early 19th century (EB 1911, "Lighting").

The arc can be started only by bringing the electrodes in contact with each other, but then the electrodes are slowly separated. Since the rods burn away, you need a mechanism to maintain the arc gap. The stability of the arc is improved by putting a ballasting resistance in series with it (which increases the power requirement).

Direct current is preferred as it causes the anode to form a crater, which gives off most of the light. The intensity is greatest at a 30–45° angle from the anode axis, and this facilitates capturing the light with the reflector (Baird 1917). High currents (130–300 amperes in 1917) are used in military searchlights, so the source must be close by.

The first carbon arc lamp emitted over 10,000 lumens (Banke 2015), and I found an ad for a 60-inch World War II carbon arc searchlight that put out 525,000 lumens (candlepowerforums 2008). Carbon arc lamps have low luminous efficacy (2–7 lumens/watt) and efficiency (0.3–1 percent). Hence, they generate a lot of heat; consideration must be given to providing proper ventilation.

The power requirement for a searchlight-scale carbon arc lamp is considerable. The U.S. Navy Model 24-G-20 24-inch searchlight used in World War II was operated at an arc current of 75–80 amperes and an arc voltage of 65–70 volts. However, the line voltage was 105–125 volts, so almost half the power was absorbed by the rheostat/ballast (General Electric 1944). That corresponds to a power draw of 7,875–10,000 watts. If we assume 80 percent efficiency in the generator and distribution system, then it would need 12,500 watts.

In the 1940s, Schulz discovered that if electricity was passed through xenon gas at high pressure, it produced a near-continuous spectrum, or white light (Maecker 2009, 1–6). In the 1950s, xenon arc lamps began to replace carbon arc for motion picture projection, and they now dominate that market.

In the early 1950s, the U.S. Bureau of Ships developed a "mercury arc lamp" that could be installed in the standard 12-inch signaling searchlight housing in place of the normal incandescent light (*Bureau of Ships Journal* 1952, 49). Later, a mercury-xenon arc lamp was used (Bureau of Naval Personnel 1966, 166). It was visible at eight to ten miles, versus five miles for the incandescent version (Feldman 1962).

Light Concentration

The "candlepower" (light intensity in the direction of the target) of a light increases if its light is more tightly focused, even though the total light output is constant. A searchlight may have millions of candlepower in its beam. Light may be concentrated by mirrors, lenses, or combinations ("catadioptric") of the two.

Reflectors. The earliest documented use of a polished metal reflector to concentrate candlelight was in 1532, at the lighthouse of Gollenberg. In 1669, Braun used a cast steel reflector with an oil lamp at the lighthouse of Landsort, Sweden (Tag 2023). The ACW searchlights used crude mirrors made of an unspecified metal that absorbed one-third to one-half of the incident light (Nerz 1893, 713).

Reflector shape. The ideal shape (figure) for a reflector is parabolic; if the light source is at the focal point, then all of the reflected rays will be parallel to the optical axis of the reflector. (Because the light source is not a point source and the focal length is finite, there will be divergence with distance—Nerz 1907, 16ff.) There were occasional experiments with spherical reflectors at lighthouses, since the spherical shape was easier to achieve. These proved to provide little concentration (Tag 2023).

Reflective material. The ideal material would be highly reflective across the visible light spectrum, easily formed into the parabolic shape, resistant to corrosion (tarnishing), easily cleaned and polished, low in density, and inexpensive. Most modern mirrors are composites—typically a metal coating on a glass or plastic substrate.

Ideally, the metal has a high reflectivity across the entire visible spectrum. This is obviously not the case for gold or copper, which have a warm cast. Silver (85–95 percent) and aluminum (91–92 percent) are the best from a purely optical standpoint, but aluminum was not available until the late 19th century, and it was more expensive than silver until the invention of the Hall-Heroult process of reducing aluminum oxide in the 1880s. Rhodium is only a little inferior optically (76–81 percent), but much more corrosion resistant. Chromium is similar (64–68 percent). Iron is markedly inferior (67–58 percent) (Weaver 1988, E-387ff).

Silver of course is expensive and so there is some advantage to combining the high reflectivity of a silver coating with a lower-cost metal. A silvered copper parabolic reflector was fitted to the La Heve lighthouse in 1781 (Marriott 2003, 25). Robert Stevenson combined an Argand lamp with a silver-clad copper parabolic reflector. Installed at the Bell Rock lighthouse in 1811, it produced 2500 candlepower (Tag 2023).

Silver however is subject to tarnishing as a result of hydrogen sulfide in the atmosphere (or in perspiration if the mirror surface is touched). The resulting silver sulfide is black. The tarnishing is more rapid if the air is humid. (This is indeed why, for consumer use, silvered glass mirrors have the silver on the back surface, so it is protected from oxidation by the glass. The downside is that there would be a ghost reflection from the front surface of the glass, but that is a serious concern only for telescope mirrors.)

Costs could be reduced further by use of speculum metal (45 percent tin, 55 percent copper). Its reflectivities are 63 percent at 0.45 nm and 75 percent at 0.65 (Tolansky 1947). Unfortunately, it too tarnishes and is also somewhat brittle.

The first telescope with a parabolic mirror was built by Hadley in 1721. It was a six-inch diameter piece of speculum metal. The Royal Society praised his achievement, but expressed the hope that someone would either figure out how to keep the metal from tarnishing, or how to make a silvered glass mirror (Pendergrast 2009, 161). This proved to be a difficult proposition, and speculum continued to be used well into the 19th century.

When a metal mirror needed to be cleaned it also had to be repolished and often refigured. The Rosse telescope (1845), the largest in the world until 1917, had two six-foot speculum mirrors; one would be in use while the other was being refigured (Pendergrast, 176–80).

For those for whom cost was an issue, Fitzmaurice invented platinum glazed porcelain reflectors. They cost one-quarter of the equivalent silvered metal reflector but were inferior in performance. They were used at Sunderland Lighthouse (1860) (Tag 2023).

Premodern glass mirrors weren't actually silvered; rather, a tin-mercury amalgam was applied to the rear surface of the glass. Modern replicas show an initial reflectivity of about 70–75 percent (Hasan 2015, 51). In the early 18th century, James Short tried and failed to use this method to make a paraboloid mirror; he switched to speculum metal (Pendergrast, 161). In 1788, Rogers made lighthouse reflectors of "silvered" glass, but they proved to be too fragile (Tag 2023).

Advances in the arts of silvering glass and of grinding glass to paraboloidal shape made the silvered glass paraboloidal mirror possible.

In 1835, von Liebig discovered how to deposit pure silver on glass by chemically reducing (with sugar) a boiling silver nitrate solution. Drayton patented several cold processes in the 1840s, but the mirrors so manufactured were unsatisfactory (they developed brownish red spots after a few weeks) (Chattaway 1907). Liebig came to the rescue in 1856 with the first truly satisfactory method, which used caustic soda and ammonia to accelerate the reduction. In 1856, Steinheil used it to silver a four-inch diameter telescope mirror (King 2003, 262).

In 1858, Foucault devised the knife-edge test, which could be used to determine how much a glass surface departed from spherical. Hence, you could make an accurate paraboloid surface by an iterative hand grind-and-check process. The same year, he made a 40 cm silvered glass paraboloid telescope mirror. The method was perfected in the 1870s by Draper, who preferred the Cimeg silvering process (Lemaitre 2008, 20).

Nonetheless, governments contented themselves in the 1880s with inferior catadioptric reflectors of the Mangin type (see below) for military searchlights (Burstyn 1893). In 1885, Schuckert "invented a machine that could accurately grind glass into a parabolic" curve (Tag 2023) and quickly put this to work in making searchlight mirrors. These Schuckert searchlights were used in 1887–8 in the Italian campaign in Ethiopia (Rey 1917, 97), and a Schuckert searchlight was exhibited at the 1893 Chicago World's Fair. Schuckert mirrors of 30-inch diameter were used to make 40-million-candlepower searchlights for the Heligoland lighthouse in 1902.

Articles in the electrical and military literature credit Schuckert with being the first to make "paraboloid glass mirrors with a sufficient degree of accuracy for searchlight work" (Murdock 1893, 359). Were they simply ignorant of the existence of telescope mirrors of that type? Or was the hand grinding done by telescope makers prohibitively expensive for military and lighthouse use?

In 1909, the mirror alone for Lowell's 42-inch reflector cost $10,800 (Cameron 2010, 117); a Model-T Ford in 1910 cost $950 (135). (It is conceivable that the high price was necessitated by the degree of accuracy demanded for astronomical work, rather than the hand grinding.)

What about tarnishing? On a telescope, the silvering must be applied to the front surface, to avoid ghost reflections from the glass. Hence, the silver is exposed to the atmosphere. It does tarnish, but it was discovered that the old coating could be removed and a new one applied without loss of the parabolic figure. On a searchlight reflector, the silvering can be applied to the rear surface, where it is better protected from the atmosphere, though it will still deteriorate with time.

With large carbon arc searchlights, the heat generated may be such that one cannot use ordinary glass, but rather must use thermal shock-resistant borosilicate glass, made by Schott in the late 19th century and mass-produced by Corning (as Pyrex® glass) in1915 (Van Helden 1984, 146). Siemens Schuckerwerke received British Patent 13,920 (1914) for a searchlight reflector made "of heat resisting material such as borosilicate glass."

Aluminum is highly reflective and only a little denser than glass. Aluminum reacts with oxygen in the atmosphere, but the resulting aluminum oxide is clear and hard, protecting the aluminum from further attack. A mirror was first aluminized in 1932 and an aluminized glass reflector was first used in a telescope in 1935. Aluminization of glass requires a high vacuum, but the film is more durable (Yoder 2015, 62). Mirrors may also be made entirely of cast aluminum (264).

A continuing concern with silvered (or aluminized) glass searchlight mirrors was vulnerability to breakage (the enemy had a tendency to shoot at searchlights). Two types of coated metal mirrors were tested in World War I; one had its coating destroyed after a few hours' exposure to the carbon arc, and the other was of inferior illuminating power to a silvered glass mirror (Baird 1917, 10–11).

In World War II, we had 60-inch, 800,000,000-candela carbon arc searchlights that used a rhodium-plated parabolic mirror (Meza 2019).

In the 1950s-era U.S. Navy, "The materials used in searchlight reflectors are Stellite (cobalt-chromium alloy), Hastelloy [nickel-molybdenum-chromium alloy], aluminum with a special Alzac [anodized] finish, chromium-plated steel, and rhodium-plated copper. Stellite … is used in all 24-inch searchlights. It has a reflectivity of about 65 percent, is relatively unaffected by arc fumes, and is highly resistant to corrosion by salt water" (U.S. Bureau Naval Personnel 1959, 184).

Segmented reflectors. Hutchinson built faceted reflectors in. Some of his designs were tin plates soldered together, but the largest, 12 feet in diameter, was of wood with pieces of mirror glass (clear glass coated with a tin-silver amalgam) attached to approximate the parabolic shape. It was coupled to an oil lamp, and reportedly could be seen 10 miles away (Tag 2023).

Another glass faceted reflector was produced by Walker (18-inch parabolic reflector for the Old Hunstanton Lighthouse, 1776). The facets were set in a parabolic plaster shell in a metal frame. Reportedly, its beam, of 1000 candlepower, was two-thirds the intensity of a one-piece parabolic reflector of the same diameter. Thomas Smith similarly built an 18-inch parabolic reflector with 350 pieces of mirror glass; used with a lamp having four rope wicks, the combination produced 1,000 candlepower at the Kinnaird Head lighthouse in 1787.

Lens. Big telescopes uses mirrors rather than lenses of the same diameter because the latter are much more expensive. However, Fresnel invented a lens composed of separate concentric annular sections, whose surfaces approximate that of a simple lens of the same focal length. Since it is only using the part of the glass that contributes to the proper refraction of the light, it much lighter and less costly than a simple lens.

The more sections, the less degradation in performance relative to a one-piece lens, but the greater the cost. The sections may have curved (better concentration) or flat (cheaper) surfaces. A Fresnel lens was first used in a lighthouse in 1823 (Pickthall 2016, 94). The largest ("hyper-radial") had a height of 148 inches and weighed 18,485 pounds. A ship's searchlight would probably be one of "third order" (62-inch height, 1,984 pounds) or smaller (Tag 2023).

Mirror-lens combinations. Robert Stevenson invented the holophotal reflector (1849). This combined a central spherical reflector, a peripheral parabolic reflector and a Fresnel lens, and the point was to capture essentially all of the light from the source (Tag 2023).

A Mangin reflector was a lens having two concave surfaces of different radii, the front surface having the shorter radius, and the back surface having a reflective coating (thus constituting a spherical mirror). The radii were chosen so the spherical aberration produced by the lens was exactly opposite to that produced by the reflective coating. Mangin reflectors were available by 1876 (Nerz 1893, 11–12).

The Mangin reflector had the disadvantage of having a longer focal length and therefore a smaller effective angle than a parabolic mirror of the same diameter; if the diameter were 60 cm, the angles would be 83° and 123° respectively, and as a result the parabolic reflector would gather 2.11 times as much light.

Weight. A 60-inch searchlight (delivering 800 million candlepower!) for military use, with a six-cylinder gasoline engine, 16.7 kW generator, carbon arc, metal mirror, protective glass, and aiming apparatus, all mounted on a small four-wheel trailer, had a combined weight of 6,000 pounds (Fort MacArthur 2013; 2016). That may seem like a lot, but it was not unusual for a mid–19th-century naval gun to weigh 150–200 times the weight of its shot (Ward 1845, 30), which would make the 60-inch searchlight equivalent to a 30–40-pounder. (And in the late 17th-century, guns were heavier, 175–250 times shot weight [Glete 2010, 516].)

Artificial light is a necessity in the ship interior, or at night. But where there is artificial light—whether provided by an open flame or electricity—there is a risk of fire, as we will see in the next chapter.

Chapter 9

Lest It Spread: Fires and Infections

This final chapter concerns itself with two disparate threats to human life, fire and infectious disease. What they have in common is the ability to spread if unchecked.

Fire Hazards

C.S. Forester's famous naval character, Horatio Hornblower, muses that "the four elements of Aristotle ...—earth, air, water, and fire—were the constant enemies of the seaman, but the less shore, the gale, and the wave, were none of them as feared in wooden ships as fire" (Forester 1950, 79).

Forester alludes to several of the fire hazards on a wooden sailing warship: "Timbers many years old and coated thick with paint burnt fiercely and readily. Sails and tarry rigging would burn like fireworks. And within the ship were tons and tons of gunpowder waiting its chance to blast the seaman into fragments."

But that is hardly a complete list of the ways in which a ship may catch fire. The fire could be attributable to the burning of combustibles to generate heat (especially for cooking), light, or (after the introduction of steam) propulsive power, to enemy use of heated shot or incendiary shells, or to lightning strikes. If the ship were in port, then a fire could spread from a dockside building, or from another docked ship. Also, some 19th-century shipboard fires were the result of arson committed by owners seeking to collect on insurance.

The 130-ton *Tijger*, carrying a cargo of oily pelts, was lost to fire within in 1614 (south of the present Times Square) (Blackmore 2004, 40), and the *Royal Sovereign* perished in 1696, its burning ascribed to an overturned candle (Fraser 1904, 201). Between 1793 and 1815, ten British warships (including eight ships-of-the-line) were lost to fire (Lavery 1987, 185).

Fire remained a substantial maritime peril even in the early 20th century. According to marine insurance companies, from March 1930 to February 1931, 688 fires aboard ships were reported (Hearings 1935, 19). Careless smoking by passengers

Night & a Ship on Fire by Pierre Charles Canot (ca. 1710–1777), after *A Ship on Fire at Night* by Peter Monamy (1681–1749) (Yale Center for British Art, Paul Mellon Collection. Accession B1995.13.129). Note also the stern windows (cf. chapter 8).

and crew caused some of the incidents. Of course, fires could be ignited not only by open flames, but also by electrical shorts and spontaneous combustion of finely divided powders. And there were new combustible materials carried by ships, as fuel or as cargo.

Fire Discipline

The best defense against fire was rigid control of any open flame illumination. In 1595, Drake's general orders included "to avoid the danger of fire, you must not bear about any candle or light in the ship, unless in a lantern … you must take the greatest care with the fire in the galley" (Maynarde 1849, 64). On Spanish galleons, dinner was served before sunset (Perez-Mallaina 1998, 143), and after it was completed, an officer would make sure that the cooking fires were extinguished (70, 180).

The night was divided into three watches, and the officer of the watch (who would be the pilot, the master, and either the captain or the master's assistant) had to police the crew to make sure that no unnecessary fires were lit. If a lantern were needed, it had to be signed out from the dispenser, an assistant officer, who also was the only person allowed to carry a light into the storeroom (Fish 2011, 408). Indeed, on some ships the rule may have been even more stringent; Perez-Mallaina (1998, 180) says that the only lights allowed at night were the compass-light and one lantern shared by the deck guard.

Similar rules were followed in the 18th-century British navy, except that there it was a lesser officer (midshipman or master-at-arms) that was on "unnecessary light patrol," and more exceptions were made. First, officers and elite passengers were permitted to use uncovered candles. Second, the crew had a horn lantern for each mess-group and there were a few large horn lanterns for lighting the gun deck (where the crew hung their hammocks) during the early evening. Finally, the navigator and the gunners had their own dark lanterns and the boatswain and carpenter had small horn lanterns (Quinn 1999, 50–1).

Dana (author of *Two Years Before the Mast*) served on the merchant brig *Pilgrim* in its voyage of 1834–36. The captain banned any light in a store room (many flammable items) and hence also in the adjoining steerage. A single swinging lamp was permitted in the forecastle, but it had to be extinguished at 8:00 p.m.

Warships faced the particular fire danger posed by the magazine and powder room. Measures were taken to avoid causing sparks, and the magazine itself was a large box hung below the deck, so it could be flooded in an emergency (Pope 1987, 60).

Lightning Protection

A wooden mast, like an isolated tree, may provide the path of least resistance for the lightning bolt. According to a review of lightning strikes on 220 Royal Navy ships during late 18th and early 19th centuries, 152 topmasts and 164 lower masts were ruined, and 300 seamen were killed or severely injured. In "about forty ... instances ... the ships were on fire in some part of the masts, sails, or rigging" (Harris 1852, 251). "Mid-nineteenth century American clippers were also affected. In 1855, a strike hit the clipper *Radiant*; her topgallant and royal masts [were] damaged, and the topgallant sail burned" (Knoblock 2014, 213).

In 1751, Benjamin Franklin had suggested "preserving houses, churches, ships, etc. from the stroke of lightning" by fixing, "on the highest parts of those edifices, upright rods of iron made sharp as a needle..., and from the foot of those rods a wire ... down round one of the shrouds of a ship, and down her side till it reaches the water" (Franklin 1751, 62).

In 1812, there was a general order that British warships, from first-rates down to 32-gun frigates, be equipped with lightning conductors (Blake 2005, 236n31). However, it appears that this took the form of chains "raised when lightning was expected," and this protection "often was not installed when lightning struck; interfered with seamen manning the rigging; and was not capable of conducting some lightning strokes without damage to itself or the ship" (Bernstein 1978). "In 1813,

[KK3]Frontispiece of W. Snow Harris, *Remarkable Instance of the Protection of Certain Ships of Her Majesty's Navy, from the Destructive Effects of Lightning* (1847), showing the lightning strike on the frigate *Fisgard* (42 guns) on September 26, 1846. The *Fisgard* was "at anchor on the Nisqually River, in the Oregon territory" (17). According to the logbook, "the main-mast was struck by lightning.... The next morning, on examining the conductor along the mast, the vane-spindle was discovered to have been fused at the point, and blackened one-third of the way down.... The electrical current having passed down the main-mast, took the direction of the branches to the bolts through the side,—one leading through the boatswain's cabin, and the other through the midshipmen's berth." From there it presumably flowed through "two bands of copper passing down externally over the ship's side." At "the point of contract with the branches and the iron knees within the ship, the metal appeared blackened." A boatswain's mate was temporarily blinded by the lightning, and both he and a midshipman were knocked down by the associated expansion of air (which is what produces the thunder). The lightning was forked, also striking the lower part of the same mast, as shown by "slight singing" in that location. The captain and the senior lieutenant were in agreement that "had it not been for the efficiency of the conductors the mainmast must have been totally destroyed," and Harris concluded by noting that "if in time of war, the *Fisgard* ... had been disabled and had lost her mainmast on a foreign station" as remote as Oregon, "the consequences might have been most serious."

nearly half of the inshore squadron ... on the blockade of Toulon, were ... struck at the same time by lightning, and more or less disabled" (Harris 1852).

One complication was that lightning didn't necessarily strike the highest point on the ship, and Harris reported "instances of lightning striking the fore-mast, a lightning conductor being applied on the main-mast" (253).

In 1820, Harris proposed "a system of permanently fixed lightning conductors," running down to the hull's copper sheathing. It was tested on several vessels, notably on HMS *Beagle* in 1831–6 (Forbes 1848, 16). The Royal Navy adopted Harris's system in 1842.

With modern ships having a metal hull and superstructure, lightning is much less of a concern, as they conduct the electricity down to the water surface. There is still a risk of damage to electronics as a result of a direct strike on topside antennae or "an electromagnetic pulse from a near-miss" (Swanson 2022).

On modern wooden sailboats, a "Faraday cage" may be created by the use of grounded conductors along the rigging as well as the mast. Other outboard metal structures, such as railings, may be incorporated into the cage. The resulting "cage" does not provide as much protection as an unbroken metal shell would, but it is better than a single lightning rod (Klopman 2007).

Automatic Fire Alarms

These alarms have two components, a detector and an annunciator. The detector may respond to heat (temperature or rate of change of temperature), smoke (by ionization, light scattering, or light absorption) or combustion gas. The annunciator provides a warning light or sound (or both), possibly in multiple remote locations, and may also indicate the location of the fire.

In 1890, Upton and Dibble patented a "portable electric fire alarm." When a "thermostatic coil" was heated, it caused the movement of a lever, closing a circuit if the temperature was high enough for it to close the gap between contacts. This caused a battery to energize an electromagnet, which brought a bell hammer against a bell dome, but also breaking the circuit. A spring retracted, returning the bell hammer to its original position. The circuit would then be closed again provided the temperature was still high enough. Thus, the alarm would continue to sound.

I believe that Upton's "thermostatic coil" was simply a conventional bimetallic strip of the kind used in thermostats of the 19th century. These were made of two metals with different coefficients of thermal expansion, and thus would bend when heated. Thus, detection was by a mechanical process, and annunciation, electromechanical.

In 1896, "fire was discovered in one of the coal bunkers of the United States warship *Indiana* which lay at her dock…. The alarm was given by the automatic fire-alarm with which the shipp is equipped, and [this] undoubtedly saved the vessel from destruction" (*Electrical World* 1896, 663). The alarm was manufactured by the Electric Heat Alarm Company. Its manner of operation is not indicated, but that company was assigned an 1892 patent (Fitzpatrick, UK576, January 12, 1892) which relied on the thermal expansion of mercury (GBPO 1898, 74).

Around 1913, the *Seeandbee* was equipped with the Aero Automatic Fire Alarm System. Its sensor was a fine copper tube. When the tube was heated by a fire, the air in the tube expanded, moving a diaphragm on one end. That movement closed an electrical circuit in a "switchboard" in the engine room, actuating an alarm gong. Since the *Seeandbee* was a large passenger ship (1,500 passengers), it had six switchboards. Its system included "nine miles of copper tubing … divided into more than fifty circuits." The provision of this fire protection was touted as being "contrary to the almost universal rule of steamship construction" (*Safety Engineering* 1913, 407). The same system was installed on the *Washington Irving* and the *Narragansett* (434–5).

The Aero alarm mechanism is described in more detail by Braidwood (1913, 76–77), together with several other alarm systems. Some of these were designed to detect a sudden temperature rise rather than a rise to a set threshold. Braidwood generally recommended that the alarm thermostats be "set to act at about 30 degrees below the fusing point of the sprinkler heads" (70).

The Fire Triangle

Modern firefighting manuals speak of the "fire triangle": fuel (any combustible material), heat (to bring the fuel to its ignition temperature), and oxygen (to combine with the fuel in the combustion reaction). For a fire to start, all three points of the triangle must be present. If any is removed, the fire will go out (MTAB 1994, 72).

Fireproofing

The potential fuels on a ship may be solid ("cordage, canvas, dunnage, furniture," wooden hulls, decks and masts, and coal in the steamship era, and solid cargoes), liquid (diesel fuel, lubricating oil, solvents, and liquid cargoes), or gaseous (typically in tanks, but possibly released by cargo in the hold).

The occasional use of incendiary missiles and heated shot in premodern naval warfare encouraged ad hoc forms of fireproofing. The deck, sails and hammocks in hammock nettings could be wetted prior to action, as was reportedly done by Nelson on the *Victory* before Trafalgar (De Steiguer 1900, 125). (Wetting sails could also increase speed, probably by sealing pores.)

A more permanent, but limited form of fireproofing was associated with the ship's stove. On Spanish ships in the 16th century, these were "built of firebricks and separated from the deck with sand in order to prevent the fire from spreading" (Perez-Mallaina 1998, 134). In the Georgian navy, the powder room and entryway featured lead or copper sheeting and mortar-plastered floors, and water-soaked, flannel "fear-nought" curtains were used if the ship went into action (Pope 1987, 59–60).

During the 19th century, there was a transition from wood to steel construction. Pitch-sealed wood is highly flammable, while steel isn't. But steel does have the disadvantage that it is an excellent conductor of heat, and hence could facilitate the spread of a fire by heating nearby combustibles. Moreover, steel loses strength when heated.

The 1929 SOLAS convention required ships "to have fire-resistive bulkheads that could withstand a temperature of 815 degrees Celsius for an hour; asbestos was almost the only material light enough for a bulkhead that could pass this test" (Maines 2013, 76). Asbestos was incorporated into other parts of ships, too, as well as in protective clothing for firefighters. More than a half-century later, asbestos was banned because of the risk of asbestosis.

As of 1999, the U.S. Navy was using "brominated vinyl-ester matrix resin with glass reinforcement for composite applications in topside surface ship structures" (Sorathia 1999).

Water vs. Fire

One may cool down the fuel by attacking it with water. "Blankets ... were soaked in water or even urine" (Perez-Mallaina 2018, 70), for use in smothering the fire as well as cooling it.

Each sailor had his firefighting station, and would help with buckets or pumps (see chapter 6) to put it out. Of course, to bring the water to where the fire was, they needed to attach a hose to the pump. Until the 19th century, this would have been made of leather or tarred canvas.

European scientists began studying natural rubber (caoutchouc) objects from Mesoamerica in the mid–18th century. While this rubber was elastic and waterproof, it was sticky, which limited its utility. Nonetheless, Edwin Chaffee received a patent (US 7946) for an "India Rubber Hose for Fire Engines, etc." in 1834. This was actually "linen or cotton duck ... coated on both sides with a solution of caoutchouc" (Jones 1834, 110).

In the 1830s, Europeans discovered that when heated, the latex polymer reacted with sulfur to form a non-sticky material, which came to be known as vulcanized rubber. Vulcanized India rubber hose was being advertised for sale by 1837 (*Mechanics Magazine* 1871, x). In the 20th century, various synthetic rubbers were developed.

The current standard U.S. Navy fire hose "has an interior lining of rubber, covered with two cotton or synthetic jackets" (Cutler 2017).

The type of nozzle is also important. A nozzle may be designed to produce either a straight stream or a spray (fog). The straight stream may be directed with greater accuracy, and has a greater reach. However, only a small fraction of the water actually comes in contact with the fire.

With a spray (fog) nozzle, accuracy and reach are limited, but the same volume of water, provided as small droplets, presents a larger total surface area than a solid stream, and thus can absorb more heat (MTAB 1994, 125–6). Moreover, the droplets

"Fire fighting aboard USS ENTERPRISE (CV-6), after a Kamikaze hit off Japan, 20 March 1945. Burning planes are F6F 'Hellcats'" (National Archives. Photo NH 80-G-274216).

flash into steam, and as a result absorb the water's heat of vaporization from the fire. Water expands 1,700 times when it evaporates into steam, so this helps dilute the oxygen in the air (Liu 2000). Combination nozzles exist that may be switched between the two modes.

A ship may be equipped with a manual (crew-controlled) or automatic sprinkler system. The problem with a manual system is that it is effective only if, first, a fire alarm is timely triggered, and second, the control system is not compromised by the fire. On the USS *Bunker Hill*, fires initiated by a 1945 kamikaze attack destroyed Hangar Deck Control (where many sprinkler controls were centralized), and many of the aluminum water main control wheels were melted by the heat.

Automatic sprinkler systems were introduced in the late 19th century for use in textile mills. The fire melted a solder, opening the valve (BAAS 1885). One of the first ships equipped with an automatic sprinkler system was "the steamship *Alabama*, built in 1910" (Hearings 1935, 25–26).

The effectiveness of a sprinkler system is also dependent on the "pressure head" to which the water is subjected. The higher the supply tanks are positioned, the greater the head, but that is also detrimental to ship stability (Hearings, 13).

There were two basic problems with the use of water in fighting shipboard fires. First, it could damage cargo and ship furnishings, although that was better than them being consumed outright. Second, and more important, if the water remained on board, it reduced the net buoyancy and, by the "free water effect" (see chapter 6) and (if on an upper deck) a shift in the center of gravity, stability (Kennedy 2009, 421; Maines 2013, 75; Hearings, 15).

The French ocean liner SS *Normandie* was detained in New York in 1939, after the Nazi invasion of France, and in 1941, it was being prepared to be refitted as an American troopship. The kapok life preservers were accidentally set on fire by a cutting torch. The docked ship was hosed down by fireboats and city fire engines. However, the *Normandie* had "low metacentric heights," so "a small shift in weight from one side to the other could cause it to list." While the fire was brought under control, the ship acquired a 10-degree list away from the pier. About six hours later the list was 35 degrees, "with water pouring in through open ports and a garbage chute." It capsized about three hours later (MTAB 44–5).

In 1963, the cargo and passenger ship *Rio Jachal* successfully avoided the twin perils of fire and water. It was tied up at a pier when a fire began in a stateroom. The fire was attacked aggressively with water. The water was "trapped on the upper decks," and the ship began to list to starboard. When it reached 15 degrees, the decision was made to limit firefighting operations and concentrate on correcting the list by discharging water from the starboard ballast tanks. Then the upper decks were freed of water by removing stateroom windows, etc. With the ship righted, the remaining fire was extinguished (MTAB 46–9).

Smothering a Fire

One may also extinguish a fire by diluting the oxygen in the surrounding air with a non-combustible gas such as carbon dioxide. The fire will go out when the oxygen concentration is less than 16 percent. This strategy works best when the space can be sealed off from entry of fresh air.

In 1859, the HMS *Excellent* tested the Phillips Fire Annihilator. Charged with saltpeter and charcoal, and also bearing water, it was supposed to produce carbonic acid, nitrogen and steam. It was found to be inferior to a "fire engine" (probably a conventional pump) in suppressing a fire started by white-hot shot (*Excellent* 1866, 164).

An 1880 *Naval Encyclopedia* article says that fire extinguishers generating "carbonic acid gas" were "designed as early as 1816." In essence, sodium or calcium carbonate reacts with sulfuric acid, forming carbon dioxide and a sulfate salt. The carbon dioxide dissolves in the water held by the extinguisher, and is forced out under pressure. The U.S. Navy was said to use extinguishers of the "Babcock pattern" (*Naval Encyclopedia* 1880, 278).

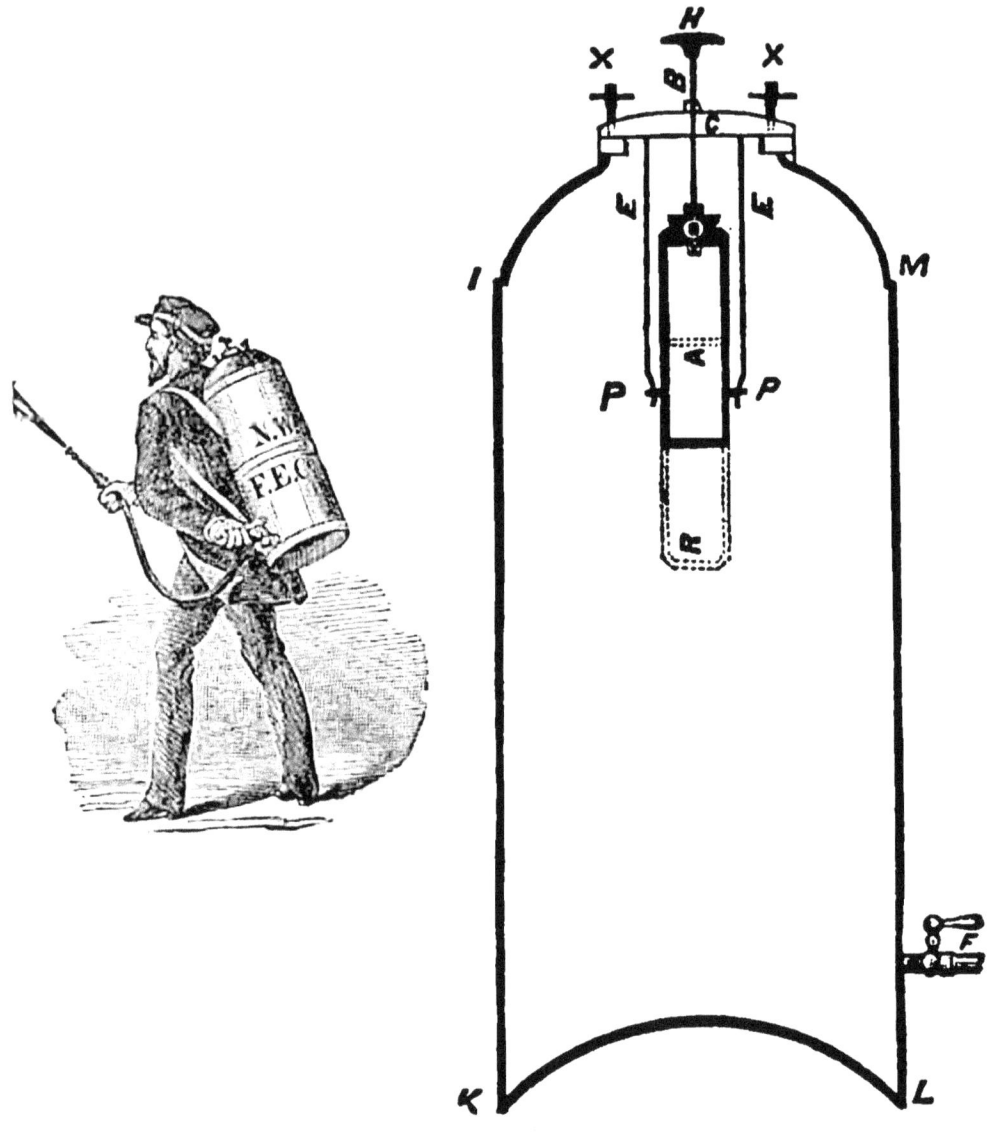

Babcock fire extinguisher, from "Implements Exhibited at State Fair, 1870," in J.M. Shaffer, *Report of the Secretary of the Iowa State Agricultural Society for the Year 1870*, page 308. The firefighter dissolves the salt in water and fills the extinguisher with the water "to within three inches of the top." He pours acid into a lead bucket, closing the latter with a lead stopper, and inserts it into the extinguisher, and secures it with screws X. "In case of fire, pull up … handle H; this draws out stopper O, and bucket A turns bottom side up, as shown by the dotted lines R," around pivots P, "thus discharging contents into the carbonated water." The products of the chemical reaction, confined by the tank, generated a pressure of 60–90 psi within ten seconds of pulling the handle. Note that the spout F is at the bottom of the tank. Note also James Babcock, "Improvement in Apparatus for Extinguishing Fires," US Patent 80,701 (granted August 4, 1868), but its construction is different. By the early 1900s, the Babcock fire extinguisher featured "acid … held in a sealed bottle," which was "crushed by a screw device at the head of the extinguisher" when put to use. See *Marine Engineering* 10: 443 (October 1905).

The chemical foam extinguisher was invented by Aleksandr Loran in 1902. Sodium bicarbonate was reacted with aluminum sulfate. The evolved carbon dioxide formed bubbles, and additives kept the bubbles stable (Hildebrand 2017).

"Firefighting foam is used to form a blanket on the surface of flaming liquids.... The blanket prevents flammable vapors from leaving the surface and prevents oxygen from reaching the fuel.... The water in the foam also has a cooling effect" (MTAB 130).

At the onset of World War I, there were foam apparatus "permanently installed in the fire rooms of oil burning destroyers," and "portable foam type extinguishers kept in fire rooms, engine rooms," etc., of naval vessels (Navy 1914, 44).

Chemical foam has been superseded by mechanical foam, in which "the bubbles are formed by the turbulent mixing of air and the foam solution." The bubbles themselves are filled with air, not carbon dioxide. Foam solutions have been based on proteins and alkyl sulfonates, "Aqueous Film-Forming Foam," "a combination of fluorocarbons, surfactants and solubilizers" (Stag 2021), was developed by the navy in the 1960s and widely adopted by the military to fight fuel fires. Its use has since been limited out of environmental concerns.

Breaking the Chain

During the combustion chain reaction, hydroxyl free radicals are generated. Halons (halogenated hydrocarbons), when heated, release halogen free radicals that react with the hydroxyl free radicals and thereby terminate the combustion reaction. Carbon tetrachloride (Halon 104) was used in a portable fire extinguisher in 1911. The U.S. Navy later employed bromotrifluoromethane (Halon 1301) and bromochlorodifluoromethane (Halon 1211) (NASEM 1997, Appendix A).

Halons are extremely effective fire extinguishing agents, but they were ultimately banned because of their effect on the ozone layer.

Firefighter Protection and Training

We have already alluded to protective clothing. In World War II, the navy began providing firefighters with rescue breathing apparatus (*Safety Review* 1947).

According to Block, "prior to our entry into World War II the subject of firefighting was basically ignored by the U.S. Navy" (Block 2009, 125). According to Edward Kehoe, after repeatedly losing warships in the Pacific to fire, the navy conducted experiments to determine the most suitable equipment and techniques for handling oil and gasoline fires, and set up "realistic fire schools." Firefighting teams were led or trained by reservists "recruited from municipal fire departments." The Coast Guard adopted a similar approach to the protection of cargo ships (*Safety Review* 1947, 21).

The USS *Forrestal* fire (1967) came at a time when the U.S. Navy confined firefighting training to specialized damage control teams. The explosion of unstable ammunition killed "nearly all of the aircraft carrier's trained firefighters," resulting in a muddled response. "All current Navy recruits receive week-long training" in firefighting, and are tested on their ability "to use portable extinguishers and charged hoses to fight fires."

Contagious Diseases

Sailors returning from leave brought pathogens on board. In the close confines of a ship, contagious diseases could spread like wildfire. There was a cholera outbreak on the troopship HMS *Apollo* in 1849. A former frigate, its normal wartime crew had been about 300 men, but it was now carrying 711 passengers from England to China. "The first case of cholera occurred in a soldier after a few days at sea." After additional cases, the ship diverted to the "Ilha Grande in Brazil for refitting and cleaning." By the time it arrived, there had been 18 deaths. The pattern of morbidity was indicative of person-to-person transmission rather than a contaminated water supply (Goodyer 2008). On June 17, 1910, there were 63 cases of typhoid fever on the Japanese battleship *Iwami*, and more than 50 on the *Mikasa* and *Suo* (PHR 1910).

Ships also played a substantial role in the dissemination of infectious diseases by carrying infected humans (or other carriers) from one port to another. For example, the SS *Shonga* brought influenza from Freetown to Accra in 1918. By the time of its arrival, "virtually the entire crew was sick.... The vessel and the crewmen were quarantined," but the infection was not contained. The disease was widespread in Accra within two weeks, and at the height of the epidemic, "it had sometimes been necessary to bury several people in a single grave" (Patterson 1983, 487–8).

Nineteenth-century statistics on shipboard disease can be somewhat frustrating to review. The disease may be merely characterized as a "fever," or a "disease of the digestive system," rather than something more specific. That said, in the British Royal Navy in 1891, the average daily sick rate was 41.33 per thousand, and the death rate from illness alone, 4.68. Cases and deaths, included, inter alia:

Table 9–1

Ailment	Cases	Deaths
connective tissue and skin*	8,193	1
digestive system*	6,455	15
venereal	4,525	4
influenza	3,527	6
rheumatism*	2,563	4
simple continued fever	1,804	1
malarial**	1,559	14

Ailment	Cases	Deaths
nervous system*	1,390	10
respiratory system*	745	72
enteric fever	206	46

* These obviously include some ailments other than infectious diseases.
** Severe yellow fever may be confused with malaria.

There were additionally cases and deaths specifically identified as measles (93 cases, 0 deaths), scarlet fever (58, 1), cholera (50, 29), dysentery (51,1), and smallpox (17, 0) (Stevenson 1898, 575).

On modern cruise ships, "over 90% of gastrointestinal illness outbreaks with a confirmed cause are due to norovirus." Diseases which are transmissible by aerosol, such as influenza, Legionnaires' disease, and COVID-19 are also of concern. There have also been outbreaks of vaccine-preventable diseases, such as varicella, among crew members, many of whom come from countries with low immunization rates. Passengers may be exposed to arthropod vector-borne diseases at ports of call, especially in the tropics (Tardivel 2023).

In the remainder of this chapter, I will focus on the role of ships in the dissemination of plague, and the corresponding efforts to reduce rat and flea populations on board.

Quarantine

Quarantine was the first moderately effective defense against the plague (and other infectious diseases). The concept of quarantining incoming vessels from foreign ports was introduced by the Venetian Ragusa in 1377. The period was originally 30 days (*trentino*) (Mackowiak 2002) but was increased to 40 (*quarantino, quarentena, quaranta*) by Marseilles in 1383. Other periods were used; Dublin in 1625 specified 20 days (Byrne 1998).

Some ports of embarkation were generally deemed safer than others. In 17th-century northern Italy, people on vessels that came directly from England or Flanders might be allowed entry after a few days, whereas those from Spain had to do 25 days of quarantine. If there were deaths or illnesses during the voyage, or during the quarantine, which weren't clearly attributable to an acceptable cause, the period of quarantine would be extended (Cipolla 1981, 111ff).

In the 16th century, a procedure was developed for more rapidly clearing passengers and goods. If their last port was free of disease, and the goods had been purified, the vessel would be given a "clean bill of health" by the local inspectors, and the captain would have this validated by the local consul for the port of destination. On arrival, if there was no reason to doubt the bill of health, the passengers and goods would enter without quarantine, or with an abbreviated one (Gensini 2004).

But it depended on circumstances. In 1652, Leghorn admitted passengers from the Levant after three days, if their vessels had "clean bills." On the other hand, Genoa would make them sit tight for 30–40 days, depending upon "information received" (Cipolla, 113).

If a vessel came from a port under ban or suspension, it would have to go through quarantine even if it had a clean bill of health. Indeed, in Genoa, the crew and passengers of a ship coming from a city under "ban" or "suspension" (a provisional ban) would not be allowed to disembark at all, and only goods considered not subject to contagion could be offloaded (Cipolla, 19).

There were attempts to game the system. Plague was brought to Messina 1743 by a merchant vessel from Morea that relied on a clean bill of health from an intermediate stop and also claimed that a plague victim was merely a sailor who had fallen overboard (Simpson 1905, 34).

The Venice Convention of 1897 classified ships as being healthy (left infected port at least ten days before yet had no cases of plague on board), suspect (had cases, but not within the last 12 days), or infected. Crew and passengers from healthy ships were just placed under surveillance for ten days (they could go about their business, but doctors checked up on them). Those from suspect ships were treated the same way, but the ship was disinfected. Only those from infected ships were quarantined, for up to ten days (Simpson, 355). The 1903 convention reduced the quarantine to five days and required that all rats on an infected vessel be destroyed (358).

A general problem with quarantining an entire vessel is that it confines those who actually have the disease with others who are merely possible contacts, thus putting those at greater risk. Of course, it is possible that on the ship, the afflicted will be isolated from everyone else. The weak link in such isolation is that it is still necessary for healthy crew members to provide care to the afflicted.

Plague was "the chief target of quarantine until yellow fever struck European and North American ports with increasing frequency in the late eighteenth century, followed by cholera after 1831" (Barnes 2014, 77). Those three are also the current "internationally quarantinable diseases."

Pest Control

One 19th-century traveler reported, "rats, mice, mosquitoes, locusts, flies, bugs, moths, cockroaches, fleas, scorpions, centipedes, and others, infest the shipping more than the shore, for having numberless places of concealment in the holds, in the interstices of the beams and timbers, and in the provision casks, it becomes quite impossible to eject the enemy after his once having made a good lodgment." They had been "frightened in our beds by the prowling of rats, … bitten by fleas, driven out of bed by bugs, and in danger of being fairly carried off by thousands of cockroaches" (Shaw 1823, 3: 251).

Daniel Ammen (1891, 35) complained vehemently about cockroaches on the sloop-of-war *Vandalia*, stationed at Pensacola, Florida. These came out when the hammocks were hung up (presumably because the lights were doused) and flew around. The crew's solution was to set out molasses-baited traps (large jars with narrow mouths).

However, Robert Stevens declared that of all vermin infesting ships, "the rat is the most injurious" (1873, 759).

Rats and Their Adverse Effects

The rat species of greatest concern are *Rattus rattus* (roof, ship or black rat) and *Rattus norvegicus* (Norway or brown rat). The former is more common on ships and weighs 4–13 ounces; the Norway rat is larger (7–17 ounces) (Szumlas 2008, 2–25). Mice (0.25 ounces) may also invade ships.

Besides contaminating or consuming provisions and cargo, rodents may "gnaw electrical insulation," causing "electrical shorts, outages and fires," and serve as vectors for communicable diseases, notably "plague, murine typhus, leptospirosis, and food-borne illnesses (e.g., salmonellosis)" (Szumlas, 2–24).

A risk specific to wooden ships was that the rats, searching for water, would gnaw a hole in the side of the ship. "Guided by the rippling of the sea, they select a plank where the sap is gone close to a seam, and by combining together, work incessantly until salt water oozes through and they find their labour useless. The weak barrier left behind gives way sooner or later, the cargo is injured, and the lives of all on board are in jeopardy." Stevens proposed giving them "a daily supply of water rather than risk such perils" (1873, 760).

The Age of Sail was also the age of rats eating the sails (aloft or stored), or tearing out chunks for use as nesting material. "Some masters ... saved their sails by supplying soft paper for the nests" (761).

A ship can harbor an amazing number of rats. In San Francisco, a 260-ton lumber carrier was fumigated, and 310 dead rats were collected. A grain-carrying vessel yielded 1,700 rats (Treasury 1910, 208). Simeon Shaw advised that by the time the 70-gun warship HMS *Valiant* returned from Havana in 1766, its rat population had increased to the point that they "devoured daily one hundredweight [112 pounds] of biscuit" (5:80).

Since shipyards are often infested with rats, a ship may become a harborage for rats while it is still under construction (Grubbs 1931, 3). Once it is launched, it may acquire a new contingent of rats whenever it visits a port. The rats may walk up the gangway or mooring lines, or be brought aboard together with the cargo.

Rats "can squeeze through a hole no larger than a quarter dollar," jump upward two feet (three with a running start) and four feet horizontally (more if jumping from a height), and climb a vertical wall if they can find a claw hold or brace their backs (such as at an inside corner). They can survive a five-story fall and "tread water for three days" (Hendrickson 1983, 87).

Rats, Fleas and Plague

Urban epidemics were often preceded by numerous rat deaths, sometimes taking the form of "rat falls" (especially in buildings with thatched roofs) (Abbott 2012, 32).

The earliest European reference I found to rat control as a plague preventative was in Defoe's fictionalized account of the Great Plague of London 1665–6: "All possible endeavors were used, also, to destroy the mice and rats, especially the latter, by laying ratsbane [arsenic] and other poisons for them" (Hendrickson 1983, 58). However, the role of the rat flea remained unsuspected.

The principal insect hosts of the plague bacillus (*Yersinia pestis*) are fleas. The flea acquires the bacterium as a result of biting an infected host, and transmits it by biting a second host. So a flea may serve to carry the infection from humans to rats and from rats to humans.

The first experimental demonstration of rat-to-rat flea-to-rat transmission was by Simond in 1898.

In general, fleas remain on a living host. Thus, exterminating flea-ridden rats may result in increased plague transmission to nearby humans, unless the fleas were previously or simultaneously eliminated, or the humans are wearing protective clothing. A plague-infested rat is more likely to come out into the open, and is also more readily killed (Simpson 1905, 363).

With individual exceptions, rats are not resistant; plague is just as much of a calamity for them as it is for us. And so infected fleas will cycle rapidly through the rat population, giving fleas more reason to jump to us.

Flea Control

Fleas were deemed a nuisance to humans and their pets long before their connection to plague was suspected.

To bite you, the flea must reach you. There are basically two approaches to preventing this: interpose a physical or chemical (repellent) barrier between the fleas and our bodies, and trap or kill the fleas beforehand by mechanical or chemical means.

It is doubtful that sleeping in hammocks was an effective defense. While the hammock itself is certainly elevated higher than the fleas can jump, they can crawl up the wall and then along the hammock rope. Moreover, sailors could go to sleep with their clothes on, which the fleas might have already boarded. An officer on the mid–18th-century East Indiaman *Argonaute* said that the berths were "infested" with fleas, despite the use of hammocks (Proulx 1984, 103).

Various hydrocarbons, and pyrethrum, an insecticide produced by the Dalmatian chrysanthemum, were in use (at least on land) to kill fleas by at least the 19th century. Munson (1920, 371) declared that "both pyrethrum powder and coal oil are effective against fleas," and if they failed to eliminate the pest, "the ship may be fumigated with sulphur dioxide or hydrocyanic acid gas." DDT was available by the 1940s; it had a more prolonged effect than pyrethrum, but it also killed more slowly (PHS 1947, 73, 383). It has since been superseded by other insecticides.

Rat Extermination

Rats may be exterminated using natural predators, traps, chemical rodenticides, and rat pathogens.

"Hieroglyphics show Egyptian seafaring traders brought cats along on trips to the Mediterranean Sea" (Sweeney 2021). The medieval "Good Customs of the Sea" provided, "if goods shall be damaged by rats on board a ship, and there be no cat in the ship, the managing owner of the ship is bound to make compensation" (Twiss 1874, 90). But 19th-century Anglo-American case law was ambivalent as to whether the presence of cats was a defense (Stevens 1873, 766). In 1946, the War Shipping Administration advised against carrying cats, expecting the effect on the rat population to be "negligible," and fearing that "the cat may act as a vehicle of disease transmission" (131).

Predators may kill rats, or they may just encourage them to move somewhere that they cannot smell the predator's presence. The latter effect is probably more useful on land than on a ship underway, where the rats will just move to someplace inaccessible to the predator, rather than jump ship. Whether a predator will stalk and kill a rat will depend on how hungry the predator is, and the risk to the predator of injury if the prey fights back. Parsons (2018) reported that feral cats showed little inclination to hunt large urban rats.

Stevens (1873, 764) warned that in 1857, "a cat was taken on board the…. *Konigin der Nederlanden* in the evening; the next morning nothing was to be seen but her skin and bones." (Probably a sailor's tall tale, since an outnumbered cat would be more likely to decide that discretion was the better part of valor.)

Rucker complained that the ordinary cat won't attack a large rat and favored dogs, especially terriers, as ratters (Treasury 1910, 159). Stevens (764) said that "bull-terriers are considered preferable to domestic cats, but they cannot follow vermin so easily." The *Mary Rose* had a dog on board, with DNA similar to that of modern Jack Russell terrier and whippet breeds ("Hatch" 2023).

Ferrets are rat predators, and "frequent attempts have been made to keep the ferret on board ship," but ferrets proved "badly suited" for shipboard life (Fennell 1841, 88).

Professional rat catchers have existed for centuries. They could be hired in port to deal with an infestation. At Woolwich, when the steam frigate *Terrible* (launched

1845) was found to be "swarming with rats," a rat-catcher was brought on board. He "caught over 400 alive, and the ship was afterwards fumigated, to suffocate the remainder" (Rodwell 1858, 163).

Traps may imprison, immobilize or kill rodents. The imprisoning (cage) traps may either drop down on the vermin or permit entry through a trap door that shuts when triggered. The immobilizing traps may have glue boards or steel jaws. The killing traps may have a spring-loaded wire (snap) to break the animal's back, or they may be barrel traps with tipping tops to drown it (Treasury 1910, 154ff; Pollitzer 1952, 460). But rats are clever and will learn to ignore traps, or to snatch the bait without getting caught.

Bait may be impregnated with a rodenticide that kills by ingestion or even skin contact. Arsenic (trioxide), sold as "ratsbane" in the 1500s, is relatively cheap and slow acting (so the poisoned rats die in their burrows, away from humans). Phosphorus was also used as a rodenticide (Stevens 1873, 764).

The bulbs of an onion-like Mediterranean plant (red squill, *Urginea maritima*) contain scilliroside, a potent rodenticide. It is one of the safest rat poisons because it causes humans and pets to vomit (rats can't vomit) (Pollitzer 1952, 407ff; Verbiscar 1986). It was still listed as a recommended rodenticide in the 1987 edition of *The Ship's Medicine Chest* (10).

Chronic (slow-killing) poisons have the advantage that the survivors are less likely to learn to avoid the bait. The bait is first left without the poison, and once it is taken the poison is added. The first chronic poison used was the anticoagulant warfarin, which was registered as a rodenticide in 1948. The rats perish as a result of internal bleeding. A 1950 report urged that "warfarin appears admirably suited for the control of rats on ships" (Hayes 1950, 1553).

Extermination, by itself, is not an effective control technique. When a substantial number of rats are killed, the survivors become more wary of traps and hunters, and they also face less competition for the available food. The decrease in rat population is temporary (Creel 1896).

Nonetheless, even in the early 1900s, "every good ship operator" had rat traps in "the storerooms and galleys" (Grubbs 1927, 513), and rat-trapping was useful as a means of monitoring the effectiveness of fumigation.

Paying the crew to trap rats helped; in 1916, 99 percent of the rats trapped on trans–Pacific liners coming to Hawaii were from liners of the Toyo Kisen Kaisha Steamship Company, since that company paid a *sen* (half a cent) for each rat trapped (Blue 1916, 167). (Hopefully, the crew didn't bring rats onto the ship to boost their earnings.)

Fumigation

A fumigant is a substance which, under expected temperature and pressure conditions, can be produced in the form of a gas at a fatal concentration for a target pest. Fumigants kill by inhalation, at least partially by asphyxiation (displacement of oxygen), and in most cases by a direct toxic mechanism, too. A gas can enter spaces that a rat-catcher cannot.

Fumigation was not strictly an anti-rat measure. It could be intended to kill insects, with the effect on rats simply a bonus. Also, in the premodern period, fumigation was considered a method of combating disease, whether or not there was any recognized relationship between the disease and an animal vector vulnerable to fumigation (Peterson 2016, 60).

Fumigating vessels with cargo is more complicated than fumigating an empty ship, as you must avoid damaging the cargo or starting a fire (Treasury 1910, 212ff). There is the option of transporting the cargo to an off-ship fumigation chamber.

Use of fumigants was potentially dangerous. Most fumigants were toxic to humans as well as to their intended targets. Sailors could be sickened or killed if they didn't respond to the order to evacuate prior to fumigation, if they remained on board during fumigation and there was a leakage of the gas, or if they entered a previously fumigated area before the gas was completely flushed out.

Moreover, if the fumigant were released by heating, there was the risk of a fire getting out of control. Elisha Kent Kane led a search for the Franklin expedition. Plagued by rats, his crew burned charcoal below deck to generate "carbonic acid gas" (this decomposes into carbon dioxide and water). The "hatches were fastened down; but by some carelessness, the ship herself caught fire and, in the suffocating agonies of the poisonous gas, it was with great danger and difficulty that it was extinguished" ("Author of Brazil" 1799, 329). Stevens (1873, 763) asserts that "some ships" have been burned as a result of fumigations at sea, "and many lives lost."

Under British law (Passenger Act 1803), passenger vessels had to be "fumigated with vinegar at least twice in every week." Vinegar is dilute acetic acid. The method of fumigation isn't specified but most likely it involved boiling the vinegar. It was later acknowledged that its efficacy as a disinfectant was very limited (Beasley 1878, 362).

Another early fumigant was smoke. In 1813, on the *Essex*, the rats were eating "not just provisions but clothes, flags, sails, and gun cartridges, even endangering the planking of the hull with their gnawing" (Budiansky 2011, 273). Commodore Porter sailed to the Marquesas and "fumigated the ship with smoke, suffocating over 1,500 rats (which were dumped into the bay)" (Cox 2021). It appears that he burned charcoal, so the rats were asphyxiated by carbon dioxide (Budiansky).

On steamships, funnel gas (a mixture of carbon monoxide and dioxide) was sometimes used for fumigation against rats. However, it was not effective against fleas, which limited its utility (Byam 1922, 1036).

Stevens (1873, 763) says that "ships are generally smoked in dock, when the cargo is out," but in "warm latitudes," it can be carried out as "all hands can sleep on deck under awnings." He warns "not to allow anyone to go below … until several hours after the hold has been well ventilated."

The earliest documented fumigant, sulfur dioxide, was obtained by burning sulfur. Odysseus ordered Euryklea, "Bring me sulfur, which cleanses all pollution, and fetch fire also that I may burn it, and purify the cloisters" (Butler 1900). At the Vasa Museum in Stockholm, a placard asserts that in 1628 Kalmar, "ships which had dead and sick men on board were fumigated with sulphur and then tarred."

The sulfur could be burned in small lots in well-distributed iron pots, or in a central furnace that was equipped with large tubes through which it would be blown throughout the ship. The former approach was favored in the United States. Sulfur dioxide may also be obtained by burning carbon disulfide, which produces sulfur dioxide more rapidly (Williams 1934, 89–90).

Sulfur dioxide will kill plenty of rats, but the smell alerts them to danger, and there are many places for them to hide where air circulation is poor (Grubbs 1931, 1267). Other issues with sulfur dioxide fumigation include that it is a relatively heavy gas, which slows its diffusion; it is rapidly absorbed by moisture, a problem in humid holds; it can damage "certain cargoes and … various ships' fittings; and there were variations in both sulfur purity and the fraction that is burnt and converted into the fumigant" (Williams 1934, 91–92).

Wooden hulls are leakier than metal ones, and thus required more intense fumigation. Havard (1914, 745) says that with sulfur dioxide, you want 48 to 72-hour exposure in wooden ships as opposed to just 24 hours in metal ones.

Nonetheless, in 1914, sulfur dioxide was virtually the sole ship fumigant in use at American ports, and even in 1932 it was used 6.5 percent of the time in the continental U.S. and 89 percent in the Philippines (Williams Feb. 1934, 192).

Even after hydrogen cyanide displaced it at major ports, sulfur dioxide remained in use at those with occasional needs for fumigation, because the fumigation crews didn't require the same stringency of training. Sulfur dioxide has the advantage that it can be smelled at concentrations much lower than those that are dangerous to humans (Pollitzer 1953, 505).

Hydrogen cyanide was first isolated in 1786, although bitter almond extracts were used as poisons at an earlier date. Its use was first authorized by the U.S. Public Health Service in 1910 (Link 1955, 67). In 1915–16 New Orleans, public health officers compared the efficiency of fumigation with either sulfur dioxide (3 pounds per 1,000 cubic feet, 6-hour exposure) or hydrogen cyanide (5 ounces per 1,000 cubic feet, 0.5 hours for superstructures and 1.25 for holds). The efficiency was the number of rats killed by fumigation as a percentage of the sum of those so killed and those trapped after fumigation. The results were as follows:

Table 9–2

Target	Sulfur Dioxide Efficiency	Hydrogen Cyanide Efficiency
entire vessel, loaded or empty	77	95
loaded holds	64	80

Target	Sulfur Dioxide Efficiency	Hydrogen Cyanide Efficiency
empty holds	96	99
above-deck compartments	55	94

(Creel 1917)

This showed that sulfur dioxide was excellent for fumigating empty holds, but otherwise inferior to cyanide.

There was a plague pandemic that began in the 1890s. "Within 10 years (1894–1903) plague entered 77 ports on five continents: Asia (31 ports), Europe (12), Africa (8), North America (4), South America (15) and Australia (7)"(WHO 2011, 26). In 1913, the surgeon general ordered that "all vessels from ports in Africa, South America, Asia ... and the West Indies were to be fumigated for rats on arrival at US ports, regardless of prior inspections" (Frierson 2012). This overwhelmed the sanitary officials at some major ports, including San Francisco. Ships had to remain at anchor or at docks longer, resulting in extra costs.

The problem was exacerbated by the time needed for sulfur dioxide fumigation, and so the authorities were pressured to use the more potent hydrogen cyanide instead. At San Francisco, hydrogen cyanide was approved for broader use in February 1916. The fumigators were provided by the shipping company, and "the Public Health Service supervised the work."

Several tragedies followed. The first was when a sailor died in October 1916, on the *Tokiwa Mura*. The reason for the crewman's death aren't clear but possibly he didn't leave the ship prior to the fumigation. In November, on the SS *Roald Amundsen*, a crewman was lowered into the hold three hours after the hatches had been opened to ventilate the fumigated vessel (the instructions then called for just a half hour of ventilation). He "signaled frantically for help," and rescuers were lowered to aid him. His life was saved, but a left-behind rescuer perished (Id.).

A diplomatic incident was caused by the November 1916 U.S. Public Health Service fumigation of the British steamship *Devonian*, which resulted in the death of three carpenters. The British ambassador protested that hydrogen cyanide was among the "most dangerous to human life of the better known fumigants and that its use requires considerable care owning to its lightness and consequent liability to become concentrated in the upper part of a ship's hold" (Vidich 2021, 463 n54).

Despite various successive safety measures—use of a caged test animal, provision of gas masks, use of power ventilators—deaths continued to occur. Cyanide fumigation was suspended for a time. Perhaps the most significant improvement was the addition of an eye irritant, chloropicrin, as an "early warning" of exposure.

Currently, the most commonly used ship fumigant is phosphine gas, generated by the reaction of a metal phosphide upon exposure to the moisture in the area. Since the phosphide is solid and no heating is needed, it is easy to apply.

There have been several illnesses and deaths attributable to human exposure to phosphine gas when it was used for in-transit fumigation of agricultural cargo. The first was on the MS *Marian Buzcek* in 1958 (Djurhuus 2021, 213). On the MV *Monika* in 2007, "a young seafarer died.... His cabin was located straight above the cargo hold, and inspection ... revealed pinholes into the cabin from the cargo hold" (212).

The length of time actually required to complete fumigation depends on the temperature and humidity in the hold. If the hold is cold and dry, a procedure that should take five days could stretch to a month (Gard 2011). Thus, if the moisture level in the cargo hold at the time of deposit of the phosphide tablets is limited, then some residue of metal phosphide may be left unreacted. "When the holds are opened..., fresh air with moisture will enter the holds and reaction will restart." This caused illness on the part of both crewmembers and port workers on the *Arklow Meadow* in 2012 (Djurhuus, 212).

Rat-Guards

Even if a ship manages to clear itself of rats, there is a risk that it will be reinfested once it comes into port, and port authorities worry that a newly arrived vessel will inadvertently disembark a horde of rodents, possibly ones infected with plague.

An 1857 article claimed that "it is common upon coming into port to fill up the hawser holes, or else to run the mooring-cable through a broom, the projecting twigs of which effectually stop the ingress of these nautical quadrupeds" (LQR 1857, 70). Given the climbing and jumping ability of rats, the actual effectiveness of these methods is doubtful. It was better to stop them before they reached the side of the ship.

Stevens (1873, 761) said that ships loading sugar at Port Louis, Mauritius, had the problem that rats at sunset would swim over, climb up the cables, and feast all night. "The ordinary prevention is a circular piece of wood, like the head of a cask, made in two parts, to fit on the cable at right angles; the outside covered with tin." (There is no reference to this device in prior editions.) In an 1887 novel by William Westall, the protagonist's ship takes a fever ship in tow, and belatedly places a "round board, studded with nails" on each of the hawsers, "to prevent an invasion of rats" (72). In both cases, the guard was an improvised device.

By 1899, New South Wales required that all lines to shore be "defended by not less than two discs of metal ... one near the ship and one near the wharf-end" (162, 488).

The earliest hawser-guard patent that this author is aware of is Conoley, British Patent 5003 (1901). This featured a single cone and thus obstructed traffic in only one direction. He later received a patent for a double cone construction (27,957; 1911).

Early 20th-century authorities required that ships be fended off six feet from the dock, gang planks raised at night, and mooring lines equipped with "rat funnels" (Link 1955, 551). Hurdy recommended that these funnels be of heavy galvanized iron, with a three-inch diameter spout and increasing to at least 36 inches diameter at the other end. But Bilderbeck (1931) said that rats can jump 30 inches vertically, so the "shield should be of a size approximating 5 feet in diameter," which he admitted was "unwieldy."

"Hodgson and Chitre found that, although the metal guards at first formed a considerable obstruction to the passage of rats, within a few days the rats learnt how to pass them" (Bilderbeck). There were indeed numerous design and installation problems with disc rat-guards, as shown by Denney's photographic studies (1937).

Taylor and Chitre proposed use of an electrified guard. The guard they designed and tested was a triangular box, 30 inches long, that could be folded around the rope. The box

A single-cone rat-guard is installed on a mooring line of the USS *Blue Ridge* (LCC-19) by Seaman E.H. Henderson. Note that this design, as installed, only blocks rats from boarding the ship, not from leaving it (photo taken March 3, 1971. Naval History and Heritage Command. Catalog K-88617).

was wired on all three sides. The interior of the triangle was filled in by "suitably shaped pieces of wood attached to the inside of each end." A 230-volt alternating current was fully effective, but that was then available only in port. Ships could provide 110 volts direct current, but that was only partially effective (Taylor 1924, 651).

Sounds promising? Bilderbeck found that when wet—a rather likely condition, especially in the monsoon season—"the current short-circuited on the wet wood and rats were able to cross unharmed" (18). Bilderbeck proposed a cylindrical rat guard, with longitudinal electrifiable strips running between hinged half-rings at each end. These were composed of both insulating and conductive layers. The rings were "designed to destroy any continuity of any water film between conductors of

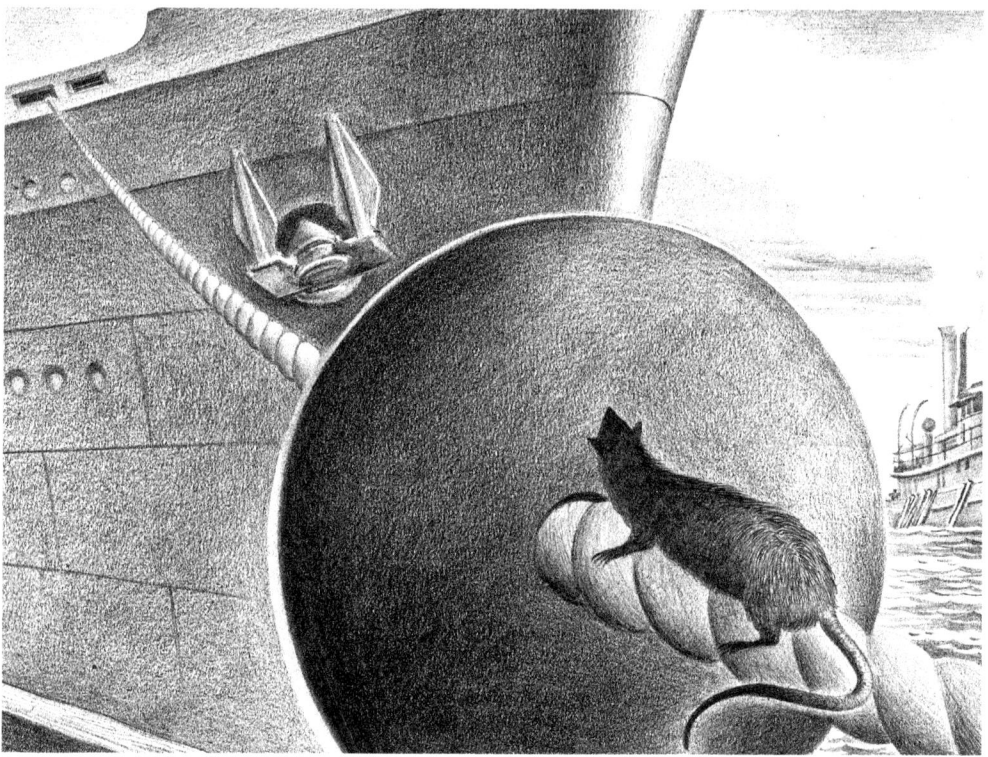

Artist's conception of rat guard in action. Drawing ("Obstructions on mooring-lines to stop rats boarding ships") by Albert Lloyd Tarter (1913–1988), in 1940s. One of several drawings "produced for an educational film" about the plague that "was never made" (Wellcome Collection, 10501i).

opposite potential." While this guard was effective in halting the rats' progress, the current used wasn't strong enough to kill, and if the rat fell into the water, it might swim to shore.

It doesn't appear that the electrified guards caught on. Perhaps they were considered too costly, or perhaps there was fear of self-electrocution. But the simple mechanical guards are still in use. While simple disc guards are commercially available, the U.S. Navy uses conical guards (Szumlas 2008, 2–28).

Rat-Proofing

Fumigation, while more effective than poisoned bait and traps, did not guarantee the death of all rats on board. According to a 1907 report, the steamship *Innanimoba* had to be fumigated with sulfur five times over the course of less than two weeks, unloading the coal bunkers before the final treatment, before she was deemed free of infestation (Grubbs 1913, 1,268). The problem was that the fumigant "will not always kill a rat if he can find any protection." This could be in cargo, dunnage, coal bunkers, pipe casings, bilges, the interstices of a double-walled hull, behind a stove, in drawers, or in loose material on deck.

The authorities therefore recommended opening up these spaces for fumigation. On some ships, which frequently visited ports where fumigation was carried out, some spaces were left open permanently (Grubbs 1927, 513). On other ships, panels were installed "which could be easily taken out or opened on hinges" (Grubbs 1913, 1268). These and similar measures, which began in 1912 (Link 1955, 67), came to be called "partial rat-proofing."

Grubbs (1925, 1,507–8) defined a "rat-proof ship" as "one on which it is impossible or difficult for a rat to hide, nest, or move about in search of food." He acknowledged that rats may board such a vessel, but they "will be confronted with an acute housing problem, high cost of living, and poor transportation between home and business (food getting).... They will breed with difficulty, and instead of multiplying will decrease or even disappear." (Note that it did not mean that rats couldn't get on board in the first place.)

He enumerated three basic principles: (1) "all foodstuffs ... should be in rat-proof containers or kept within rooms ... into which rats cannot come from without"; (2) "all small enclosed spaces should either be abolished or if this is not possible, should be blocked off by material impenetrable to rats"; and (3) all "openings should be closed by screening or by doors or windows that fit and are impenetrable to rats at gnawing levels" (1,509). The openings to be considered included gaps where pipes pierce a deck or wall (to be closed by sheet metal collars). Large openings could be covered with metal screens rather than solid sheets.

It is obvious that it would have been more difficult to rat-proof a wooden ship than one of metal construction. "The jaw muscles of a rat can exert up to 12 tons (or 24,000 pounds) per square inch. By comparison, a great white shark bites with a force of 1.8 to 2 tons per square inch." Moreover, "rat teeth are ranked 5.5 on [the Mohs hardness] scale, which means their teeth are harder than copper and iron," as well as aluminum. So they can easily chew through wood (Terminix 2023). However, hardened steel, like that used in a steel file, is about 6.5 (Peters 2001, 35).

Ship owners had several strong financial incentives for rat-proofing their vessels (or buying ships built to be rat-proof). Rat-proofing reduced damage to cargo, and rat-proofed ships were less frequently subjected to fumigation (and its direct and indirect costs) (Link 1955, 68). Grubbs (1927, 509) asserted that fumigation costs were "often several dollars per rat," and Sherrard (1943, 1889) estimated $600–1,000 per fumigation as the direct cost for an average-sized freight ship.

In 1929, the American Marine Standards Committee published Standard H41-1929, "Rat Proofing of Ships." In 1931, Grubbs updated his popular rat-proofing guide to reflect the standard and made some interesting points: (1) conventional extermination and fumigation rarely eliminate all rats, because of the many protected places on board; (2) the survivors will multiply and rapidly replace the colony's losses; and (3) the real limits on shipboard rat population are harborage (nesting sites) and food, but given rats' feeding habits, it is difficult to protect all possible food sources. Hence, rat-proofing should be aimed primarily at eliminating harborage and "runs" between possible harborage and food or water sources. And he opined that "rat proofing is more effective on ships than on shore" (12).

Out of 4,418 ships entering American Atlantic ports in a seven-month period in 1936–37, 8.4 percent were infested by rats. In contrast, in 1925–27, half the ships arriving at the port of New York were so afflicted. Rat-proofing of vessels was given partial credit (PHS 1937, 413).

The following statistics are even more persuasive:

Table 9–3: Vessels Undergoing Sanitary Inspection, New York, FY Ending June 30, 1943

	Rat-Proofed	Non-Rat-Proofed
Total	763	3004
Rat-Infested	15	983
… and fumigated	0	157

(Sherrard 1943, 1889–90).

Conclusion

The discomforts and perils of life at sea were considerable, but they were (and had to be) endured. There was increased interest in improving life at sea from the late 18th century on. The twin engines for these improvements were science and humanism. Scientists studied aspects of life at sea and engineers devised ways to make it better. Social reformers also examined the plight of seafarers and legislated health and safety standards that addressed their needs. The improvements sometimes came haltingly—there was resistance to changes that increased the cost of the navy and merchant marine, and sometimes even those who were to benefit directly preferred to do things as they always had—but they came.

References

Abbreviations

The abbreviations listed here appear in the parenthetical citations in the body of the book.

ABS   American Bureau of Shipping
Aide-Memoire   British Army, Royal Engineers. Aide-mémoire to the Military Sciences
AJ   *Acetylene Journal*
AMS   Agricultural Marketing Service
Army   U.S. Army
ASHRAE   American Society of Heating, Refrigerating and Air-Conditioning Engineers
BAAS   British Association for the Advancement of Science
BMJ   *British Medical Journal*
BNP   Bureau of Naval Personnel
BOSJ   *Bureau of Ships Journal*
BSA   Bureau of Supplies and Accounts
CAO   Cruise Arabia Online
DOE   Department of Energy
EB   *Encyclopaedia Britannica*
EPA   Environmental Protection Agency
ETB   The Engineering Toolbox
EW   Electrical World
FAA   Federal Aviation Administration
FLPM   *Frank Leslie's Popular Monthly*
GBPO   Great Britain, Patent Office
HHS   Department of Health and Human Services
IMO   International Maritime Organization
LJA   Life Jacket Association
LQR   *London Quarterly Review*
ME   Marine Engineering
MHCPT   Museum of Historical Chamber Pots and Toilets
MM   *Mechanics Magazine*
MR   *Medical Record*
MT   Museum of Technology
MTAB   Maritime Training Advisory Board
NASEM   National Academies of Sciences, Engineering, and Medicine
NATO   North Atlantic Treaty Organization
NAVEDTRA   Naval Education and Training Command
NAVMED   Navy Medicine
NAVPERS   Navy Personnel Command
NAVSHIPS   Naval Ship Systems Command
NAVSUP (Naval Supply Systems Command
navycs   Navy Cyberspace
NCA   National Candle Association
NHHC   Naval History and Heritage Command
NIH   National Institutes of Health
NIMA   National Imagery and Mapping Agency
NIOSH   National Institute for Occupational Safety & Health
NM   *Nautical Magazine*
NMAH   National Museum of American History
NMNC   *Nautical Magazine and Naval Chronicle*
NPS   Naval Postgraduate School
OSHA   Occupational Safety and Health Administration
PHR   *Public Health Reports*
PHS   Public Health Service
RMG   Royal Museums Greenwich
RNLI   Royal National Lifeboat Institution
SMG   Science Museum Group
SMNJ   *The Sailor's Magazine, and Naval Journal*
SNR   Society for Naval Research
SOLAS   The International Convention for the Safety of Life at Sea
SR   *Safety Review*
SSTV   Safe and Sound, Tees Valley
SSVLB   South Shields Volunteer Life Brigade
TFB   The Firearm Blog

Treasury   U.S. Treasury Department
USArmyPJ   *Professional Journal of the U.S. Army*
USBNP   United States Bureau Naval Personnel
USMM   United States Merchant Marines
USN   U.S. Navy
USNDM   *U.S. Navy Diving Manual*
USPS   United States Power Squadron
WCFM   Worshipful Company of Fan Makers
WCT   *West Coast Times*
WHO   World Health Organization
WLM   Wood Library Museum
WWH   What-When-How

Aarnio, Markus. 2022. *Cruise Ship Handbook*. Springer International.
Abbott, Rachel C., and Tonie E. Rocke. 2012. "Plague." USGS Circular 1372. pubs.usgs.gov/circ/1372/pdf/C1372_Plague.pdf.
Abbott, Zack. 2019, Dec. 6. "Does Alcohol Really Cause Dehydration." https://zbiotics.com/blogs/journal/alcohol-and-dehydration.
Acetylene Journal. 1906. "Annual Meeting of the British Acetylene Association." *Acetylene J.*, 7(11) (May 1906); 7(12) (June 1906).
Adams, Henry. 1999 [1944]. *The War of 1812*. Cooper Square Press.
Adkins, Douglas, and Llewellyn Howand III. 2012. *Dorade: The History of an Ocean Racing Yacht*. David R. Godine.
Adkins, Roy. 2006. *Nelson's Trafalgar: The Battle That Changed the World*. Penguin Books.
Admiralty. 1883. *Manual of Seamanship for Boys' Training Ships of the Royal Navy*.
Admiralty. 1901. "Specifications for Fitting a Ship for the Conveyance of Men." In *Regulations (of 1893 as Corrected by Errata to End of 1899) for His Majesty's Transport Service (with Specifications for Fitting)*... HM Stationery Office.
Admiralty. 1964. *Admiralty Manual of Seamanship*, Vol. 1. HM Stationery Office.
AgeofSail. 2008, Dec. 31. "Sailors and Swimming." https://ageofsail.wordpress.com/2008/12/31/sailors-and-swimming/.
Aggarwal, Roshan L., and Kambiz Alavi. 2018. *Introduction to Optical Components*. CRC Press.
Agius, Dionisius A. 2012. *Seafaring in the Arabian Gulf and Oman: People of the Dhow*. Routledge.
Agricola, Gnaeus Julius. 1912 (Hoover transl.) *De Re Metallica*. http://www.hellenicaworld.com/Germany/Literature/GeorgiusAgricola/en/DeReMetallica.html#Page_200.
Agricultural Marketing Service. 2012. *Sulfuric Acid*. https://www.ams.usda.gov/sites/default/files/media/Sulfuric acid report.pdf.
Ainsworth, B.E., et al. 2011. "The Compendium of Physical Activities." *Medicine and Science in Sports and Exercise*, 43(8): 1575–1581. https://sites.google.com/site/compendiumofphysicalactivities/home.
Al-Gobiasi, M.K. 2010. *Physical, Chemical and Biological Aspects of Water*, Vol. I. Eoiss Publishers.
Allan, Philip K. 2023, April. "The Evolution of Frigates in the Age of Sail." *Naval History Magazine*. https://www.usni.org/magazines/naval-history-magazine/2023/april/evolution-frigates-age-sail.
AllHands. 1974, July. "A New Look at Shipboard Berthing." *All Hands* 690: 20–21.
AllHands. Undated. "USS *Constitution*." https://allhands.navy.mil/Features/Constitution/.
Alpinist. 2008, June 1. "The First Modern Climbing Harness." http://www.alpinist.com/doc/ALP24/tool-users-the-first-modern-climbing-harness.
American Bureau of Shipping. 2014, Sept. *Guide for Compliance with the ILO Maritime Labour Convention, 2006 Title 3 Requirements*.
American Society of Heating, Refrigerating and Air-Conditioning Engineers. 2019. *Ventilation for Acceptable Indoor Air Quality*, ANSI/ASHRAE Standard 62.1–2019.
American State Papers. 1831. Vol. 1. Naval Papers No. 2. 3d Congress, 2d Session. "Construction of Frigates Under the Act of March 27, 1794." Gates and Seaton.
Ammen, Rear Admiral Daniel. 1891. *The Old Navy and the New*, Vol. 1. J.B. Lippincott Co.
Amy, Commander John V., Jr. 2002. "Considerations in the Design of Naval Electric Power Systems." Power Engineering Society Summer Meeting, IEEE.
Anderson, Nathan J., et al. 2006 (Spring/Summer). "Peak Inspiratory Flows of Adults Exercising at Light, Moderate and Heavy Work Loads." *Journal of the International Society for Respiratory Protection*, 23: 53–62.
Anderson, R.C. 1994 [1927]. *The Rigging of Ships in the Days of the Spritsail Topmast, 1600–1720*. Dover.
Andrews, Frank. 2005. "A Short History of Electric Light." http://www.debook.com/Bulbs/LB26efficient.htm.
Anund, Anna, et al. 2015. *Countermeasures for Fatigue in Transportation: A Review of Existing Methods for Drivers on Road, Rail, Sea and in Aviation*. Swedish National Road and Transport Research Institute.
Aonghais, Clinton Mhic. 2014. *The Baker Boys*. Trafford Publishing.
Arnott, Neil. 1831. *Elements of Physics*, Vol. 2, Part I. Carey & Lea.
Atkinson, James, et al. 2009. "Natural Ventilation for Infection Control in Health-Care Settings." World Health Organization.
Attwood, Edward Lewis. 1904 [2009, Applewood]. *War-Ships: A Textbook on the Construction, Protection, Stability, Turning*. Longmans, Green.
Atwood, George. 1796. "V. the Construction and Analysis of Geometrical Propositions, Determining the Positions Assumed by Homogeneal Bodies Which Float Freely, and at Rest, on a Fluid's Surface; Also Determining the Stability of Ships, and of Other Floating Bodies."

Philosophical Transactions of the Royal Society of London, 86: 46–130.
Atwood, George. 1798. "X. a Disquisition on the Stability of Ships." *Philosophical Transactions of the Royal Society of London*, 88: 201–307.
"Author of Brazil." 1799. *Arctic Discovery and Adventure*. Religious Tract Society.
Azevedo, Sara, et al. 2016. "Microbiologically, Wine Is a Low Food Safety Risk Consumer Product." *BIO Web of Conferences* 7, 04003.
Baird, First Lieutenant Clair W., and Edward P. Noyes. 2017, Jan.-Feb. "Searchlights." *J. U.S. Artillery*, 47 (143): 1.
Baker, M.N., 1999. *The Quest for Pure Water: The History of Water Purification from the Earliest Records to the Twentieth Century*. American Water Works Association.
Baker, William A. 1958. *The New Mayflower: Her Design and Construction*. Barre Gazette.
Baker, William A. 1966. *Sloops and Shallops*. Barre Publishing Co.
Bander, James, And J.D. Davies. 2014. *Dutch Warships in the Age of Sail 1600–1714*. Seaforth Publishing.
Banke, Jessica. 2015, April 23. "The Evolution of Artificial Lighting, Part 2." http://blog.1000bulbs.com/home/artificial-lighting-part-2.
Barker, Richard. 1987. "John Wilkinson and the Early Iron Barges." *The Journal of the Wilkinson Society*, No. 15. https://www.broseley.org.uk/Archive/Broseley/15.htm.
Barksdale, Nate. 2015, May 19. "Who Invented the Flush Toilet?" https://www.history.com/news/who-invented-the-flush-toilet.
Barnaby, Sir Nathaniel. 1890. "The Protection of Iron and Steel Ships Against Foundering from Injury to Their Shells, Including the Use of Armour." *The Journal of the Iron & Steel Institute* 37: 438.
Barnes, David S. 2014, Spring. "Cargo, 'Infection,' and the Logic of Quarantine in the Nineteenth Century." *Bulletin of the History of Medicine*, 88(1): 75–101.
Baron, Jeremy Hugh. 2009. "Sailors' Scurvy Before and After James Lind—A Reassessment." *Nutrition Reviews* 67(6): 315–332. https://doi.org/10.1111/j.1753-4887.2009.00205.x.
Barratt, Peter. 2004. *Bahama Saga: The Epic Story of the Bahama Islands*. AuthorHouse.
battleshipbean. 2020, July 12. "Naval Rations Part 2." https://www.navalgazing.net/Naval-Rations-Part-2.
battleshipbean. 2020, July 22. "Naval Rations Part 3." https://www.navalgazing.net/Naval-Rations-Part-3.
Bauer, K. Jack, and Stephen S. Roberts. 1991. *Register of Ships of the U.S. Navy, 1775–1990: Major Combatants*. ABC-CLIO.
Bay, Alexander R. 2008. "Beriberi, Military Medicine, and Medical Authority in Prewar Japan." *Japan Review*, 20: 111–156.
Bayer, Laurence, et al. 2011, June. "Rocking Synchronizes Brain Waves During a Short Nap." *Current Biology*, 21(12): 8461–8462.

Bearss, Edwin C. 1984. *Charleston Navy Yard 1800–1842*. Historic Resource Study, Boston National Historic Park, Vol. 2. National Park Service.
Beasley, Henry. 1878. *The Druggist's General Receipt Book*. Lindsay and Blakiston.
Beeden, John. 2015, June 7. "The Cure for Salt Sores Is Exposure." http://solopacificrow.com/2015/06/07/the-cure-for-salt-sores-is-exposure/.
Beekman, E.M. 1988. *Fugitive Dreams: An Anthology of Dutch Colonial Literature*. University of Massachusetts Press.
Begley, Sharon, 2017, Dec. 20. "Does Exercise Burn More Calories in the Cold Than in Warm Weather?" https://www.statnews.com/2017/12/20/exercising-cold-calories-burned/.
Bender, Robert. 2011. "The McCloskey Ships of the Second World War." https://www.concreteships.org/ships/ww2/.
Bennett, Michael H. 2005. *Union Jacks: Yankee Sailors in the Civil War*. University of North Carolina Press.
Bentham, Samuel. 1802. "Method of Preserving Fresh Water Sweet During Long Voyages." *Repertory of Arts and Manufactures*, 16: 238–240.
Bernan, Walter. 1845. *On the History and Art of Warming and Ventilating Rooms and Buildings*, Vol. 2. George Bell. [Google Books misidentifies the author as Robert Stuart Meikleham.]
Bernstein, Theodore, and Terry S. Reynolds. 1978, Feb. "Protecting the Royal Navy from Lightning—William Snow Harris and His Struggle with the British Admiralty for Fixed Lightning Conductors." *IEEE Transactions on Education*, 21(1): 7–14.
Beyer, Henry G. 1905, Dec. 16. "The Water Supply in Ships from Its Beginning to the Present Time." *The Journal of the American Medical Association*, 45(25): 1846–1852.
Bhatia, A. 2014. "HVAC—Natural Ventilation Principles." Course No: M04-038, CED Engineering.
Bierens, Joost J.L.M. 2006. *Handbook on Drowning: Prevention, Rescue, Treatment*. Springer.
Biesty, Stephen. 2011. "Daily Calorie Intake of an 1812 Sailor and a Modern Combat Ration." USS *Constitution* Museum. https://ussconstitutionmuseum.org/wp-content/uploads/2018/09/Daily-Calorie-Intake-of-an-1812-Sailor-Dinnertime-Sailors-Eating.pdf.
Bilderbeck C.L., et al. 1931, Jan. "Rat-Guards for Ships' Hawsers." *Indian Medical Gazette*, 17–20.
Bishop[?], Kellene. 2013, June 12. "Tips and Tricks for Using Oil Lamps." http://www.preparednesspro.com/tips-tricks-oil-lamps.
Black & Veatch Corporation. 2011. *White's Handbook of Chlorination and Alternative Disinfectants, 5th ed*. Wiley.
Blackmore, David. 2004. *Blunders and Disasters at Sea*. Pen & Sword.
Blake, Captain John. 1758. *A Plan, for Regulating the Marine System of Great Britain*. A. Millar.
Blake, Nicholas. 2005. *Steering to Glory: A Day in the Life of a Ship of the Line*. Chatham.

Blake, Nicholas, and Richard Lawrence. 2005. *The Illustrated Companion to Nelson's Navy*. Stackpole Books.

Blefeld, Lawrence P. 1953. U.S. Patent 2,660,736. "Flotation Equipment."

Bloch-Dano, Evelyne. 2012. *Vegetables: A Biography*. University of Chicago Press.

Block, Leo. 2009. *Aboard the Farragut Class Destroyers in World War II: A History with First-Person Accounts of Enlisted Men*. McFarland.

Blomfield, Rear-Admiral Sir R. Massie. 1911a. "Hammocks and Their Accessories." *The Mariner's Mirror*, 1: 144–7.

Blomfield, Rear-Admiral Sir R. Massie. 1911b. "Man-of-War Boats." *The Mariner's Mirror*, 1: 235–240.

Blue, Surgeon General Rupert. 1916. *Annual Report of the Surgeon General of the Public Health Service of the United States for the Fiscal Year 1916*. Government Printing Office.

Bobe, Leonid, et al., 2016. "Design and Operation of Water Recovery Systems for Space Stations." ICE-2016-28 (46th International Conference on Environmental Systems, 10–14 July 2016, Vienna, Austria.). https://ttu-ir.tdl.org/bitstream/handle/2346/67479/ICES_2016_28.pdf.

Boehlert, C. Richard, and Raymond C. Coulter. 1971. *Legislative History: Saline Water Conversion Act, Volume 1*.

Boles, Frank. 2017. *Sailing Into History: Great Lakes Bulk Carriers of the Twentieth Century and the Crews Who Sailed Them*. Michigan State University Press.

Boreham, Ian. 2023. "January–March 1773." http://www.captaincooksociety.com/home/detail/225-years-ago-january-march-1773. [Originally published in Cook's Log, page 23, vol. 46, no. 1 (2023).]

Borgenstam, Curt, and Anders Sandstrom. 1984. *Why Vasa Capsized*. Stockholm: Statens Sjöhistoriska Museum (National Maritime Museum).

Bosquet, Abraham 1800, Feb. 25. "Structures and Observations on the Various Ways and Means That Have Been Proposed for the Preservation of Shipwrecked Mariners." Bosquet, Abraham. 1818. *A Series of Essays on Several Most Important New Systems and Inventions*, 5–59. London.

Boulware, Captain Jonathan, and Captain Rick Miller. 2015, Jan. *Tall Ships America Guidelines for Safety Aloft*. https://tallships.files.wordpress.com/2015/02/007-tsa-safety-aloft-draft-jan-20151.pdf.

Bowen, Brian H., and Marty W. Irwin. 2008, Oct. *Coal Characteristics*. CCTR Basic Facts File #8. Indiana Center for Coal Technology Research (CCTR). https://www.purdue.edu/discoverypark/energy/assets/pdfs/cctr/outreach/Basics8-CoalCharacteristics-Oct08.pdf.

Bown, Stephen R. 2005. *Scurvy: How a Surgeon, a Mariner, and a Gentlemen Solved the Greatest Medical Mystery of the Age of Sail*. St. Martin's Griffin.

Boyle, James. 1831. *A Practical Medico-historical Account of the Western Coast of Africa*. Highley.

Braddon, W. Leonard. 1907. *The Cause and Prevention of Beri-Beri*. Rebman.

Braidwood, John S. 1913. *Fire, Its Prevention & Extinguishing*. Charles & Edwin Layton.

Braisted, W.C. 1916. *Annual Report of the Surgeon General, US Navy... for the Fiscal Year 1916*. Government Printing Office.

Braithwaite, R. 2009, Oct. *32 Gun Frigate HMS Southampton Stability Analysis*. https://web.archive.org/web/20150417225544/http://richardsmodelboats.webs.com/pdf/Southampton%20Stability%20Report%20Rev%2001.pdf.

Brammer, Ulf. 2007, Sept. "Kulsejlet." *Traeskibs Sammenslutningen*, 36(3): 11–16.

Brandan, Maria Alejandra Menchaca, et al. 2018. "Modeling Natural Ventilation in Early and Late Design Stages: Developing the Right Simulation Workflow with the Right Inputs." 2018 Building Performance Analysis Conference and SimBuild co-organized by ASHRAE and IBPSA-USA, Chicago, Sept. 26–28, 2018.

Brenckle, Matthew. 2016. "Clothing the Royal Navy Sailor, 1765 to 1775." Newport Historical Society. https://newporthistory.org/wp-content/uploads/2018/10/clothingstandards-1765-1775royalnavy.pdf.

Brenckle, Matthew. 2018. "Food and Drink in the U.S. Navy, 1794 to 1820." USS *Constitution* Museum. https://www.usscm.org/publications/food-and-drink-in-the-us-navy-1794-to-1820.pdf.

Brigadier, S.R. 2002. *The Artifact Assemblage from the Pepper Wreck: An Early Seventeenth Century Portuguese East-Indiaman That Wrecked in the Tagus River*. M.A. thesis, anthropology, Texas A&M.

British Army. 1811. *General Regulations and Orders for the Army*. W. Clowes.

British Army, Royal Engineers. 1862. "Penetration of Projectiles." *Aide-mémoire to the Military Sciences*: Vol. 3, Paleontology. Zig-zag. Lockwood.

British Association for the Advancement of Science. 1885. Woodbury, " Automatic Sprinklers for Fire Extinction." In *Report of the 54th Meeting of the British Association for the Advancement of Science*, Aug.-Sept. 1884. John Murray.

British Medical Journal. 1948. Oct. 9. "Habitability of Ships." *British Medical Journal*, 2(4579): 685–6.

British Museum. Undated, but artifact acquired 1848. "Wall Panel; Relief." https://www.britishmuseum.org/collection/object/W_1848-1104-8.

Brodie, Alexander. 1780, Dec. 8. British Patent 1271. "Ship Stove, Kitchen or Hearth, with a Smoke-Jack and Iron Boilers."

Brooks, Chris J. 2001, Aug. 24. *Survival in Cold Water: A Report Prepared for Transport Canada*. https://www.ntsb.gov/news/events/Documents/fishing_vessel-5-Hiscock-CBrooks-Survival-In-Cold-Water-2001.pdf.

Brooks, Chris J. 2008. "Chapter 9B—All You Need

to Know About Life Jackets: A Tribute to Edgar Pask." In . *Survival at Sea for Mariners, Aviators and Search and Rescue Personnel*. RTO-AG-HFM-152. http://citeseerx.ist.psu.edu/viewdoc/download?doi=10.1.1.214.8436&rep=rep1&type=pdf

Brown, D.K. 2006. *Nelson to Vanguard: Warship Design and Development 1923–1945*. Naval Institute Press.

Brown, Stephanie A.T. 2012, June. *Maritime Platform Sleep and Performance Study: Evaluating the SAFTE Model for Maritime Workplace Application*. Thesis, Naval Postgraduate School.

Brown, Steven. 2022, July 6. "From Waxlight to Moonlight: 21st Century Standard Candles at NIST." https://www.nist.gov/blogs/taking-measure/waxlight-moonlight-21st-century-standard-candles-nist.

Brychta, R.J., and K.Y. Chen. 2017, March. "Cold-Induced Thermogenesis in Humans." *European Journal of Clinical Nutrition*, 71(3): 345–352. https://www.ncbi.nlm.nih.gov/pmc/articles/PMC6449850/.

Buchanan, Angus. 2018. *The Engineering Revolution: How the Modern World Was Changed by Technology*. Pen & Sword.

Budiansky, Stephen. 2011. *Perilous Flight*. Knopf Doubleday.

Bureau of Naval Personnel. 1966. *Shipboard Electrical Systems*.

Bureau of Ships Journal. 1952, July. "Mercury Arc Lamp Increases Power of Standard Searchlight." *Bureau of Ships Journal*, 1(3): 49.

Bureau of Supplies and Accounts. 1944. *The Cook Book of the United States Navy*. NAVSANDA Publication 7.

Burney, William. 1815. *A New Universal Dictionary of the Marine*. Cambridge University Press.

Burstyn, M. 1893, Mar. 22. "Search Light Projectors—An Experimental Comparison of the Mangin and Schuckert Projectors." *The Electrical Engineer*, 289.

Burton, David A. 2022, Jan. 15. "Atmospheric Carbon Dioxide (CO_2) and Methane (CH_4) Levels, 1800–Present." https://www.sealevel.info/co2_and_ch4.html.

Butler, Samuel. 1900. *Homer, the Odyssey*. Perseus Digital Library 22.139. http://www.perseus.tufts.edu/hopper/text?doc=Hom.%20Od.%2022.139&lang=original.

Butts, Ed. 2019, Oct. 14. "The Cautionary Tale of Whale Oil." *The Globe and Mail*. https://www.theglobeandmail.com/opinion/article-the-cautionary-tale-of-whale-oil/.

Byam, William, and Robert George Archibald. 1922. *The Practice of Medicine in the Tropics*, Vol. 2. Henry Frowde.

Byrne, William R., et al. 1998, March. "Antibiotic Treatment of Experimental Pneumonic Plague in Mice." *Antimicrobial Agents and Chemotherapy*, 42(3): 675–681. http://www.ncbi.nlm.nih.gov/pmc/articles/PMC105516/.

Cai, Wei, Li Cheng and Xi Longfei. 2010. "Watertight Bulkheads and Limber Holes in Ancient Chinese Boats." In Jun Kimura, *Shipwreck ASIA: Thematic Studies in East Asian Maritime Archaeology*. https://www.shipwreckasia.org/wp-content/uploads/Chapter2.pdf.

Cameron, G. 2010. *Public Skies: Telescopes and the Popularization of Astronomy in the Twentieth Century*. Ph.D. dissertation, history of science and technology, Iowa State University.

candlepowerforums.com. 2008, Nov. 30. "For Sale: 525,000 Lumens Monster 60" Carbon Arc Anti-Aircraft Light from WWII." http://www.candlepowerforums.com/vb/showthread.php?214515-For-Sale-525-000-Lumens-Monster-60-quot-Carbon-Arc-Anti-Aircraft-light-from-WWII.

Capello, Captain Lenny. 2000, Aug. "Ready Around the Clock." *U.S. Naval Institute Proceedings*, 126(8): 1, 170. https://www.usni.org/magazines/proceedings/2000/august/ready-around-clock.

"Captain in the Royal Navy." 1804. *Observations and Instructions for the Use of the Commissioned, the Junior, and Other Officers of the Royal Navy on All the Material Points of Professional Duty*. P. Steel.

Carlisle, Rodney P. 1998. *Where the Fleet Begins: A History of the David Taylor Research Center, 1898–1998*. Naval Historical Center.

Carpenter, Kenneth J. 1986. *The History of Scurvy and Vitamin C*. Cambridge University Press.

Carpenter, Kenneth J. 2000. *Beriberi, White Rice, and Vitamin B: A Disease, a Cause, and a Cure*. University of California Press.

Cassady, Michael. 2000, Aug. 18. *Risk Analysis of Shipboard Drinking Water Chemical Contaminants*. http://www.oldbluewater.com/harbors/waterproject Navy ships1.pdf.

Cengel, Yunus A., and Michael A. Boles. 2002. *Thermodynamics: An Engineering Approach*, 4th ed. McGraw-Hill.

Cerezo, Alvaro. 2012, May 29. "I'm Dying of Thirst. Can I Start Drinking Seawater?" http://paradise.docastaway.com/drinking-sea-water/.

Chaline, Eric. 2018, July 16. "How Europe Learnt to Swim." https://www.historytoday.com/miscellanies/how-europe-learnt-swim.

Chalmers, David. 2012. *Sir Samuel Bentham—Naval Architect—1757–1831 an Extraordinary Career*. portsmouthdockyard.org.uk/Sir Samual Bentham Biography.pdf.

Chamier, Captain Frederick. 1850. *The Life of a Sailor*. R. Bentley.

Chappelle, Howard I. 1949. *The History of the American Sailing Navy*. W.W. Norton & Co.

Chappelle, Howard I. 1967. *The Search for Speed Under Sail, 1700–1855*. W.W. Norton & Co.

Chattaway, F.D. 2007, Sept. 27. "The Deposit of Metallic Silver on Glass." *The Chemical News*, 96 (2496): 151.

Chester Energy and Policy, LLC. 2017, Oct. 16. "What Is the Most Climate Friendly Way to Light Your Jack-O'-Lantern?" http://chesterenergyandpolicy.com/2017/10/16/what-is-the-most-climate-friendly-way-to-light-your-jack-o-lantern/.

Childs, David. 2009. *Tudor Sea Power: The Foundation of Greatness*. Pen & Sword.

Chisholm, H.W. 1873. *Seventh Annual Report of the Warden of the Standards on the Proceedings and Business of the Standard Weights and Measures Department of the Board of Trade for 1872–73*. George E. Eyre and William Spottiswoode.

Chourpiliadis, Charilaos, and Abhishek Bhardwaj. 2021, Sept. 20. "Physiology, Respiratory Rate." National Library of Medicine. https://www.ncbi.nlm.nih.gov/books/NBK537306/.

Cipolla, Carlo M. 1981. *Fighting the Plague in Seventeenth-Century Italy*. University of Wisconsin Press.

Clarke, Francis Gedney. 1838. *The American Ship-Master's Guide, and Commercial Assistant*. Allen & Co.

Cleborne, C.J. 1879. "Life-Saving at Sea." In Joseph B. Parker, *Hygienic and Medical Reports by Medical Officers of the US Navy*, 1029–1048. Government Printing Office.

Clow, Bertha, and Abby L. Marlatt, 1929, Dec. 15. "The Vitamin-C Content of Fresh Sauerkraut and Sauerkraut Juice." *Journal of Agricultural Research*, 39(12): 963–971.

Coats, Ann Veronica, and Philip MacDougall. 2011. *The Naval Mutinies of 1797: Unity and Perseverance*. Boydell Press.

Colomb, Philip Howard. 1898. *Memoirs of Admiral the Right Honorable Sir Astley Cooper Key*. Methuen & Co.

Combs, Gerald F., Jr., and James P. McClung. 2022. *The Vitamins: Fundamental Aspects in Nutrition and Health*. Elsevier Science.

Compton, Nic. 2014. *The Anatomy of Sail: The Yacht Dissected and Explained*. Bloomsbury.

USS *Constitution* Museum. 2011. "Berth Deck Area Activity." https://ussconstitutionmuseum.org/wp-content/uploads/2018/09/Berth-Deck-Area-Activity-1.pdf.

USS *Constitution* Museum. 2014, April 25. "Water, Water Everywhere." https://ussconstitutionmuseum.org/2014/04/25/water-water-everywhere/.

Cook, Captain James. 1776. "XXII. the Method Taken for Preserving the Health of the Crew of His Majesty's Ship the Resolution During Her Late Voyage Round the World." *Philosophical Transactions of the Royal Society of London*, 66(66): 402–6. https://royalsocietypublishing.org/doi/epdf/10.1098/rstl.1776.0023.

Cook, Captain James. 1813. *The Voyages of Captain James Cook Round the World*, Vol. 4. Sherwood, Neely & Jones.

Cooke, Brian Douglas. 2014. *Australia's War Against Rabbits: The Story of Rabbit Haemorrhagic Disease*. CSIRO Publishing.

Cope, L.F. 1910, Aug. 20. "Air and Ventilation in Modern Warships." *British Medical Journal*, 2: 443–5.

Corbin, Thomas W. 1917. *Marvels of Scientific Invention*. J.B. Lippincott.

Cordle, Captain John, and Nita Shattuck. 2013, Jan. "A Sea Change in Standing Watch." *US Naval Institute Proceedings*, 139: 1, 319.

Coris, Eric E., et al. 2004. "Heat Illness in Athletes: The Dangerous Combination of Heat, Humidity and Exercise." *Sports Medicine*, 34(1): 9–16. https://doi.org/10.2165/00007256-200434010-00002.

Coursey, Todd. 2013, Sept. 5. "Closing the Loop: Boosting Operational Sustainability and Quality of Life Through Wastewater Reclamation." https://cimsec.org/closing-the-loop-boosting-operational-sustainability-and-quality-of-life-through-wastewater-reclamation/.

Cowan, J.M. 2010, June. "The Relationship of Normal Body Temperature, End-expired Breath Temperature, and Bac/brac Ratio in 98 Physically Fit Human Test Subjects." *Journal of Analytical Toxicology*, 34(5): 238–42. https://www.ncbi.nlm.nih.gov/pubmed/20529457.

Cox, Samuel J. 2021, June. "H-062-1: 'Battles You've Never Heard Of.'" Naval History and Heritage Command. https://www.history.navy.mil/about-us/leadership/director/directors-corner/h-grams/h-gram-062/h-062-1.html.

Crawford, Richard W. 2013. *San Diego Yesterday*. Arcadia Publishing.

Creel, R.H. 1917, Sept. 7. "Rodent Destruction on Ships: A Report on the Relative Efficiency of Fumigants as Determined by Subsequent Intensive Trapping Over a Period of One Year." *Public Health Reports* 32(36): 1445–50. https://www.jstor.org/stable/4574619.

Cringle, Tom. 1863. *Tom Cringle's Letters on Practical Subjects, Suggested by Experiences in Bombay*. Bombay: Education Society's Press, Byculla.

Crisman, K. 1999. "Angra B: The Lead-Sheathed Wreck at Porto Novo." *Revista Portuguesa Arqueologia* 2(1): 255–62. http://www.patrimoniocultural.gov.pt/media/uploads/revistaportuguesadearqueologia/2_1/14.pdf.

Critchley, Macdonald. 1945, Aug. 4. "Problems of Naval Warfare Under Climatic Extremes: Part I. Lecture I." *British Medical Journal*, 2(4413): 145–8.

Critchell, James Troubridge, and Joseph Raymond. 1912. *A History of the Frozen Meat Trade*. Constable.

Cruise Arabia Online. 2013. "History of Air Conditioning at Sea." http://www.cruisearabiaonline.com/Historical/2013/08/25/Historical—History-of-air-conditioning-at-sea.

Cutler, Thomas J. 2017. *The Bluejacket's Manual*. Naval Institute Press.

D'Albe, Edmund Edward Fournier. 1924. *The Life of Sir William Crookes*. D. Appleton & Co.

Dalton, Kyle. 2017, May 31. "Canvas Hats, British Tars 1740–1790." https://www.britishtars.com/2017/05/canvas-hats.html.

Dana, R.H. [Richard Henry], Jr. 1841. *The Seaman's Friend*. Boston.

Dana, Richard Henry, Jr. 1907. *Two Years Before the Mast: A Personal Narrative of Life at Sea*. T.Y Crowell & Co. (originally published in 1840).

Danielski, John. 2023, March 27. "Downtime on the High Seas—Inside the Curious Sleep Habits of 18th Century Sailors." https://militaryhistorynow.com/2023/03/27/downtime-on-the-high-seas-inside-the-curious-sleep-habits-of-18th-century-sailors/.

Davis, Charles G. 2012 [1926]. *The Ship Model Builder's Assistant*. Dover.

Davis, Charles G., and Irving R. Wiles. 1929. *Ships of the Past*. Bonanza Books.

Dean, James Seay. 2014. *Tropic Suns: Seadogs Aboard an English Galleon*. History Press.

De Castro, Filipe Vieira, et al., eds. 2022. *Heritage and the Sea, Volume 2: Maritime History and Archaeology of the Global Iberian World (15th–18th Centuries)*. Springer International.

De Graauw, Arthur. 2017, Feb. 2. "From Amphora to TEU: Journey of a Container—An Engineer's Perspective." Conference: Portus Limen Project, Rome. https://www.academia.edu/31201988/From_Amphora_to_TEU_Journey_of_a_container_An_engineers_perspective.

Delanghe, Joris R. 2011, Nov. "Vitamin C Deficiency: More Than Just a Nutritional Disorder." *Genes & Nutrition*, 6(4): 341–346. http://www.ncbi.nlm.nih.gov/pmc/articles/PMC3197848/.

Delanghe, Joris R., et al. 2007, Aug. "Vitamin C Deficiency and Scurvy Are Not Only a Dietary Problem but Are Codetermined by the Haptoglobin Polymorphism." *Clinical Chemistry*, 53(8): 1397–400. http://www.ncbi.nlm.nih.gov/pubmed/17644791.

Delgado, James P. 2019. *War at Sea: A Shipwrecked History from Antiquity to the Twentieth Century*. Oxford University Press.

Delyannis, Anthony A., and Eurydike A. Delyannis. 1974. *Sauerstoff: Anhangband Water Desalting Wasser-Entsalzung*. Springer.

Dempster, Daniel E., and Todd R. Berger. 2002. *Lighthouses of the Great Lakes*. Voyageur Press.

Denney, O.E. 1937, June 4. "Some Experiments with Rats and Rat Guards." *Public Health Reports*, 52(23): 723–6.

Denny, Mark. 2008. *Float Your Boat! The Evolution and Science of Sailing*. Johns Hopkins University Press.

Department of Health and Human Services. 2020, Dec. *2020–2025 Dietary Guidelines for Americans* (9th ed.).

Department of Energy. 2013, Nov. 22. "The History of the Light Bulb." https://www.energy.gov/articles/history-light-bulb.

De Paula, Leandro, et al. 2007. "A Contribution for the Construction of Parabolic Mirrors." *Caderno Brasileiro de Ensino de Física*, 24: 338–352. English abstract. http://arxiv.org/abs/0810.4165.

De Queriros, Pedro Fernandes. 1904. *The Voyages of Pedro Fernandez De Quiros, 1595–1606*. University of California.

De Steiguer, Lieutenant Louis R. 1900, July. "Electrical Notes." Office of Naval Intelligence, *Notes on Naval Progress*.

Digby, Sir Everard, and Christofer Middleton. 1595. *A Short Introduction for Learne to Swimme*. https://quod.lib.umich.edu/cgi/t/text/text-idx?c=eebo;idno=A20436.0001.001.

Dilek, Yildirim, and Sally Newcomb. 2003. *Ophiolite Concept and the Evolution of Geological Thought*. Geological Society of America.

Dill, Gregory J. 2006. *Myth, Fact, and Navigators' Secrets: Incredible Tales of the Sea and Sailors*. Lyons Press.

Djurhuus, Rune. 2021. "Fumigation on Bulk Cargo Ships: A Chemical Threat to Seafarers." *International Maritime Health*, 72(3): 206–216.

Dobbs, Christopher. 2012, April 6. "The Mary Rose: Celebrating Three Decades of Research." https://the-past.com/feature/the-mary-rose-celebrating-three-decades-of-research/.

Dodds, James, and James Moore. 2022. *Building the Wooden Fighting Ship*. Greenhill Books.

Donoghue E.R., and S.C. Minnigerode. 1977, July. "Human Body Buoyancy: A Study of 98 Men," *J Forensic Sci*, 22(3): 573–9. https://pubmed.ncbi.nlm.nih.gov/617991/.

Duly. 1938, Mar. 25, "Condensation on Board Ship." *Journal of the Royal Society of Arts*, 86: 439–55.

Duly. 1950, July 14. "The Ventilation of Ships' Holds." *Journal of the Royal Society of Arts*, 98: 710–25.

Du Monceau, Duhamel. 1754. *The Elements of Naval Architecture; Or, a Practical Treatise on Ship-Building*. D. Henry and R. Cave.

Dunmore, John (ed.) 2022. *The Pacific Journal of Louis-Antoine de Bougainville, 1767–1768*. Taylor & Francis.

Eastland, Jonathan, and Ian Ballantyne. 2011. *HMS Victory: First Rate 1765*. Pen & Sword.

Edwards, Frederick. 1868. *The Ventilation of Dwelling Houses and the Utilization of Waste Heat from Open Fire-Places*. Longmans, Green & Co.

Ekirch, A. Roger. 2001, April. "Sleep We Have Lost: Pre-Industrial Slumber in the British Isles." *American Historical Review*, 106(2): 343–86.

Ekirch, A. Roger. 2015, Feb. "The Modernization of Western Sleep: Or, Does Insomnia Have a History." *Past & Present*, 226: 149–192. See also Ekirch, Roger, "Additional Historical References to 'Segmented Sleep.'" https://sites.google.com/vt.edu/roger-ekirch/sleep-research/segmented-sleep.

Ekirch, A. Roger. 2016. "Segmented Sleep in Preindustrial Societies." *Sleep*, 39(3): 715–716.

Electrical World. 1896, May 30. "News of the Week." *Electrical World*, 27(22).

Electricity. 2018, June 7. "Some Interesting Points on Good Illumination as a Factor of Safety." *Electricity: A Practical Trade Journal* 32.

Ellicott, W.M. 1896. "The Plimsoll Mark." *St. Nicholas: A Monthly Magazine for Boys and Girls*, 24 (Part 1): 45–47.

Elliott, F.S., and D.A. Kobus. 1992, June 16. *Standardizing Shipboard Lighting: Light Fixtures and Light Bulbs on U.S. Navy Ships*, Naval Health Research Center Report No. 92–22. https://apps.dtic.mil/sti/citations/ADA273682.

Ellis, F.P. 1948, Oct. "Victuals and Ventilation and the Health and Efficiency of Seamen." *British Journal of Industrial Medicine*, 5(4): 185–197.

Encyclopaedia Britannica. 1911. (11th ed.). (Various entries cited in form EB1911/Article Name.)

The Engineering Toolbox. 2023. "Water—Saturation Pressure Vs. Temperature." Online water saturation pressure calculator. https://www.engineeringtoolbox.com/water-vapor-saturation-pressure-d_599.html.

English, Andrew Ramsey. 2016, April. *The Laird Rams: Warships in Transition 1862–1885*. Ph.D dissertation, maritime history, University of Exeter.

Environmental Protection Agency. 2011, Nov. *Graywater Discharges from Vessels*.

Erichsen-Brown, Charlotte. 2013. *Medicinal and Other Uses of North American Plants*. Dover.

Esparza, Bill. 2017, Sept. 13. "Drink Like a Ryukyu Kingdom Royal." https://explorepartsunknown.com/okinawa/drink-like-a-ryukyu-kingdom-royal/.

Estimates and Accounts; Army, Navy, Ordnance, Civil List; Miscellaneous Services, Session 27 January–10 June, 1818. 1818. "I. Copy of the Correspondence Between the Master General of the Ordnance and Mr. Mallison, 21 July 1817–13 Feb. 1818, II. Copy of the Correspondence Between the Board of Ordnance and Mr. Mallison, 28 Oct. 1817–6 Feb 1818; Copy of the Correspondence Between the Board of Admiralty and Mr. Mallison, 17 July 1817–14 Jan. 1818."

Estrella, Jesus Moo, and Gloria Arankosky Sandoval. 2021, Feb. "Comparison and Prediction of Sleep Quality in Users of Bed or Hammock as Sleeping Device." *Sleep Health*, 7(1): 93–97. https://doi.org/10.1016/j.sleh.2020.06.002.

Evans, David Stanley. 2014. *Herschel at the Cape: Diaries and Correspondence of Sir John Herschel, 1834–1838*. University of Texas Press.

Ewbank, Anne. 2018, Feb. 22. "How Killer Rice Crippled Tokyo and the Japanese Navy." https://www.atlasobscura.com/articles/rice-disease-mystery-edo-tokyo-navy-beriberi.

Ewbank. Thomas. 1876. *A Descriptive and Historical Account of Hydraulic and Other Machines for Raising Water, Ancient and Modern*. Scribner Armstrong.

Excellent. 1866. *Experiments with Naval Ordnance, HMS Excellent, 1866*. Harrison & Sons.

Fairley, R. 2007. "Why the Vasa Sank: 10 Lessons Learned." http://www.cse.ogi.edu/~dfairley/The_Vasa.pdf.

Falconer, William. 1780. *Falconer's Marine Dictionary*. T. Cadell.

Falconer, William, and William Burney. 1815. *An Universal Dictionary of the Marine*. T. Cadell.

Farrer, Keith. 2005. *To Feed a Nation: A History of Australian Food Science and Technology*. CSIRO Publishing.

Fatsecret.com, Undated. "Vitamins in Seaweed." http://www.fatsecret.com/calories-nutrition/food/seaweed/vitamins.

Federal Aviation Administration. 2012, Oct. 11. "Fitness for Duty." Advisory Circular 117–3.

Feldman, S., and T.I. Monahan. 1962, Feb. "A New Signaling Beacon." *Naval Research Revs*. 12–14.

Fennell, James Hamilton. 1841. *Natural History of British and Foreign Quadrupeds*. Joseph Thomas.

Ferreiro, Larrie D. 2007. *Ships and Science: The Birth of Naval Architecture in the Scientific Revolution, 1600–1800*. MIT Press.

Fields, Nic. 2007. *Ancient Greek Warship, 500–322 BC*. Osprey Publishing.

Fincham, John. 1825. *An Introductory Outline of the Practice of Ship-building*. William Woodward.

The Firearm Blog. 2018, Sept. 27. "POTD: The Coast Guard's 80 Year Old Line Throwers" https://www.thefirearmblog.com/blog/2018/09/27/coast-guard-line-throwers/.

Fish, Shirley. 2011. *The Manila-Acapulco Galleons: The Treasure Ships of the Pacific*. AuthorHouse UK.

Forbes, R.B. 1848. *Protection of Ships from Lightning*. Skeeper & Rogers.

Forbes, R.B. 1871, Nov. "Life Saving Hammocks." *The Nautical Magazine (New Series)*, 40: 794–7.

Forbes, R.B. 1902. "Bamboo, the Chinaman's Vade Mecum." *Modern Culture*, 14: 485, 487.

Forester, C.S. 1950. *Mr. Midshipman Hornblower*. Pinnacle Books.

Foster, Jo. 2019, May 10. "The Mary Rose." https://www.solentplastics.co.uk/news/the-mary-rose/.

Foster, Russell G. 2020. "Sleep, Circadian Rhythms and Health." *Interface Focus*, 10(3): 20190098. https://doi.org/10.1098%2Frsfs.2019.0098.

Fougner, Nicolay Knudtzon. 1922. *Seagoing and Other Concrete Ships*. Henry Frowde.

Francescutto, Alberto, and A.D. Papanikolaou. 2011, Feb. "Buoyancy, Stability, and Subdivision: From Archimedes to Solas 2009 and the Way Ahead." *Proceedings of the Institution of Mechanical Engineers, Part M: Journal of Engineering for the Maritime Environment*, 225: 17–32. https://doi.org/10.1177/14750902JEME238.

Frank Leslie's Popular Monthly. 1880, Aug. 10(2): 254.

Franklin, Benjamin, and Peter Collinson. 1751. *Experiments and Observations on Electricity*. E. Cave. https://library.si.edu/digital-library/book/experimentsobser00fran.

Fraser, Edward. 1904. *Famous Fighters of the Fleet: Glimpses Through the Cannon Smoke in the Days of the Old Navy*. Macmillan and Co.

Freemantle, Sir Edmund Robert. 1904. *The Navy as I Have Known It, 1849–1899*. Cassell.

Friedman, Norman. 2015. *The British Battleship 1906–1946*. Naval Institute Press.

Frierson, J. Gordon. 2022. *Guarding the Golden Gate: A History of the U.S. Quarantine Station in San Francisco Bay*. University of Nevada Press.

Fuentes, Gidget. 2021, Feb. 15. "Fatigue at Sea: Researching How Lack of Sleep Can Affect Sailors." https://seapowermagazine.org/fatigue-at-sea-researching-how-lack-of-sleep-can-affect-sailors/.

Gard. 2011. "Fumigation of Cargo on Board Ships: The Invisible Killer." *Gard News* 204. http://www.gard.no/web/updates/content/20650371/fumigation-of-cargo-on-board-ships-the-invisible-killer.

Gardiner, Robert. 1989. "Frigate Design in the 18th Century, Part 3." *Warship*, 12: 269.

Gardiner, Robert. 2006. *Frigates of the Napoleonic Wars*. Naval Institute Press.

Gardiner, Robert. 2011. *Warships of the Napoleonic Era: Design, Development and Deployment*. Naval Institute Press.

Gates, Manley F. 1907, April. "A Sanitary Scuttle-Butt." *The Military Surgeon*, 20(4): 313–6.

Gatewood, J.D. [James Duncan]. 1909. *Naval Hygiene*. P. Blakiston's Son & Co.

Gatewood, J.D. 1909, Jan. "The Artificial Illumination of Naval Vessels—A Study in Naval Hygiene." *United States Naval Bulletin*, 3(1): 1–22.

Gelder, Paul. 2013. *Total Loss: Dramatic First-Hand Accounts of Yacht Losses at Sea*. Bloomsbury.

General Electric. 1944, April. "Instruction Book for 24-inch Searchlight Model 24-G-20." http://www.maritime.org/doc/searchlight24/index.htm.

Gensini, Gian Francisco, et al. 2004, Nov. "The Concept of Quarantine in History: From Plague to SARS." *Journal of. Infection*, 49: 257–61.

Gihon, Albert Leary. 1871. *Practical Suggestions in Naval Hygiene*. Government Printing Office.

Giot, Claire, et al. 2023. "A Survey Exploring How Watch Officers Manage Effects of Sleep Restrictions During Maritime Navigation." *International Journal of Environmental Research and Public Health*, 20(2): 986. https://doi.org/10.3390/ijerph20020986.

Girbayedoff, Valerien. 1893. "The Sailors of the Czar." *American Illustrated Magazine*, 36: 115–125.

Glascock, Captain W.N. [William Nugent]. 1829. *Sailors and Saints; Or, Matrimonial Maneuvers*. Vol. 2. Henry Colburn.

Glascock, Captain W.N. 1848. *The Naval Officers Manual*, 2d. ed, London.

Glass, Roger I., et al. 1980. "Deaths from Asphyxia Among Fishermen." *Journal of the American Medical Association*, 244(19): 2193–2194.

Gleick, P.H. 1996. "Basic Water Requirements for Human Activities: Meeting Basic Needs." *Water International*, 21: 83–92.

Glete, Jan. 2010. *Swedish Naval Administration, 1521–1721*. Brill.

Goethe, W.H.G., et al. 2012. *Handbook of Nautical Medicine*. Springer Berlin Heidelberg.

Goldsmith, Elizabeth C. 2014, May. "Famous Sailors Who Couldn't Swim, Wonders & Marvels." http://www.wondersandmarvels.com/2014/05/famous-sailors-who-couldnt-swim.html.

Goodrich, Lieutenant-Commander Caspar Frederick. 1883. *Report on the British Naval and Military Operations in Egypt 1882*. Government Printing Office.

Goodyer, Bronwen E.J. 2008, Sept. "An Assistant Ship Surgeon's Account of Cholera at Sea." *Journal of Public Health*, 30(3): 332–338. https://doi.org/10.1093/pubmed/fdn031.

Gougeon, Meade, and Ty Knoy. 1973. *Evolution of Modern Sailboat Design*. Winchester Press.

Grandjean, A.C. 2004, Aug. *Water Requirements, Impinging Factors, and Recommended Intakes*. WHO. http://www.mayoclinic.org/healthy-lifestyle/nutrition-and-healthy-eating/in-depth/water/art-20044256.

Great Britain, Patent Office. 1898. Patents for Inventions. Abridgments of Specifications, Vol. 7.

Greene, Arthur Maurice. 1919. *Pumping Machinery: A Treatise on the History, Design, Construction and Operation of Various Forms of Pumps*. John Wiley & Sons.

Griffiths, Captain Anselm John. 1828. *Observations on Some Points of Seamanship*, 2d ed.

Griffiths, Major Frederick Augustus. 1859. *The Artillerist's Manual*. London.

Groak, Irwin D. 1918, Sept. "Recent Development in Equipment Used in Ozone Method of Water Treatment." *Municipal & County Engineering*, 55(3): 109–111.

Gross, Harry. 2013, Feb. "Air Sea Rescue Dumb Barges—called the 'Cuckoo.'" https://www.readkong.com/page/air-sea-rescue-dumb-barges-called-6000887.

Grossman, Daniel. 2016, July 27. "High CO_2 Levels Inside & Out: Double Whammy." https://www.yaleclimateconnections.org/2016/07/indoor-co2-dumb-and-dumber/.

Grosvenor. 1921, Sept. "Red, Green and White Lights on Vessels." *Grosvenor Library Bulletin*, 69.

Grove, C.E. 1900, April 5. "The Electrical Equipment of Ships of War." *Proceedings of the. Institution of Electrical Engineers*, 29: 530–626.

Grubbs, S.B., and B.E. Holsendorf. 1913, June 20. "Fumigation of Vessels for the Destruction of Rats." *Public Health Reports*, 28: 1266–1274.

Grubbs, S.B., and B.E. Holsendorf. 1925, July 17. "The Rat-Proofing of Vessels." *Public Health Reports*, 40(29): 1507–1515.

Grubbs, S.B., and B.E. Holsendorf. 1927. "Rat Proofing of Vessels." *The Nautical Gazette*, May 7: 509, 513; May 14: 537, 541.

Grubbs, S.B., and B.E. Holsendorf. 1931. *Rat Proofing of Vessels (Third Edition)*, Supplement No. 93 to the Public Health Reports. Government Printing Office.

Guildbaud, Julie, et al. 2010, Dec. "Laundry Water Recycling in Ship by Direct Nanofiltration with Tubular Membranes." *Resources, Conservation & Recycling*, 55(2): 148–54. https://www.sciencedirect.com/science/article/abs/pii/S0921344910002016#!

Gunsteren, Lex A. 2013. *Quality in Design and Execution of Engineering Practice*. IOS Press.

Halacy, D.S. 1974, Sept/Oct. "How to Build a Solar Still." *Mother Earth News*. http://www.motherearthnews.com/renewable-energy/solar-stillzmaz74sozraw.aspx (The article was originally published as a chapter in Halacy's book, *Fun with the Sun* [1959]).

Hales, Stephen. 1743. *A Description of Ventilators...* London.

Hallett, Sasha, et al. 2022, May 8. "Physiology, Tidal Volume." National Library of Medicine. https://www.ncbi.nlm.nih.gov/books/NBK482502/.

Halperin, Samuel. 2007. "Titanic's Masthead Light." http://www.glts.org/articles/halpern/masthead_light.html.

Halsey, Taylor. 2023. "Our Story: Satisfying Thirst Since 1912." https://www.halseytaylor.com/us/en/about.html.

Hargis, Robert. 2012. *US Submarine Crewman 1941–45*. Bloomsbury.

Harland, John. 1985. *Seamanship in the Age of Sail*. Naval Institute Press.

Harrelson, M.L. 1992. *Basic Military Requirements*. NAVEDTRA 12043. Government Printing Office.

Harris, Fraser D. 1916, "Stephen Hales, the Pioneer in the Hygiene of Ventilation." *The Scientific Monthly*, 3: 440–454.

Harris, Mike. 1985. "A Matter of Survival." *Fathom*, 17(2): 30–31.

Harris, Tom, et al. 2021, Feb. 11. "How Light Emitting Diodes (LEDs) Work." https://electronics.howstuffworks.com/led.htm.

Harris, Sir William Snow. 1852, May. "Review of the History and Progress of a System of Permanently Fixed Lightning Conductors for H.M. Ships..." *The Nautical Magazine and Naval Chronicle for 1852*. Cambridge University Press.

Harrower, A.B. 1922, Aug. 24., "Acetylene on Board Ship." letter to the editor, Aug. 21, 1922, in *Shipbuilding and Shipping Record* 20(8) 234.

Harvey, Robert. 2008. *Maverick Military Leaders: The Extraordinary Battles of Washington, Nelson, Patton, Rommel, and Others*. Skyhorse.

Hasan, Dalia, et al. 2015, Dec. "An Experimental Study for the Evaluation of Materials and Methods Used in the Treatment of Tin Amalgam Mirrors." *Shedet (Fayoum University)*, 2(2): 40–53. https://www.researchgate.net/publication/337037130_An_Experimental_Study_For_The_Evaluation_Of_Materials_And_Methods_Used_In_The_Treatment_Of_Tin_Amalgam_Mirrors.

"Hatch." 2023. "Hatch, the Mary Rose Dog." https://maryrose.org/hatch/.

Havard, Valery. 1914. *Manual of Military Hygiene for the Military Services of the United States, 2d ed*. William Wood and Company.

Hawkins, Richard. 1847 [1622]. *The Observations of Sir Richard Hawkins, Knt. in His Voyage Into the South Sea in the Year 1593*. Hakluyt Society.

Hayes, Wayland J., Jr., and Thomas B Gaines. 1950, Nov. 24. "Control of Norway Rats with Residual Rodenticide Warfarin." *Public Health Reports*, 65(47): 1537–1555.

Hayward, Arthur H. 1962 [1923]. *Colonial and Early American Lighting, Vol. 2*. Dover.

Hearings Before the Committee on Merchant Marine and Fisheries, House of Representatives. 1935. *Safety of Life and Property at Sea* Seventy-fourth Congress, First Session. Government Printing Office.

Hearings Before the Subcommittee on Irrigation and Reclamation of the Committee on Interior and Insular Affairs. 1965, May 18 and 19. *Saline Water Conversion*, United States Senate, 89th Congress, First Session. Government Printing Office.

Heath, T.L. 1953. *The Works of Archimedes Edited in Modern Notation*. Dover (unabridged reissue, originally published in 1897, with 1912 supplement, by Cambridge University Press).

Heck, J.G., and Spencer F. Baird. 1851. "Plate 21: Scenes from Shipboard Life." In Plates, Vol. 2, Division VI, "Naval Sciences." *Iconographic Encyclopedia of Science, Literature and Art*. Rudolph Garrigue. https://library.si.edu/digital-library/book/iconographicencplathecka.

Heck, J.G., and Spencer F. Baird. 1851. *Iconographic Encyclopedia of Science, Literature, and Art*. Rudolph Garrigue. https://library.si.edu/digital-library/book/iconographicencplathecka

Hemingway, James Peter. 2002, May. *The Work of the Surveyors of the Navy During the Period of the Establishments: A Comparative Study of Naval Architecture Between 1672 and 1755*. Ph.D. dissertation, University of Bristol.

Henderson, Wilfred. 1907. *Seamanship*. J Griffin.

Hendrickson, Robert. 1983. *More Cunning Than Man: A Social History of Rats and Men*. Stein and Day.

Herschel, Sir J.F.W. 1861, Jan. 4. "Swinging Cot for Sea Voyages." *J. Society Arts*, 97–98.

Herwig, R.P., et al. 2006, Oct. "Ozone Treatment of Ballast Water on the Oil Tanker S/t Tonsina: Chemistry, Biology and Toxicity." *Marine Ecology Progress Series*, 324: 37–55.

Heymsfield, S.B., et al. 1989, Dec. "Chemical Determination of Human Body Density in Vivo: Relevance to Hydrodensitometry." *American Journal of Clinical Nutrition*, 50(6): 1282–9. https://doi.org/10.1093/ajcn/50.6.1282.

Hickox, Rex. 2007. *All You Wanted to Know About 18th Century Royal Navy*. Self-published.

Hildebrand, Gregory C., et al. 2017. *Above Ground Bulk Storage Tank Emergencies*. Jones & Bartlett Learning.

Hilditch, Cassie J., and Andrew W. McHill. 2019. "Sleep Inertia: Current Insights." *Nature and Science of Sleep*, 11: 155–165.

Hill, Ninian. 1838. "Health of the British Navy." *The Lancet (London)*, 2: 865–868.

Hiscocks, Richard. 2021, Aug. 7. "The Boyne Is Destroyed by Fire at Spithead—1 May 1795." https://morethannelson.com/the-boyne-is-destroyed-by-fire-at-spithead-1-may-1795/.

Hobdy, William C. 1910. "The Rat in Relation to Shipping." In U.S. Treasury Department, 207–214.

Holmes, George C.V. 1906. *Ancient and Modern Ships, Part II*. HM Stationery Office.

Holms, A. Campbell. 1918. *Practical Shipbuilding: A Treatise on the Structural Design and Building of Modern Steel Vessels*. Longmans.

Hong, B.D., and E.R. Slatick. Undated. "Carbon Dioxide Emission Factors for Coal." https://www.eia.gov/coal/production/quarterly/co2_article/co2.html. Originally published in Energy Information Administration, Quarterly Coal Report, January–April 1994, DOE/EIA-0121(94/Q1) (Washington, D.C.: Aug. 1994), pp. 1–8.

Honsberg, Christiana, and Stuart Bowden. 2019. "Calculation of Solar Insolation." https://www.pveducation.org/pvcdrom/properties-of-sunlight/calculation-of-solar-insolation.

House Committee on Naval Affairs. 1914. *Hearings, Estimates Submitted by the Secretary of the Navy*. Government Printing Office.

Hoving, A.J., and Diederick Wildeman. 2012. *Nicolaes Witsen and Shipbuilding in the Dutch Golden Age*. Texas A&M University Press.

Howard, Frank. 2002. *Sailing Ships of War, 1400–1860*. Bloomsbury.

Howe, Henry. 1870. *Life and Death on the Ocean*. Henry Howe.

Hubbard, R.W. 1990. "The Role of Exercise in the Etiology of Exertional Heatstroke." *Medicine & Science in Sports & Exercise*, 22(1): 2–5.

Hume, Martin A.S., ed. 1899, *Calendar of Letters and State Papers Relating to English Affairs, Vol. IV, Elizabeth 1587–1603*.

Hunt, William H. 1881, Nov. 28. *Report of the Secretary of the Navy for the Year 1880*. Government Printing Office.

Hutchinson, Joseph Chrisman. 1872. *A Treatise on Physiology and Hygiene for Educational Institutions and General Readers*. Clark & Maynard.

Igwemmar, N.C. 2013, Nov. "Effect of Heating on Vitamin C Content of Some Selected Vegetables." *International Journal of Science and Technology Research*, 2(11): 209.

Inglis-Arkell, Esther. 2011, Nov. 23. "How Long Would You Survive in an Airlock?" https://io9.gizmodo.com/how-long-would-you-survive-in-an-airlock-5861679.

The International Convention for the Safety of Life at Sea. 2014, Jan. 20. http://www.archive.org/stream/textofconvention00inte#page/108/mode/2up.

International Maritime Organization. 2008. *Officer in Charge of a Navigational Watch on a Fishing Vessel*. IMO.

International Maritime Organization. 2023. "Part B—Guidance on Individual SOLAS Chapter II-1 Subdivision and Damage Stability Regulations." https://www.imorules.com/MSCRES_281.85_ANN_PTB.html.

Irwin, Emily. 2012. The Spermaceti Candle and the American Whaling Industry. https://www.eiu.edu/historia/2012Irwin.pdf.

Jaksic, Fabian M., and Sergio A. Castro. 2021. *Biological Invasions in the South American Anthropocene: Global Causes and Local Impacts*. Springer International.

Jane, Frederick Thomas. 1899. *Imperial Russian Navy: Its Past, Present, and Future*. W. Thacker & Co.

Japan, Bureau of Medical Affairs, Navy. 1911. *The Surgical & Medical History of the Naval War Between Japan & Russia During 1904–1905*. Tokyo: Toyo Printing Co.

Jefferson, Thomas. 1791, Nov. 21. "Report on Desalination of Sea Water." https://founders.archives.gov/documents/Jefferson/01-22-02-0296.

Jepsen, Jorgen Riis, et al. 2015. "Seafarer Fatigue: A Review of Risk Factors, Consequences for Seafarers' Health and Safety and Options for Mitigation." *International Maritime Health*, 66(2): 106–117.

Jing, X.K., and Y.Q. Li. 2012, Sept. 2–6. "Wind Pressure Acting on Flat Roofed Buildings." The Seventh International Colloquium on Bluff Body Aerodynamics and Applications (BBAA7) Shanghai, China.

Johns, A.W. 1922, Aug. 4. "The Stability of the Sailing Warships, II." *The Engineer*, 134: 108–110.

Johnson, Bill. 2012, May 17. "S.T. Betty Johnson FD168." http://www.fleetwood-trawlers.info/index.php/2012/05/s-t-betty-johnson-fd168/.

Johnson, Caleb. 2020. "John Howland." http://mayflowerhistory.com/howland.

Johnson, Doug. 2018, April 26. "The Chemical Composition of Exhaled Air from Human Lungs." https://sciencing.com/chemical-composition-exhaled-air-human-lungs-11795.html.

Johnson, J. 1807. *The Oriental Voyager* (describes voyage to India and China on HMS *Caroline* in 1803–6). J. Asperne.

Johnson, Steven. 2006. *The Ghost Map*. Penguin.

Johnston, Paul Forsythe. 1985. *Ship and Boat Models in Ancient Greece*. Naval Institute Press.

Jones, George. 1829. *Sketches of Naval Life*. Vol. 2. H. Howe.

Jones, James Rees. 1996. *The Anglo-Dutch Wars of the Seventeenth Century*. Longman.

Jones, Thomas P. 1834. *Journal of the Franklin Institute of the State of Pennsylvania*, Vol. 14 (new series). Franklin Institute.

Jordan, Borimir. 2000. "The Crew of Athenian Triremes." *L'Antiquité Classique*, 69: 81–101.

Kalloniatis, Michael, and Charles Luu. 2007, July 9. "Light and Dark Adaptation." In H. Kolb, E. Fernandez, and R. Nelson, eds., *Webvision: The Organization of the Retina and Visual System [Internet]*. University of Utah Health Sciences Center. https://www.ncbi.nlm.nih.gov/books/NBK11525/.

Kamler, Kenneth. 2004. *Surviving the Extremes: A Doctor's Journey to the Limits of Human Endurance*. St Martin's Press.

Kari, Alexander. 1927. *The Design of Merchant Ships and Cost Estimating: A Treatise on Ship Design and Cost Estimating*. C. Lockwood and Son.

Keevil, J.J. 1957. *Medicine and the Navy 1200–1900*. Vol. 1 (1200–1649). E&S Livingstone.

Kennedy, Maxwell Taylor. 2009. *Danger's Hour: The Story of the USS Bunker Hill and the Kamikaze Pilot Who Crippled Her.* Simon & Schuster.

Kent, Barry C. 2004. *Jacob My Friend: His 17th Century Account of the Susquehannock Indians.* XLibris US.

Kindleberger, C.P. 2018, April. "A Substitute for the Sailor's Hammock." *United States Naval Medical Bulletin*, 12(2): 251–3.

King, Ellis G., and Edward R. Frederick. 1954, Nov. 2 (issue date). U.S. Patent 2,692,994 (1954)]. "Fibrous Glass Life Preserver."

King, Henry C. 2003 [1955]. *The History of the Telescope.* Dover.

Kingston, William Henry Giles, et al. 1868. *The Boy's Own Book of Boats.* Sampson Low.

Klopman, Jonathan. 1997. "Lightning Ground Systems." https://www.woodenboat.com/lightning-ground-systems.

Knight, Edward Henry. 1884. *American Mechanical Dictionary.* Houghton Mifflin.

Knight, Roger, and Martin Wilcox. 2010. *Sustaining the Fleet, 1793–1815: War, the British Navy and the Contractor State.* Boydell Press.

Knoblock, Glenn A. 2014. *The American Clipper Ship, 1845–1920: A Comprehensive History, with a Listing of Builders and Their Ships.* McFarland.

Koennecke, Erik. Undated post. "What Is the Mass of the Carbon Dioxide Exhaled in a Breath." https://www.quora.com/What-is-the-mass-of-the-carbon-dioxide-exhaled-in-a-breath.

Koletzko, B. 2012. *The Vitamin a Story: Lifting the Shadow of Death.* S. Karger A.G.

Krulder, Joseph J. 2021, Aug. 19. "Dr. Hale's Ventilator and the Seven Years' War." https://www.joehistorian.com/blog/2021/8/19/dr-hales-ventilator-and-the-seven-years-war.

Kumar, Omraj. 2009, Dec. "Ascorbic Acid Contents in Chili Peppers (Capsicum L)." *Notulae Scientia Biologicae*, 1(1): 50–52. https://www.researchgate.net/publication/40832861_Ascorbic_Acid_Contents_in_Chili_Peppers_Capsicum_L

Kumar, Vinay. 2015, June. *Buoyancy Materials for Marine Instrumentation.* https://www.researchgate.net/publication/277994229_Buoyancy_Materials_for_Marine_Instrumentation.

Lamb, Thomas. 2000. *Exploration and Exchange: A South Seas Anthology, 1680–1900.* University of Chicago Press.

Lanbourn, Elizabeth. 2018. *Abraham's Luggage: A Social Life of Things in the Medieval Indian Ocean World.* Cambridge University Press.

Lance, Rachel. 2020. *In the Waves: My Quest to Solve the Mystery of a Civil War Submarine.* Dutton.

Langer, G.G., and N.T. Launert. 2018, Oct 2–4. "Lighting Future Naval Ships—Mission Optimized and Human Centric." *Proceedings of the International Ship Control Systems Symposium (iSCSS).* http://doi.org/10.24868/issn.2631-8741.2018.022.

Langstrom, Bjorn. 1983. *The Ship: An Illustrated History.* Doubleday.

Lardas, Mark. 2012. *Ships of the American Revolutionary Navy.* Bloomsbury.

Laughton, Leonard George Carr. 2012 [1925]. *Old Ship Figure-Heads and Sterns.* Dover.

Lavery, Brian. 1987. *The Arming and Fitting of English Ships of War 1600–1815.* Conway Maritime Press.

Lavery, Brian. 1988. *The Colonial Merchantman: Susan Constant, 1605.* Conway Maritime Press.

Lavery, Brian. 2020. *Shipboard Life and Organisation, 1731–1815.* Taylor & Francis.

Lavery, Brian. 2021. *Anson's Navy: Building a Fleet for Empire 1744–1763.* Pen and Sword.

Lavie, Peretz. 1996. *The Enchanted World of Sleep.* Yale University Press.

Layman, William. 1813. *Precursor to an Exposé on Forest Trees and Timber...* Asperne.

Lazaridis, Giorgos. 2011, May 30. "How Gas Discharging Lamps Work (Cold Cathode CCFL and Hot Cathodes)." http://pcbheaven.com/wikipages/How_Gas_Discharging_Lamps_work/.

Lemaitre, Gérard René. 2008. *Astronomical Optics and Elasticity Theory: Active Optics Methods.* Springer Berlin Heidelberg.

Lenihan, Daniel. 2010. *Submerged: Adventures of America's Most Elite Underwater Archaeology Team.* University of Virginia.

Leslie, Robert Charles. 1886. *A Sea-Painter's Log.* Chapman & Hall.

Leslie, Robert Charles. 1894. *A Waterbiography.* Chapman & Hall.

Lewis, W.H. 1986. *Underground Coal Mine Lighting Handbook, 1. Background.* Bureau of Mines Information Circular 9073. United States Bureau of Mines, Dept. Interior.

Licuanan, Virginia Benitez, and José Llavador Mira. 1991. *The Philippines Under Spain: The Legazpi Expedition, Conquest, and Colonization.* Manila: National Trust for Historical and Cultural Preservation of the Philippines.

Lienhard, John H. 1998. "Engines of Our Ingenuity No. 1344, the Monitor's Flush Toilet." https://uh.edu/engines/epi1344.htm.

Lienhard, John H., et al. 2012. "Chapter 9, Solar Desalination." In *Ann Rev Heat Transfer*, Vol. 15. Begell House.

Life Jacket Association. 2022, "Cold Water Survival." https://www.lifejacketassociation.org/life-jackets/choosing-a-life-jacket/cold-water-survival/.

Lifeline. 1973. *Lifeline: The Naval Safety Journal* 2(2).

Lim, Kyung-Sum. 1994. *How Products Are Made: An Illustrated Guide to Product Manufacturing · Vol. 2.* Pennsylvania State University.

Lind, James. 1788. *An Essay on Diseases Incidental to Europeans in Hot Climates*, 4th ed. J. Murray.

Link, Vernon B. 1955. *A History of Plague in the United States of America.* Public Health Monograph 26. Government Printing Office.

Liss, Carolin. 2011. *Oceans of Crime: Maritime Piracy and Transnational Security in Southeast Asia and Bangladesh*. Institute of Southeast Asian Studies.

Little, Benerson. 2005. *The Sea Rover's Practice: Pirate Tactics and Techniques, 1630–1730*. Potomac Books.

Liu, Zhigang, and Andrew K. Kim. 2000. "A Review of Water Mist Fire Suppression Systems—Fundamental Studies." *Journal of Fire Protection Engineering*, 10(3): 32–50.

London Quarterly Review. 1857, Jan. "Rats." *London Quarterly Review*, 101: 68–78.

Long, Philip E., and Nicola J. Palmer. 2007. *Royal Tourism: Excursions Around Monarchy*. Channel View Publications.

Luce, Stephen Bleeker, and Aaron Ward. 1884. *Text-book of Seamanship*. D. Van Nostrand.

Lugar, Catherine. 2011. *The History of the Manila Galleon Trade*. http://pacificsearesources.com/Introduction_pp.1-81.pdf.

Lukin, James, and Henry George Howe. 1852. *The History of a Ship: From Her Cradle to Her Grave*. Simpkin Marshall and Co.

Lurting, Thomas. 1710. *The Fighting Sailor, Turn'd Peaceable Christian*. J. Sowle.

Luther, Mark B., and Hora, Peter. 2014. "Investigating and Understanding CO^2 Concentrations in School Classrooms." In F. Madeo and M.A. Schnabel, eds., *Across: Architectural Research Through to Practice: International Conference of the Architectural Science Association*, pp. 631–641. Genova University Press.

Lyle, David Alexander. 1878. *Report on Life-saving Ordnance and Appurtenances*. Government Printing Office.

Macdonald, Janet. 2006. *Feeding Nelson's Navy: The True Story of Food at Sea in the Georgian Era*. Chatham Publishing.

Macdonald, Janet. 2010. *The British Navy's Victualling Board, 1793–1815*. Boydell Press.

Macdonald, John Denis. 1881, *Outline of Naval Hygiene*. Smith, Elder.

Macintosh, R.R. 1957. "The Testing of Life-Jackets." *British Journal of Industrial Medicine*, 14: 168–176.

Mackowiak, Philip A., and Paul S. Sehdev. 2002, Nov. 1. "The Origin of Quarantine." *Clinical Infectious Diseases*, 35(9): 1071–2. http://cid.oxfordjournals.org/content/35/9/1071.full.

Macky, Ian. 2023. "Deck Prisms on the Charles W. Morgan." https://glassian.org/Prism/Deck/Charles_W_Morgan/index.html.

MacPherson, R. 1783. *A Dissertation on the Preservative from Drowning; and Swimmer's Assistant*. J. Murray (London).

Maecker, Heinz. 2009. *The Electric Arc*. Popp.

Magoun, F. Alexander. 1928. *The Frigate Constitution and Other Historic Ships*. Bonanza Books.

Maier, Julius. 1886. *Arc and Glow Lamps: A Practical Handbook on Electric Lighting*. Whittaker.

Maines, Rachel. *Asbestos and Fire: Technological Tradeoffs and the Body at Risk*. Rutgers University Press.

Mair, Michael, and Joy Waldron. 2014. *Kaiten: Japan's Secret Manned Suicide Submarine and the First American Ship It Sank in WWII*. Penguin.

Maisel, Albert Q. 1944, March. "War Leaves Its Mark on the Signposts of the Sea." *Popular Science*, 144(3): 105–8, 191.

Malinowska, Katarzyna. 2020. "Comparative Study of Damage Stability Calculations for SOLAS 2009 and SOLAS 2020 of RO-RO Passanger [sic] Vessel." https://www.topkorab.org.pl/wp-content/uploads/2020/12/Malinowska_Katarzyna-paper.pdf.

Mamabear (pseud.). 2004, April 8. "Lamp Oil and Wick Burn Time." http://www.alpharubicon.com/primitive/lampoilburntime.htm.

Mannix, Daniel P., IV. 2014. *The Old Navy: From the Personal Records of Rear Admiral Daniel P Mannix III*. eNet Press.

Manser, Martin H., et al. 2007. *The Facts on File Dictionary of Proverbs*. Facts on File.

Marine Engineering. 1921, Aug. "Acetylene for Ship's Lighting." *Marine Engineering*, 26(8): 587.

Maritime Training Advisory Board (U.S.). 1994. *Marine Fire Prevention, Firefighting and Fire Safety: A Comprehensive Training and Reference Manual*. DIANE Publishing Co.

Markert, John W. 1944. "Evolution of Ship Ventilation Systems." *The Log*, 39: 202–18.

Marriott, Leo. 2003. *Lighthouses*. PRC Publishing.

MARS Reports. 2016, Jan. 4. "Real Life Accident: Chief Mate Dies from Asphyxiation." *Marine Insight*. https://www.marineinsight.com/case-studies/real-life-accident-zinc-concentrate-kills-by-asphyxiation/.

Marsden, Peter. 2019. *1545: Who Sank the Mary Rose?* Pen & Sword.

Martin, J. 2010, Sept. 9. "Shipboard Berthing: Athwartship and Longitudinal." *IEEE Xplore*. https://ieeexplore.ieee.org/document/1154176 (originally presented at OCEAN 75 Conference, Sept. 22–25, 1975).

Martin, Sir Thomas Bryan. 1803. *Letters and Papers of Admiral of the Fleet Sir Thos. Byam Martin, G.C.B.* Navy Records Society.

Martin, Commander Tyrone G. 2003. *Constitution Close Up: Minutiae for the Modeler and Artist*. https://ussconstitutionmuseum.org//wp-content/uploads/dlm_uploads/2020/01/closeup_martin.pdf.

Masefield, John. 1905. *Sea Life in Nelson's Time*. Methuen & Co.

Matthews, R., et al., 2021, Oct. "O034 Rockabye Sailor: Investigating the Impact of Simulated Motion on Sleep and Cognitive Performance." *SLEEP Advances*, 2(Suppl. 1): A15–A16. https://doi.org/10.1093/sleepadvances/zpab014.033.

Maxim, Dan. 2015, Jan. 12. "Summary of Relevant Life Jacket Research." https://cdn.ymaws.com/www.americancanoe.org/resource/resmgr/spp-documents/life_jacket_report_Dan_Maxim.pdf.

May, Brian, et al. 2020, July 21. "Ballistic Protective Properties of Material Representative of English Civil War Buff-coats and Clothing." *International Journal of Legal Medicine*, 134: 1949–1956. https://doi.org/10.1007/s00414-020-02378-x.

Maynarde, Thomas. 1849. *Sir Francis Drake His Voyage, 1595*. Hakluyt.

McCollum, Elmer Verner. 1918. *The Newer Knowledge of Nutrition*. Macmillan.

McDonald, W.F. 1957. *Notes on the Problems of Cargo Ventilation*, Tech. Note 17. World Meteorological Organization.

McDowell, Lee R. 2013. *Vitamin History: The Early Years*. University of Florida Press.

McElvogue, Douglas. 2020. *Tudor Warship Mary Rose*. Bloomsbury Publishing.

McKay, John. 2020. *Sovereign of the Seas, 1637: A Reconstruction of the Most Powerful Warship of Its Day*. Pen & Sword.

McKay, Richard C. 2011. Donald McKay and His Famous Sailing Ships. Dover.

McKee, Christopher. 2002. *Sober Men and True: Sailor Lives in the Royal Navy, 1900–1945*. Harvard University Press.

McManamon, John M. 2016. *Caligula's Barges and the Renaissance Origins of Nautical Archaeology Under Water*. Texas A&M University Press.

Mechanics Magazine. 1849, Jan. 13. "Brig.-Gen Sir Samuel Bentham's Model Vessels." *Mechanics Magazine*, 50: 38.

Mechanics Magazine. 1871, Nov. 25. North British Rubber Company advertisement. *Mechanics Magazine*, 95.

Medical Record. 1898, June 25. "Sunstroke and Its Varied Causes." *Medical Record* 941.

Melville, Herman. 1902 [1851]. *Moby Dick, or the Whale*. Scribner.

Melville, Herman. 1922 [1850]. *White Jacket, Or, the World in a Man-of-War*. In *The Works of Herman Melville*, Standard Ed., Vol. VI. Constable and Company.

Menz, Garry, et al. 2011."Growth and Survival of Foodborne Pathogens in Beer." *Journal of. Food Protection*, 74 (10): 1670–5.

Mercieca, Simon. 2014. "Hazards at Sea: A Case-Study of Two Ex-Voto Paintings from the Church of Karmelitani Skalzi in Bormla, Malta." https://www.um.edu.mt/library/oar/bitstream/123456789/25591/1/Ships_Saints_and_Sealore_Cultural_Herita (1).pdf.

Meulstee, Louis. Undated. "Gibson Girl Part 1. Air-Sea Rescue: Long Wave and Short Wave." Wireless for the Warrior. http://www.wftw.nl/gibsongirl.html.

Meza. Bob. 2019. "World War II Searchlight History." https://www.victorysearchlights.com/victory2.html.

Michel, Wolfgang, and Elke Werger-Klein. 2004. "Drop by Drop—The Introduction of Western Distillation Techniques Into Seventeenth-Century Japan." *Journal of the Japan Society of Medical History*, 50(4): 463–492. http://wolfgangmichel.web.fc2.com/publ/aufs/72/072.htm.

Militaryhistorynow.com 2013, Sept. 30. http://militaryhistorynow.com/2013/09/30/fighting-spirits-three-centuries-of-rum-in-the-royal-navy/.

Mills, Evan. 2003. "Technical and Economic Performance Analysis of Kerosene Lamps and Alternative Approaches to Illumination in Developing Countries." https://www.researchgate.net/publication/228503671_Technical_and_Economic_Performance_Analysis_of_Kerosene_Lamps_and_Alternative_Approaches_to_Illumination_in_Developing_Countries.

Milton, James H., and Roy M. Leach. 2013. *Marine Steam Boilers*. Elsevier Science.

Mittleider, Megan C. 2020, March. *Improving U.S. Navy Shipboard Habitability: Effects of Light and Temperature in Berthing Compartments*. Thesis, Naval Postgraduate School.

ModernSurvivalBlog. Undated. "How to Make Own Olive Oil Lamp." http://www.instructables.com/id/How-To-Make-Your-Own-Olive-Oil-Lamp/.

Mollahan, Lt. Cmdr. Sean D. 2012, Feb. *United States Navy Nutrition Culture and How Best to Select Food While Underway*. Thesis, m. military art & science, U.S. Army Command and General Staff College. https://apps.dtic.mil/sti/citations/ADA599135.

Mondfeld, Wolfram zu. 2005. *Historic Ship Models*. Sterling.

Moore, John Hamilton. 1784. *The Practical Navigator and Seamans Daily Assistant*. B. Law.

Moorhead, Robert. 2002, Nov. "William Budd and Typhoid Fever." *Journal of the Royal Society of Medicine*, 95(11): 561–4.

Morel, Jen. 2022, Feb. 24. "4 Floating Positions That Could Save Your Life." https://www.livestrong.com/article/468210-different-floating-positions-in-swimming/.

Morison, Samuel Eliot. 1947. *History of United States Naval Operations in World War II: The Battle of the Atlantic, September 1939-May 1943*. Little, Brown.

Morison, Samuel Eliot. 1974. *The European Discovery of America: The Southern Voyages, 1492–1616*. Oxford University Press.

Motture, Peta. 2011. *Re-thinking Renaissance Objects: Design, Function and Meaning*. Wiley.

Muckle W. (revised by D.A. Taylor). 2013. *Muckle's Naval Architecture, 2d ed.* Elsevier Science.

Muir, Margaret. 2012, Nov. 21. "HMS Warrior—Staying Clean on an 1860 Fighting Ship." http://margaretmuirauthor.blogspot.com/2012/11/hms-warrior-staying-clean-on-1860.html.

Mungan, Carl E., and John D. Emery. 2011, May. "Rolling the Black Pearl Over: Analyzing the Physics of a Movie Clip." *Physics Teacher*, 49: 266–71.

Munson, Francis Merton. 1920. *Hygiene of Communicable Diseases*. P.B. Hoeber.

Murdock, Lieutenant J.B. 1893. "Notes on Naval Dynamo Machinery." *Naval Institute Proceedings*, 19(4): 345.

Museum of Historical Chamber Pots and Toilets. Undated. "Chamber Pots." https://muzeumnocniku.cz/en/chamber-pots.

Museum of Technology: The History of Gadget and Gizmos. 2007. "WWII Signal Corps Balloon Aerial, 1943." http://www.museumoftechnology.org.uk/objects/_expand.php?key=532.

Musset, Lucien. 2005. *The Bayeux Tapestry*. Boydell Press.

Nance, R. Morton. 2012. *Classic Sailing-Ship Models in Photographs*. Dover.

Nansen, Fridtjof, and Otto Neumann Sverdrup. *Farthest North*, Vol. 1. Harper & Bros.

Naranjo, Ralph. 1986, March. "The Solar Alternative." *Cruising World* 51.

Nathanson, Andrew. 2019, Feb. "Sailing Injuries: A Review of the Literature." *Rhode Island Med. J.*, 23–7. http://www.rimed.org/rimedicaljournal/2019/02/2019-02-23-wilderness-nathanson.pdf.

National Academies of Sciences, Engineering, and Medicine. 1997. *Fire Suppression Substitutes and Alternatives to Halon for U.S. Navy Applications*. The National Academies Press. https://doi.org/10.17226/5744.

National Academies of Sciences, Engineering, and Medicine. 2011. *The Effects of Commuting on Pilot Fatigue*. National Academies Press. https://doi.org/10.17226/13201.

National Candle Association. 2020. "Elements of a Candle: Wicks." http://candles.org/elements-of-a-candle/wicks/.

National Imagery and Mapping Agency. 2003 [rev. of 1969 ed.]. *International Code of Signals for Visual, Sound and Radio Communications*. United States Edition. NIMA.

National Institute for Occupational Safety & Health. 2020, April 1. "Interim NIOSH Training for Emergency Responders: Reducing Risks Associated with Long Work Hours." https://www.cdc.gov/niosh/emres/longhourstraining/default.html.

National Institutes of Health. 2021. "Vitamin C Fact Sheet for Health Professionals." https://ods.od.nih.gov/factsheets/VitaminC-HealthProfessional/.

National Institutes of Health. 2022, June 15. "Vitamin A and Carotenoids: Fact Sheet for Health Professionals." https://ods.od.nih.gov/factsheets/VitaminA-HealthProfessional/.

National Museum of American History. Undated. Edison type Z dynamo from SS Columbia, ID EM 181820, Accession 33703, Catalog 181830. https://americanhistory.si.edu/collections/search/object/nmah_712864.

Nautical Magazine and Naval Chronicle. 1841.

Nautical Magazine. 1835, March. *The Nautical Magazine*, 4: 168–9.

Naval Economy 1811. *Naval Economy: Exemplified in Conversations Between a Member of Parliament and Officers of a Man of War During a Winter's Cruize...* William Lindsell.

Naval Education and Training Command. 1979. *Electronics Technician 3&2*. Vol. 1.

Naval Encyclopedia 1880. *Naval Encyclopedia*. L.R. Hamersly & Co.

Naval History and Heritage Command. Undated. "Flushing Toilet, CSS Alabama, #91–217-AG." https://www.history.navy.mil/content/history/nhhc/research/underwater-archaeology/conservation-and-curation/ua-artifact-collections/css-alabama-artifact-collection/css-alabama-artifact-photo-collection/css-alabama-toilet.html.

Naval History and Heritage Command. Undated. "Uniforms and Personal Equipment." https://www.history.navy.mil/browse-by-topic/heritage/uniforms-and-personal-equipment/uniforms-1776-1783.html; https://www.history.navy.mil/browse-by-topic/heritage/uniforms-and-personal-equipment/uniforms-1797.html; https://www.history.navy.mil/browse-by-topic/heritage/uniforms-and-personal-equipment/uniforms-1802.html.

Naval History and Heritage Command. Undated. "USS Monitor Coal (1 of 2)." https://www.history.navy.mil/news-and-events/multimedia-gallery/news-photos/uss-monitor-coal.html.

Naval History and Heritage Command. 2019, Dec 18. "First Naval Vessel with Electric Lights—Trenton." https://www.history.navy.mil/browse-by-topic/exploration-and-innovation/electricity-and-uss-trenton.html.

Naval Postgraduate School. 2017, Aug. 1. *Crew Endurance Handbook: A Guide to Applying Circadian-Based Watchbills*. http://my.nps.edu/web/crewendurance.

Naval Ship Systems Command. 1967. *Boats of the United States Navy*, Manual 250–452. https://maritime.org/doc/boatcat/index.htm#toc.

Naval Supply Systems Command. 2016, Aug. *Food Service Management General Messes*, NAVSUP P-486, Vol. II, Rev. 8.

Navy Cyberspace. Undated. "1842 Navy Food and Spirit Ration." https://www.navycs.com/public-law/naval-ration-1842.html.

Navy Cyberspace. Undated. "1861 Navy Food and Spirit Ration." https://www.navycs.com/public-law/naval-ration-1861.html.

Navy Cyberspace. Undated. "Navy and Marine Corps Rations 1794." https://www.navycs.com/charts/1794-navy-rations.html.

Navy Cyberspace. Undated. "Navy and Marine Corps Rations 1801." https://www.navycs.com/charts/1801-navy-rations.html.

Navy Cyberspace. Undated. "Navy Food & 'Midrats' Ration, 1902." https://www.navycs.com/charts/1902-navy-rations.html.

Navy Medicine. 2022. *Manual of Naval Preventive Medicine*, NAVMED P-5010–6, Chapter 6, Water Quality Afloat (Rev. 1–2022).

Navy Personnel Command. 1955, Jan. *Submarine Distilling Systems*, NAVPERS 16170.

Needham, Joseph, et al. 1971. *Science and Civilization in China*, Vol. 4, Physics and Physical Technology, Part III, Civil Engineering and Nautics. Cambridge University Press.

Nerz, F. 1893. "The Construction and Use of Projectors." *The Electrician*, 30: 713.

New Scientist 2005. *Does Anything Eat Wasps?* Profile Books.

New South Wales. 1899. *Rules, Regulations, and By-laws*, Vol. 1. Law Book Company of Australasia.

Nicholson, John. 1825. *The Operative Mechanic, and British Machinist*, Vol. 1. Knight and Lacey.

Njoku, P.C., et al. 2011. "Temperature Effects on Vitamin C Content in Citrus Fruits." *Pakistan Journal of Nutrition*, 10(12): 1168–9.

Noakes, T.D. 1998, June. "Fluid and Electrolyte Disturbances in Heat Illness." *International Journal of Sports Medicine*, 19 (Suppl 2): S146-0). https://doi.org/10.1055/s-2007-971982.

Noakes, T.D. 2008, Jan. "A Modern Classification of the Exercise-related Illnesses." *Journal of Science and Medicine in Sport*, 11(1): 33–9; https://doi.org/10.1016/j.jsams.2007.02.009.

Normandy, Frank. 1909. *A Practical Manual on Sea Water Distillation, Etc.* Charles Griffin & Co.

North Atlantic Treaty Organization. 2010, March. *Nutrition Science and Food Standards for Military Operations*, RTO Technical Report TR-HFM-154.

Nortum, Reece. 2016, Sept. 30. "In the Limelight: A Civil War Military Innovation." https://hamptonroadsnavalmuseum.blogspot.com/2016/09/in-limelight-civil-war-military.html.

Notter, James Lane, and Sir Robert Hammill Firth. 1896. *The Theory and Practice of Hygiene*. Churchill.

Notter, James Lane, and Sir Robert Hammill Firth. 1908. *The Theory and Practice of Hygiene*. Churchill.

Occupational Safety and Health Administration 2015. Fall Protection in Construction. https://www.osha.gov/sites/default/files/publications/OSHA3146.pdf.

Occupational Safety and Health Administration. 2018, Feb. 7. "Carbon Dioxide Health Hazard Information Sheet."https://www.fsis.usda.gov/sites/default/files/media_file/2020-08/Carbon-Dioxide.pdf.

Oertling, Thomas James. 1984, May. *The History and Development of Ships' Bilge Pumps, 1500–1840*. M.A. thesis, anthropology, Texas A&M University.

Olsen, Olav Kjellevold. 2013. "The Impact of Partial Sleep Deprivation on Military Naval Officers' Ability to Anticipate Moral and Tactical Problems in a Simulated Maritime Combat Operation." *International Maritime Health*, 64(2): 61–65. https://journals.viamedica.pl/international_maritime_health/article/view/34686/25358.

Open University. 1997. *Seawater: Its Composition, Properties and Behavior.* Butterworth-Heinemann.

Oppenheim, Michael. 1896. *A History of the Administration of the Royal Navy and of Merchant Shipping in Relation to the Navy*, Vol. 1. J. Lane.

Ordnance Department, U.S. Army. 1920. *History of Military Pyrotechnics in World War.* Government Printing Office.

Orme, Nicholas. 1983. *Early British Swimming, 55 BC–AD 1719.* University of Florida.

Osborne, R. 1857, Jan. 30. Circular No. 283, *Regulations, for a Uniform Dress for Petty Officers, Seamen and Boys in the Royal Navy*. Quoted in https://freepages.rootsweb.com/~pbtyc/genealogy/RN/Pay_and_Condns/Uniform_1857.html.

Owen, Hugh. 1906. *The Tonnage and Freeboard of Merchant Ships.* J. Brown and Son.

Panico, Michele. 2018, Nov. *Hammocks: A Maritime Tool*. M.A. Thesis, maritime studies, East Carolina University.

Paris, John Ayrton. 1825, *The Elements of Medicinal Chemistry.* W. Phillips.

Parker, William Harwar. 1883. *Recollections of Naval Officer, 1841–1865.* Charles Scriber's Sons. https://scholarship.rice.edu/jsp/xml/1911/27246/1/aa00404.tei.html.

Parkes, Edmund Alexander. 1866. *Manual of Practical Hygiene.* John Churchill & Sons.

Parkes, Edmund Alexander. 1883. *Manual of Practical Hygiene*, Vol. 1. Wood.

Parsons, Michael H., et al. 2018, Sept. 27. "Temporal and Space-Use Changes by Rats in Response to Predation by Feral Cats in an Urban Ecosystem." *Frontiers in Ecology and Evolution*, 6: 146. https://doi.org/10.3389/fevo.2018.00146.

Passenger Act 1803. 43 Geo III c56. Cited at 429 in Steel, David. 1803. *The Ship-master's Assistant and Owner's Manual.* P. Steel.

Patel, Hiran, et al. 2022, Feb. 15. "Physiology, Respiratory Quotient." National Library of Medicine. https://www.ncbi.nlm.nih.gov/books/NBK531494/.

Patowary, Kaushik. 2020, April 13. "Rettungsbojen: The Floating Rescue Buoys of the Luftwaffe." https://www.amusingplanet.com/2020/04/rettungsbojen-floating-rescue-buoys-of.html.

Patterson, K. David. 1983. "The Influenza Epidemic of 1918-19 in the Gold Coast." *Journal of African History*, 24(4): 485–502.

Pearce, D. 1979. *Ship Bridge Lighting: Red or White*. Defence and Civil Institute of Environmental Medicine. https://apps.dtic.mil/sti/citations/ADA087320.

Pendergrast, Mark. 2009. *Mirror Mirror: A History of the Human Love Affair with Reflection.* Basic Books.

Pereira, Helena. 2011. *Cork: Biology, Production and Uses.* Elsevier Science.

Perez-Mallaina, Pablo E. 1998. *Spain's Men of the Sea: Daily Life on the Indies Fleets in the Sixteenth Century.* Johns Hopkins University Press.

Permentier, Kim, et al. 2017, April 4. "Carbon Dioxide Poisoning: A Literature Review of an Often Forgotten Cause of Intoxication in the Emergency Department." *International Journal of Emergency Medicine*, 10: 14. https://www.ncbi.nlm.nih.gov/pmc/articles/PMC5380556/.

Persily A., and L. De Jonge. 2017. "Carbon Dioxide Generation Rates for Building Occupants." *Indoor Air*, 27: 868–879.

Person, Jennifer. 2008, July. *NAVSUP Initiatives for Health & Radius.* https://slideplayer.com/slide/8659023/.

Peters, E. Kirsten, and Larry E. Davis. 2001. *Geology from Experience: Hands-On Labs and Problems in Physical Geology.* W.H. Freeman.

Peterson, Alyssa A. 2016. *"We Live in the Midst of Death": Medical Theory, Public Health, and the 1793 Yellow Fever Epidemic.* M.A. thesis, history, Eastern Illinois University.

Phillips, Lawrence. 1970, July. "The Abolition of the Rum Ration." *U.S. Naval Institute Proceedings*, 96(7): 809. https://www.usni.org/magazines/proceedings/1970/july/abolition-rum-ration.

Pickthall, Barry. 2016. *A History of Sailing in 100 Objects.* Bloomsbury.

The Pilot. 1839, Nov. "Narrative of the Wreck of the Ship 'Protector.'" *The Pilot, or Sailors Magazine* 341.

Pinwell, George. 1866, Aug. 16. "Death's Dispensary." *Fun*, 10: 233. https://ufdc.ufl.edu/UF00078627/00010/zoom/205.

Pitassi, Michael. 2012. *The Roman Navy: Ships, Men & Warfare 350 BC–AD 475.* Seaforth Publishing.

Platt, Richard. *Stephen Biesty's Cross-Sections Man-of-War.* D.K. Publishing.

Plummer, Katherine. 1991. *The Shogun's Reluctant Ambassadors: Japanese Sea Drifters in the North Pacific, 3rd ed. rev.* Oregon Historical Society Press.

Pollitzer, R. 1952. "Plague Studies: 6. Hosts of the Infection." *Bulletin of the World Health Organization*, 6: 381–465.

Pollitzer, R. 1953. "Plague Studies: 10. Control and Prevention." *Bulletin of the World Health Organization*, 9: 457–551.

Pope, Dudley. 2013. *Life in Nelson's Navy.* House of Stratus.

Popular Mechanics. 1915, April. "Life Preservers Made of Kapok Fiber." *Popular Mechanics* 542.

Porter, David. 1815. *Journal of a Cruise Made to the Pacific Ocean in the United States Frigate Essex in the Years 1812, 1813, and 1814.* Bradford and Inskeep.

Porter, Commodore David Dixon. 1875. *Memoir of Commodore David Porter of the United States Navy.* J. Munsell.

Pressly, Andrew. 2020, Dec. 14. "Taming the Depths: A History of Diving." https://www.floridamaritimemuseum.org/post/taming-the-depths-a-history-of-diving.

Professional Journal of the U.S. Army. 1964, Dec. "Military Notes." *Professional Journal of the U.S. Army*, 44: 98–106.

Prothero, David. 2010, July 29. "Flying Flags Upside Down." https://www.crwflags.com/fotw/flags/xf-flip.html.

Proulx, Gilles. 1984. *Between France and New France: Life Aboard the Tall Sailing Ships.* Dundum Press.

Pryor, James Chamber. 1918. *Naval Hygiene.* P. Blakston.

Public Health Reports. 1910, July 22. "Typhoid Fever on Japanese Battle Ships." *Public Health Reports*, 25(29): 1023.

Public Health Service. 1937, April 2. "Overseas Transmission of Bubonic Plague: A Danger Almost Eliminated." *Public Health Reports*, 52(14): 412–4.

Public Health Service. 1947. *The Ship's Medicine Chest and First Aid at Sea.* Misc. Pub. 9. Government Printing Office.

Queensland Government Emigration Office. 1869. *Instructions to Surgeon-Superintendants of Queensland Ships.*

Quinn, Katrina J., et al. 2021. *Adventure Journalism in the Gilded Age: Essays on Reporting from the Arctic to the Orient.* McFarland.

Quinn, Kendra Leeanne. 1999, Dec. *Shipboard Lighting: A.D. 400–1900.* M.A. thesis, anthropology, Texas A&M University.

Ramani, K.V. 2015. *Readings in Personnel Management.* Mittal Publications.

Ramirez, Ainissa. 2020, March-April. "Tungsten's Brilliant, Hidden History." *American Scientist*, 108(2): 88. https://doi.org/10.1511/2020.108.2.88. https://www.americanscientist.org/article/tungstens-brilliant-hidden-history.

Rao, Narashima Desirazu. 2011, Aug. *Distributional Impacts of Energy Policies in India: Implications for Equity in International Climate Change Agreements.* Ph.D. dissertation, Emmett interdisciplinary program in environment and resources, Stanford University.

Rawson, F.L. 1906, April. "Acetylene Association Abroad." *Light, Heat and Power*, 6(2): 52–3.

Raymond, John O. 1866, Oct. 30 (issue date). U.S. Patent 59,264. "Improved Wind-Sail."

Reed, Edward J. 1858. "On the Stability of Monitors Under Canvas." *Transactions of the Institution Naval Architects*, 9: 198–217.

Reed, Edward J. 1885. *A Treatise on the Stability of Ships.* Charles Griffin and Company.

Rees, Abraham. 1819. "Ventilator." *The Cyclopædia, Or, Universal Dictionary of Arts, Sciences, and Literature*, Vol. 36. Longman.

Registrar-General. 1875. *Thirty-Sixth Annual Report of the Registrar-General of Births, Deaths, and Marriages in England.* HM Stationery Office.

Reid, D.B. 1844. *Illustrations of the Theory and Practice of Ventilation.* Longman, Brown, Green & Longmans.

Reiss, Oscar. 1998. *Medicine and the American Revolution: How Diseases and Their Treatments Affected the Colonial Army.* McFarland.

The Repertory of Arts, Manufactures and Agriculture, etc. 1807. Vol. XI, 2d Ser.

Rey, Jean. 1917. *The Range of Electric Searchlight Projectors.* D Van Nostrand Company.

Richards, Phil, and John J. Banigan. 1942, June. *How to Abandon Ship.* http://www.ibiblio.org/hyperwar/USN/ref/AbandonShip/index.html.

Riess, Warren Curtis. 1980. *The History of, and Search for, the Seventeenth Century Bristol Merchantman Angel Gabriel.* M.A. thesis, anthropology, Texas A&M

Ritchie, Robert. 1843. "Suggestions for the Better Ventilation of Sailing and Steam-Vessels." *Mechanics' Magazine,* 39: 418.

Robinson, Hercules. 1860. *Harry Evelyn; Or, Romance of the Atlantic (A Naval Novel, Founded on Facts).* J. Blackwood.

Rodwell, James. 1858. *The Rat: Its History & Destructive Character.* G. Routledge & Company.

Rogers, Stanley. 1936. *Freak Ships.* John Lane.

Roman, Gustavo C. 2013. "Beriberi Neuropathy." In Chapter 30, "Nutritional Disorders in Tropical Neurology." *Handbook of Clinical Neurology,* 113: 381–404. https://www.sciencedirect.com/topics/medicine-and-dentistry/beriberi.

Romm, Joe. 2016, Oct. 26. "Exclusive: Elevated CO_2 Levels Directly Affect Human Cognition, New Harvard Study Shows." https://archive.thinkprogress.org/exclusive-elevated-co2-levels-directly-affect-human-cognition-new-harvard-study-shows-2748e7378941/.

Roscoe, Henry Enfield. 2007. *Lessons in Elementary Chemistry: Inorganic and Organic.* Macmillan.

Rose, Lisle A. 2007. *Power at Sea, Volume 2: The Breaking Storm, 1919–1945.* University of Missouri Press.

Rossum, J.R. 2000. "Fundamentals of Metallic Corrosion in Fresh Water." https:/roscoemoss.com/wp-content/uploads/publications/fmcf.pdf.

Rouppe, Lewis. 1772. *Observations on Diseases Incidental to Seamen.* London.

Royal Museums Greenwich. https://collections.rmg.co.uk/mediaLib/497/media-497813/large.jpg.

Royal National Lifeboat Institution. "1785: The First Lifeboats." https://rnli.org/about-us/our-history/timeline/1785-the-first-lifeboats.

Royal National Lifeboat Institution. "1854: First Lifejackets." https://rnli.org/about-us/our-history/timeline/1854-first-lifejackets. [This site says Ward invented the cork lifejacket. Not true. Another RNLI page says that the only survivor of the 1861 Whitby lifeboat disaster was the one who was wearing the new jacket. That is true, but his companions weren't wearing any kind of cork jacket (SSTV).]

Rucker, William Colby. 1910. "Rodent Extermination." In U.S. Treasury Department, 153–162.

Ryder, Rear-Admiral A.P. 1871, Sept. "The Naval Hammock—Its Buoyancy and Use in Saving Life at Sea—In Cases of Collision, Etc." *The Nautical Magazine (New Series),* 40: 636–642.

Safe and Sound, Tees Valley. 2015, June 15. "The Life Jacket." In "Safe and Sound, Stories of Emergency Response in the Tees Valley." https://safeandsoundteesvalley.wordpress.com/2015/06/15/the-cork-life-jacket/comment-page-1/.

Safety Engineering. 1913, Nov. and Dec. issues. *Safety Engineering,* Vol. 26.

Safety Review. 1947, Nov. "Marine Section." *Safety Review,* 4(11): 20–22.

The Sailor's Magazine, and Naval Journal. 1835, April. "Largest Ship in the World." *The Sailor's Magazine, and Naval Journal,* 7: 243.

Satran, Joe. 2017, Dec. 6. "13 Weird Moments in the History of Water Fountains" https://www.huffpost.com/entry/history-of-water-fountains_n_6357064.

Scanship. Undated. "Water Reuse Unit." https://www.scanship.no/solutions/water-reuse.

Schama, Simon. 1997. *The Embarrassment of Riches: An Interpretation of Dutch Culture in the Golden Age.* Vintage Books.

Schat, Ane P. 1929, May 21 (issue date). U.S. Patent 1,714,452. "Self-Contained Automatic Tackle."

Schat, Ane P. 1930, Feb. 18 (issue date). U.S. Patent 1,747,795. "Means for Handling Lifeboats."

Schat, Ane P. 1935, Oct. 8 (issue date). U.S. Patent 2,016,838. "Installation for Handling a Lifeboat."

Schotte, Margaret E. 2019. *Sailing School: Navigating Science and Skill, 1550–1800.* Johns Hopkins University Press.

Schultz, Charles R. 1999. *Forty-niners 'Round the Horn.* University of South Carolina Press.

Schurz, William. 1939. *The Manila Galleon.* E.P. Dutton.

Science Museum Group. Undated. "The Colt Blackout Ventilator," Object 2019–97. https://collection.sciencemuseumgroup.org.uk/objects/co8600128/the-colt-blackout-ventilator-blackout-ventilator.

Sedgwick, William Thompson. 1902. *Principles of Sanitary Science and the Public Health.* Macmillan.

Semeco, Arlene. 2023, Jan. 4. "What Happens If You Drink Too Much Water." https://www.medicalnewstoday.com/articles/318619.

Sephton, James. 2011. *Sovereign of the Seas: The Seventeenth-Century Warship.* Amberley.

Septer, Dirk. 2021. "Ane Pieter Schat's Life Boat Davits and Life Boats Improved Shipwreck Survival Worldwide." http://nauticapedia.ca/Gallery/Schat_Patent.php.

Shavinina, Laris V. 2003. *The International Handbook on Innovation.* Elsevier Science.

Shaw, Simeon. 1823. *Nature Displayed in the Heavens, and on the Earth, According to the Latest Observations and Discoveries,* Six Volumes. G. & W.B. Whitaker.

Sherrard, G.C. 1943, Dec. 24. "The Benefits Accruing from the Ratproof Construction of Vessels." *Public Health Reports,* 58(52): 1888–1890.

Simmons, Joe J. 1985, Dec. *The Development of External Sanitary Facilities Aboard Ships of the Fifteenth to Nineteenth Centuries.* M.A. thesis, anthropology, Texas A&M University.

Simpson, William John. 1905. *A Treatise on Plague Dealing with the Historical, Epidemiological, Clinical, Therapeutic and Preventive Aspects of the Disease.* Cambridge University Press.

Škrovánková, Soňa. 2011. "Chapter 28, Seaweed Vitamins as Nutraceuticals." In Se-Kwon Skim, ed., *Marine Medicinal Foods: Implications and Applications; Macro and Microalgae*, 357–370 (Vol. 64 in Advances in Food and Nutrition Research). Elsevier.

Skylighter. 2018, Feb. 15. "Realgar & Orpiment." http://www.skylighter.com/fireworks/orpiment-realgar.asp.

Slism.com. Undated. "Komatsuna." http://slism.com/calorie/106086/.

Smit, Andrea N., et al. 2019. "Sleep Timing and Duration in Indigenous Villages with and Without Electric Lighting on Tanna Island, Vanuatu." *Scientific Reports* 9: 17278. https://doi.org/10.1038/s41598-019-53635-y.

Smith, Edgar C. 2013 [1938]. *A Short History of Naval and Marine Engineering*. Cambridge University Press.

Smith, John. 1627. *The Sea-Mans' Grammar*. https://onlinebooks.library.upenn.edu/webbin/book/lookupid?key=uma12469.

Smith, Paul. 2007, July-Aug. "Sailor, Supplier, Service, and the Navy Standard Core Menu." *The Navy Supply Corps Newsletter*.

Smyth, Richard. 2012. *Bum Fodder: An Absorbing History of Toilet Paper*. Souvenir Press.

Society for Naval Research 2015. "Collecting Rain Water at Sea." https://snr.org.uk/snr-forum/topic/collecting-rain-water-at-sea/.

Solem, Berge. 2007. "Steerage Passengers—Emigrants Between Decks." In *The Transatlantic Crossing*, Chapter 2, Norway Heritage. http://www.norwayheritage.com/steerage.htm.

Sorathia, Usman. 1999, May. "Fire Safety of Composites in the U.S. Navy." *Composites Part A: Applied Science and Manufacturing*, 30(5): 707–713.

South Shields Volunteer Life Brigade. 2015. *Coast Rescue Equipment: Line Throwing from Early 1800s to 1948*. http://www.thessvlbhistory.co.uk/pdf/History%20line%20throwing%20from%20the%20shore%20presentation%202.pdf.

Spalding, Simon. 2015. *Food at Sea: Shipboard Cuisine from Ancient to Modern Times*. Rowman & Littlefield.

Sparks, Mike. 2005. "Abandon Ship! Prepare to Die! How the U.S. Navy Lives in Denial and Is Not Ready for Combat." http://www.combatreform.org/USNAVYINDANGER/abandonshippreparetodie.htm.

Squires, Harold. 1992. *S.S. Eagle, the Secret Mission, 1944–45*. Jesperson Press.

Stag, Michael G. 2021, Jan. 7. History and Development of Firefighting Foams. https://stagliuzza.com/news/history-and-development-of-firefighting-foams/.

Standards of Training, Certification, and Watchkeeping (STCW_ Handbook. Undated. Commentary on Chapter VIII: Watchkeeping. http://www.navit.fo/stcw_konventiontekstur/s-handbk12.htm.

Stare, Frederick J., and Margaret McWilliams. 1984. *Living Nutrition*, 4th ed. John Wiley & Sons

Stayton, William H. 1895. *Navy Militaman's Handbook*. A.R. Pope.

Stedman, Captain John Gabriel. 1813. *Narrative. of a Five Years Expedition; Against the Revolted Negroes of Surinam*, Vol. 1. J. Johnson & J. Edwards.

Steel, David. 1794. *The Elements and Practice of Rigging and Seamanship*. San Francisco Maritime National Park Association (2013). https://maritime.org/doc/steel/.

Steel, David. 1805. *The Shipwright's Vade-Mecum*. London.

Steele, Brett D., and Tamera Dorland. 2005. *The Heirs of Archimedes: Science and the Art of War Through the Age of Enlightenment*. MIT Press.

Stein, Stephen H. 2017. *The Sea in World History: Exploration, Travel and Trade*. ABC-CLIO.

Sterling, Christopher H. 2008. *Military Communications: From Ancient Times to the 21st Century*. ABC-CLIO.

Stevens, Robert White. 1871 and 1873. *On the Stowage of Ships and Their Cargos*. Longmans.

Stevenson, David. 2014 [1878]. *Life of Robert Stevenson: Civil Engineer*. Cambridge University Press.

Stevenson, Thomas, and Shirley F. Murphy. 1892. *A Treatise on Hygiene and Public Health*, Vol. 2. J. & A. Churchill.

Stewart, Alexander. 1798. *Medical Discipline, or Rules and Regulations for the More Effectual Preservation of Health on Board the Honourable East India Company's Ships*. Murray and Highley.

Stockwin, Julian. 2011. *Stockwin's Maritime Miscellany: A Ditty Bag of Wonders from the Golden Age of Sail*. Ebury Publishing.

Stromsted, Alf J. 1950, May. "Electrical Installations on the New APL Passenger Ships." *The Log*, 45: 31, 48–49.

Stubbs, Brett J. 2003. "Captain Cook's Beer: The Antiscorbutic Use of Malt and Beer in Late 18th Century Sea Voyages." *Asia Pacific Journal of Clinical Nutrition*, 12(2): 129–37.

Sugden, John. 2012. *Nelson: A Dream of Glory, 1758–1797*. Pimlico.

Sundal, Marjani Kjetland, et al. 2017, Oct. "Asphyxiation Death Caused by Oxygen-depleting Cargo on a Ship." *Forensic Science International*, 279: e7–e9. https://pubmed.ncbi.nlm.nih.gov/28890242/.

Swanson, Peter. "Proven Lightning Protection That Gets Little Respect (Except from Adopters)." https://loosecannon.substack.com/p/proven-lightning-protection-that.

Sweeney, Kelly. 2021. "No Kitten: Time Again to Welcome Cats Aboard Merchant Ships." https://professionalmariner.com/article/no-kitten-time-again-to-welcome-cats-aboard-merchant-ships/.

Swinburne, Layinka. 1996. "Dancing with the Mermaids: Ship's Biscuit and Portable Soup." In Harlan Walker, ed., *Food on the Move; Proceedings of the Oxford Symposium on Food and Cookery*. Prospect Books.

Symondson, F.W.H. 1876. *Two Years Abast the Mast: Or, Life as a Sea Apprentice.* William Blackwood and Sons.

Szumlas, Commander Daniel E. 2008, May. *United States Navy Shipboard Pest Management Manual.*

Tabaczek-Bejster, Iwona, and Stanislaw Zaborniak. 2019, April. "The Beginning of Swimming Teaching Methodology Before 1939." *Scientific Review of Physical Culture*, 4(4): 162. https://www.researchgate.net/publication/332468351_the_beginning_of_swimming_teaching_methodology_before_1939.

Tag, Thomas. 2023. "Reflectors"; "The Fresnel Lens Makers"; "Fresnel Lens Orders, Sizes, Weights, Quantities and Costs"; "Glass Facet Reflectors." United States Lighthouse Society. http://uslhs.org.

Talbert, S.G., et al. 1970, April. *Manual on Solar Distillation of Saline Water*, Office of Saline Water, Research and Development Progress Report No. 546. United States Department of the Interior.

Talcott, George, et al. 1841. *Ordnance Manual for the Use of the Officers of the United States Army.* J. & G.S. Gideon.

Tardivel, Kara, et al. 2023. "Cruise Ship Travel." *CDC Yellow Book 2024: Travel by Air, Land & Sea.* https://wwwnc.cdc.gov/travel/yellowbook/2024/air-land-sea/cruise-ship-travel.

Taylor, Major J., and G.D. Sitre. 1924. "Note on an Electrical Rat-Guard for Ships Hawsers." *Indian Journal of Medical Research*, 11: 643–652.

Telkes, Maria. 1953, May. "Fresh Water from Sea Water by Solar Distillation." *Industrial & Engineering Chemistry Research*, 45(5): 1108–14.

Terminix. 2023 (accessed). "What Can Rats Chew Through." https://www.terminix.com/rodents/rats/behavior/what-can-rats-chew-through/.

Thomson, Sir William. 1879, May 23. "Minutes of Evidence Taken Before the Select Committee on Lighting." Parliament, House of Commons, Reports from Committees: 22: 176.

Tolansky, S. 1947. "The Reflectivity of Speculum Metal." *Journal of Scientific Instruments*, 24(9): 248.

Toll, Ian W. 2008. *Six Frigates.* W.W. Norton.

Tomlinson, Charles. 1886. *A Rudimentary Treatise on Warming and Ventilation.* C. Lockwood and Company.

Torchia-Núñez, Juan Cristóbal, et al. 2014, Sept. "Thermodynamics of a Shallow Solar Still." *Energy and Power Engineering*, 6: 246–265. http://dx.doi.org/10.4236/epe.2014.69022.

Torck, Mathieu. 2009. *Avoiding the Dire Straits: An Inquiry Into Food Provisions and Scurvy in the Maritime and Military History of China and Wider East Asia.* Harrasowitz.

Totten, Benjamin J. 1862. *Naval Text-book, and Dictionary: For the Use of the Midshipmen of the U.S. Navy.* D Van Nostrand.

Traill, Henry Duff, and James Saumarez Mann. 1902. *Social England, etc.* Vol. II. Cassell and Company, Limited.

TransportCanada. 2003. "Chapter 5 : Interrelationship Between the Immersion Suit and the Lifejacket, in Survival in Cold Waters." In *Survival in Cold Waters*, Transport Publication TP 13822E. https://www.tc.gc.ca/eng/marinesafety/tp-tp13822-section7-2168.htm.

Trevithick, Francis 1872. *Life of Richard Trevithick: With an Account of His Inventions*, Vol. 1. E. & F.N. Spon.

Trowbridge, John. 1874, Oct. "The Electric Light for Steamships." *Pop. Sci. Monthly*, 720.

Tucker, Spencer C. 2013. *American Civil War: The Definitive Encyclopedia and Document Collection.* ABC-CLIO.

Tudor, Frederick, and Timothy T Sawyer. 1859. "Report of the Committee of the Boston Board of Trade on the Subject of the Ice Trade, January, 1857." In Frederick M. Kelley, *Union of the Oceans by Ship-Canal Without Locks Vis the Atrato Valley*, Appendix. Harper & Bros.

Twiss, Travers. 1874. *The Black Book of the Admiralty*, Appendix—Part III. Longman & Co.

United States Army. 1950. *Survival at Sea.* Field Manual FM 21–22, AFM 64–26.

United States Army. 1958, Dec. 15. *Harbor Craft Crewman's Handbook*, Tech. Manual 55–501.

United States Bureau Naval Personnel. 1956. *Curriculum for Naval Reserve Training FPG3, Pipefitters G, Shipboard Pipefitters, Class A, Part 1.*

United States Bureau of Naval Personnel. 1958. *Boatswain's Mate 1 and Chief,* Navy Training Courses, NAVPERS 10122-B (1958).

United States Bureau of Naval Personnel. 1959. *Electrician's Mate 3 & 2.* Navy Training Courses, NAVPERS 10546. Government Printing Office.

U.S. Coast Guard. Undated. "Lyle Gun, Heritage Asset Collection." Photograph by H. Farley in "The Coast Guard Heritage Asset Collection of Art and Artifacts." https://www.history.uscg.mil/ourcollections/artifacts/igphoto/2001752213/.

U.S. Congress. 1852, Aug. 30. "Steamboat Act of 1852." 32nd Congress, 1st Sess. https://www.loc.gov/law/help/statutes-at-large/32nd-congress/session-1/c32s1ch106.pdf

U.S. Department of Agriculture. 2018, April (SR Legacy). "Horseradish, Prepared." NDB 2055. https://fdc.nal.usda.gov/fdc-app.html#/food-details/173472/nutrients.

U.S. Department of Agriculture. 2018, April (SR Legacy). "Lemon Juice, Raw." NDB 9152. https://fdc.nal.usda.gov/fdc-app.html#/food-details/167747/nutrients.

U.S. Department of Agriculture. 2018, April (SR Legacy). "Lime Juice, Raw." NDB 9160. https://fdc.nal.usda.gov/fdc-app.html#/food-details/168156/nutrients.

U.S. Department of Agriculture. 2018, April (SR Legacy). "Sauerkraut, Canned, Solids and Liquids." NDB 11439. https://fdc.nal.usda.gov/fdc-app.html#/food-details/169279/nutrients.

U.S. Department of Agriculture. 2020, April 1. "Onions, Red, Raw." NDB 100252, FoodData

Central. https://fdc.nal.usda.gov/fdc-app.html#/food-details/790577/nutrients.

U.S. Department of Agriculture. 2020, April 1. "Onions, Yellow, Raw." NDB 100253, FoodData Central. https://fdc.nal.usda.gov/fdc-app.html#/food-details/790646/nutrients.

U.S. Department of Agriculture. 2020, Oct. 30. "Garlic, Raw." NDB 11215, FoodData Central. https://fdc.nal.usda.gov/fdc-app.html#/food-details/1104647/nutrients.

U.S. Department of Agriculture. 2022, Oct. 28. "Cabbage, Green, Raw." NDB 11109, FoodData Central. https://fdc.nal.usda.gov/fdc-app.html#/food-details/2346407/nutrients.

U.S. Department of Agriculture. 2022, Oct. 28. "Cabbage, Red, Raw." NDB 11112, FoodData Central. https://fdc.nal.usda.gov/fdc-app.html#/food-details/2346408/nutrients.

U.S. Department of Agriculture. 2022, Oct. 28, "Sauerkraut." Food Code 75230000, FoodData Central. https://fdc.nal.usda.gov/fdc-app.html#/food-details/2345506/nutrients.

United States Merchant Marines. 1944. "United States Coast Guard Regulations Applicable to Certain Vessels and Shipping During Emergency." 46 Code of Federal Regulations Part 153. http://www.usmm.org/lifeboat.html.

United States Merchant Marines. 1944. *Lifeboats, Lifeboat Equipment and Rafts [As taught to U.S. Maritime Service Trainees in World War II—from United States Maritime Service Training Manual. War Shipping Administration Training Organization.* Published for United States Maritime Service by Cornell Maritime Press, 1943, 1944. http://www.usmm.org/lifeboat2.html.

United States Merchant Marines. 1954, June. "History of the Development of the Life Preserver." *Proceedings of the Merchant Marine Council,* U.S. Coast Guard, 1(6): 91–93.

U.S. Navy. Undated. "Annual Report Surgeon General U.S. Navy." *Annual Reports of the Navy Department for the Fiscal Year 1932.* Government Printing Office.

U.S. Navy. 1841, Feb. 19. *Regulations for the Uniform and Dress of the Navy of the United States.* J. & G.S. Gideon, Printers.

U.S. Navy. 1865. *Allowances Established for Vessels of the United States Navy 1864.* Government Printing Office.

U.S. Navy. 1916. *The Ship and Gun Drills, U.S. Navy 1914.* Military Publishing Co.

U.S. Navy. 1925. Navy Department (Bureau of Construction and Repair). *Instructions for Painting and Cementing Vessels of the United States Navy.* Government Printing Office.

U.S. Navy. 1932. *The Cook Book of the United States Navy.* Government Printing Office. [USPHS] United States Public Health Service. 1929. *The Ship's Medicine Chest and First Aid at Sea.* Government Printing Office.

U.S. Navy. 1969. Navy Department (Naval History Division). *Dictionary of American Naval Fighting Ships,* Vol. IV. Government Printing Office.

U.S. Navy. 1979, Jan. *U.S. Navy Diving Manual,* Vol. 1 Air Diving, NAVSEA-0884-LP-001-9010.

United States Power Squadron. 2007, Jan. 23. "Hypothermia Safety." *Compass* 1(1).

U.S. Treasury Department. 1910. *The Rat and Its Relation to Public Health.* Government Printing Office.

United States V. Holmes, 1 Wall. Jr.1, Case No. 15, 383 (E.D.Pa. 1842). https://law.resource.org/pub/us/case/reporter/F.Cas/0026.f.cas/0026.f.cas.0360.pdf.

Valdez, Pablo. 2019. "Circadian Rhythms in Attention." *Yale Journal of Biology and Medicine,* 92(1): 81–92 (2019).

Van der Kooij, B.J.G. 2015, April. *The Invention of the Electric Light.* Ph.D. dissertation, University of Technology, Delft, The Netherlands.

Van der Vliet, J. 2016. "Anonymous, Ship Lantern, Netherlands, 1650-1700." In J. van der Vliet and A. Lemmers, eds., *Navy Models in the Rijksmuseum,* online coll. cat. Amsterdam: hdl.handle.net/10934/RM0001.COLLECT.244867.

Van Helden, Albet. 1984. "Building Large Telescopes, 1900–1950." In *The General History of Astronomy,* Vol. 4, Part A, 134–152. Cambridge University Press.

Van Leeuwen, Wessel M.A., et al. 2013, Nov. "Sleep, Sleepiness, and Neurobehavioral Performance While on Watch in a Simulated 4 Hours On/8 Hours off Maritime Watch System," *Chronobiology International,* 30(9): 1108–15.

Van Leeuwen, Wessel M.A., et al. 2021. "Mathematical Modelling of Sleep and Sleepiness Under Various Watch Schedules in the Maritime Industry." *Marine Policy,* 130: 104277.

Van Riper, A Bowdoin. 2007. *Rockets and Missiles: The Life Story of a Technology.* Johns Hopkins University Press.

Vantine, Wilbur H. 2011. *Some Nautical Tales.* AuthorHouse.

Vaucher, Jean. 2018, Aug. "History of Ships: Animal Skin Floats." http://www.iro.umontreal.ca/~vaucher/History/Prehistoric_Craft/Float.html.

Vaucher, Jean. 2023, Jan. "History of Ships: Prehistoric Craft." https://www-labs.iro.umontreal.ca/~vaucher/History/Ships/Prehistoric_Craft/.

Vegetius, *De Re Militari,* Book III. https://facultystaff.richmond.edu/~wstevens/history331texts/Vegetius.html.

Verbiscar, A.J., et al. 1986. "Recent Research on Red Squill as a Rodenticide." Proc. 12th Vertebrate Pest Conf. https://digitalcommons.unl.edu/vpc12/26/.

Vidich, Charles. 2021. *Germs at Bay: Politics, Public Health, and American Quarantine.* ABC-CLIO.

Vinagre-Rios, Jan, et al. 2021, Sept. "The Effect of Circadian Rhythms on Shipping Accidents." *Journal of Navigation,* 74(5): 1190–1199. https://doi.org/10.1017/S0373463321000333.

Vincent, Charles W., and Mason, James. 1875. *The Year-Book of Facts in Science and Art.* Simpkin, Marshall & Co.

Wade, J.A. 1876. "Windmill Pumps" (Letter). *Nautical Magazine* 45: 1027.

Wagner, J. Robert. 2004. *VS-931 Antisubmarine Squadron*. AuthorHouse.

Wagner, Jonathan. 2006. *A History of Migration from Germany to Canada, 1850–1939*. University of British Columbia Press.

Wagstaffe, Cate. 2010, Oct. *Furring in the Light of 16th Century Ship Design*. M.A. thesis, maritime archaeology, University of Southern Denmark.

Wales, Philip S. 1881. *Sanitary and Statistical Report of the Surgeon General of the Navy for the Year 1879*. Government Printing Office.

Walker, Ian. Undated. "SCR-578 Gibson Girl." https://musickpointradio.org/musick-memorial-radio-station-today/vintage-transmitters/scr-578-gibson-girl/.

Walker, Commander William. 1863. *The Magnetism of Ships, and the Mariner's Compass, 2d ed., rev*. Virtue Brothers.

Walling, Burns Tracy, and Julius Martin. 1907. *Electrical Installations of the United States Navy*. U.S. Naval Institute.

Walton, Thomas. 1899. *Know Your Own Ship: A Simple Explanation of the Stability, Construction, Tonnage and Freeboard of Ships*. Charles Griffin & Co.

Wang, L., et al. 2018. "Transient Bacteria Removal by Concentrated Sulfuric Acid for Cell Pollution." *Journal of Biochemistry and Biophysics*, 2(1): 103.

War Shipping Administration. 1946. *United States Maritime Service Hospital Corps School Manual*. Government Printing Office.

Ward, James Harmon. 1845. *An Elementary Course of Instruction on Ordnance and Gunnery*. Carey and Hart.

Waterhouse, A.G. 1900, Feb. 20 (issue date). U.S. Patent 643,702. "Method of Distilling and Evaporating Water."

Waterhouse, Addison G. 1897, July 6 (issue date). U.S. Patent 585,943. "Process of and Apparatus for Distilling Liquids."

Watson, Kathryn, 2018, Oct. 27. "What's an Average Shoulder Width?" https://www.healthline.com/health/average-shoulder-width#average-shoulder-width.

Watson, Nigel. 2015. *Maritime Science and Technology: Changing Our World*. Lloyd's Register.

Weaver, J.H. 1988. "Optical Property of Metals." In *CRC Handbook of Chemistry and Physics, 69th edition, 1988–1989*. CRC Press.

Webb, Silas. 1856. *The Yankee Enterprise; Or, the Two Millionaires and Other Tales*. Wentworth & Co.

Webster, Francis. 2016. *Oats: Chemistry and Technology*. Elsevier Science.

Webster, Robert W., and William Marcellino. 1954, Jan. "Defense Needs Speed Textile Research." *Bureau of Ships Journal*, 2(9): 23.

Weinmann, Michael. 2003. "Hot on the Inside." *Emergency Medical Services*, 32(7): 34.

Welin, Axel. 1909, Oct. 26 (issue date). U.S. Patent 938,448. "Device for Hoisting and Lowering Boats."

West Coast Times. 1871, Dec. 9. "A Useful Invention on Ship-Board." *West Coast Times*, Issue 1633, p. 2.

Westall, William. 1887. *A Queer Race; The Story of a Strange People*. Cassell.

Westick, Peter, and Peter Neushul. 2013. *The World in a Curl: An Unconventional History of Surfing*. Crown.

Wexler, Randell K. 2002, June. "Evaluation and Treatment of Heat-related Illnesses." *American Family Physician*, 65(11); 2307–14.

What-When-How. Undated. "Tungsten Filament Inventions." http://what-when-how.com/inventions/tungsten-filament-inventions/.

Whelan, M. 2013. "The Fluorescent Lamp"; "Incandescent Lighting." http://www.edisontechcenter.org.

White, George Clifford. 1986. *The Handbook of Chlorination*. Van Nostrand Reinhold Company.

Wilkinson, John. 1765, May 25. English Patent 824. "Floating Baths and Cork Jackets or Floats."

Wilkinson, John. 1765, Dec. 31. "XVI. a Course of Experiments to Ascertain the Specific Buoyancy of Cork in Different Waters..." *Philosophical Transactions of the Royal Society of London*, 55: 95–105. https://royalsocietypublishing.org/doi/10.1098/rstl.1765.0016.

Wilkinson, John. 1766. *Tutamen Nauticum: Or the Seaman's Preservation from Shipwreck, Diseases, and Othe Calamities Incident to Mariners, 4th ed*. Dodsley.

Williams, C.L. 1934, Jan. 19. "Sulphur Dioxide for the Fumigation of Ships: Methods of Use and Prospects of Improvement," *Public Health Reports*, 49(3): 89–100.

Williams, C.L. 1934, Feb. 9. "Liquid Sulphur Dioxide as a Fumigant for Ships." *Public Health Reports*, 49(6): 192–208.

Williams, Ian. 2006. *Rum: A Social and Sociable History of the Real Spirit of 1776*. Nation Books.

Williams, Thomas. 2012. *American Honor: The Story of Admiral Charles Stewart*. AuthorHouse.

Williamson, Jill P., et al. 2020. "Upgrades to the International Space Station Urine Processor Assembly." ICES-2020-391. https://ttu-ir.tdl.org/bitstream/handle/2346/86369/ICES-2020-391.pdf.

Willis, Sam. *Fighting at Sea in the Eighteenth Century: The Art of Sailing Warfare*. Boydell Press.

Wilson, Andrew Ian. 2011. "Developments in Mediterranean Shipping and Maritime Trade from 200 BC to AD 1000." In Damian Robinson and Andrew Wilson, eds., *Maritime Archaeology and Ancient Trade in the Mediterranean*, 33–59. Oxford Centre for Maritime Archaeology.

Wilson, Joseph. 1879. "Naval Hygiene: Human Health and the Means of Preventing Disease." *The American Journal of the Medical Sciences*, 77: 532.

Wilson, Joseph, and Gorgas, Albert. 1870. *Naval Hygiene*. Government Printing Office.

Wilson, Theodore Delavan, et al. 1873. *An Outline of Ship Building, Theoretical and Practical*. John Wiley & Son.

Winfeld, Rif. 2010. *British Warships in the Age of Sail 1603–1714: Design, Construction, Careers and Fates*. Pen & Sword.

Wood, Alan Douglas, et al. 1977, June. *Pumps and Water Lifters for Rural Development*. Colorado State University. http://hdl.handle.net/10217/180912.

Wood, Michael S. 2009, March. *Literary Subjects Adrift: A Cultural History of Early Modern Japanese Castaway Narratives, ca. 1780–1880*. Ph.D. dissertation, Department of East Asian Languages and Literatures, University of Oregon.

Wood Library Museum. Undated. "Haldane Gas Analysis Apparatus." https://www.woodlibrarymuseum.org/museum/item/980/haldane-gas-analysis-apparatus.

Woodbury, Walter E. 1896. *The Encyclopaedic Dictionary of Photography*. Scovill & Adams.

Woodward, Joseph Janver. 1879. *The Medical and Surgical History of the War of the Rebellion, Part II, Volume 1, Medical History*. Government Printing Office.

World Health Organization. 2011. *Guide to Ship Sanitation*, 3d ed. https://www.ncbi.nlm.nih.gov/books/NBK310823/.

Worshipful Company of Fan Makers. Undated. "A Short History of Mechanical Fans." https://fanmakers.com/history-of-mechanical-fans/.

Wyman, Merrill. 1846. *A Practical Treatise on Ventilation*. Munroe.

Wynter, Andrew. 1870. *Curiosities of Toil and Other Papers*. Vol. 1. Chapman and Hall.

Xu, Zhenyuan, et al. 2020. "Ultrahigh-efficiency Desalination Via a Thermally-localized Multistage Solar Still." *Energy & Environmental Science*, 13: 830–839.

Yannopoulos, Stavros, et al. 2017, Feb. "History of Sanitation and Hygiene Technologies in the Hellenic World." *J. Water, Sanitation and Hygiene for Development*. https://doi.org/10.2166/washdev.2017.178.

Yannopoulos, Stavros I., et al. 2015. "Evolution of Water Lifting Devices (Pumps) Over the Centuries Worldwide." *Water*, 7: 5031–5060. https://www.mdpi.com/2073-4441/7/9/5031.

Yeo, John. 1894. *Steam and the Marine Steam-Engine*. Macmillan.

Yetish, G., et al. 2015. "Natural Sleep and Its Seasonal Variations in Three Pre-industrial Societies." *Current Biology*, 25: 2862–8.

Yoder, Paul R., Jr., and Daniel Vukobratovich. 2015. *Opto-Mechanical Systems Design*, 4th ed. Vol. 2. Taylor & Francis.

Zaltzman, Julia. 2020, March 2. "Taking a Deep Dive Into Cruise's Eco Credentials." https://cruiseblondes.com/features/deep-dive-cruises-eco-credentials/.

Zaragoza, G., et al, 2012. "Use of Passive Solar Thermal Energy for Freshwater Production." In Jan Joinkis and Jochen Bundschuh, eds., *Renewable Energy Applications for Freshwater Production*. Taylor & Francis.

Zuckerman, Arnold. 1977. "Scurvy and the Ventilation of Ships in the Royal Navy: Samuel Sutton's Contribution." *Eighteenth-Century Studies*, 10(2): 222–234.

Zupko, Ronald Edward. 1985. *A Dictionary of Weights and Measures for the British Isles the Middle Ages to the Twentieth Century*. American Philosophical Society.

Zuurbier, Moniek, et al. 2009. "Minute Ventilation of Cyclists, Car and Bus Passengers: An Experimental Study." *Environmental Health*, 8: 48. http://www.ehjournal.net/content/8/1/48.

Author's Preexisting Work

As noted, portions of this book were previously published in the *Grantville Gazette*, which ceased publication in July 2022. That online magazine presented fiction set in and nonfiction relating to the fictional literary universe created by the late Eric Flint's alternate-history sci-fi novel *1632*. In that novel, a new timeline is created when a fictional West Virginia town (Grantville) is moved from the year 2000 to Thuringia, Germany, during the Thirty Years' War. The nonfiction considered the limited knowledge and resources that would have been available to the townspeople, and proposed how they might cope. Naturally, the fictional aspects have been omitted from the present work.

Cooper, Iver P. 2011. "The Wind Is Free; Sailing Ship Design, Part 1: Propulsion." *Grantville Gazette* 21.

———. 2011. "The Wind Is Free; Sailing Ship Design, Part 2: Seaworthiness." *Grantville Gazette* 22.

———. 2012, "The Multihull and the Mariner." *Grantville Gazette* 30.

———. 2013. "Infectious Pestilence: Part 1, Coping with Plague in Early Modern Europe." *Grantville Gazette* 50.

———. 2014. "Infectious Pestilence: Part 2, Fighting the Plague After the Ring of Fire." *Grantville Gazette* 51.

———. 2016, March. "Life at Sea in the Old and New Time Lines: Part 1, Providing Nourishment." *Grantville Gazette* 64.

———. 2016, May. "Life at Sea in the Old and New Times Lines: Part 2: Keeping Dry (and Afloat)." *Grantville Gazette* 65.

———. 2016, Sept. "Life at Sea in the Old and New Time Lines, Part 3: Shipboard Lighting and Fire Prevention." *Grantville Gazette* 67.

———. 2017, May. "Life at Sea in the Old and New Time Lines, Part 4: Lights Across the Waters." *Grantville Gazette* 71.

_____. 2020, Sept. "Life at Sea in the Old and New Time Lines, Part 5: Creature Comforts." *Grantville Gazette* 91.

_____. 2021, Jan. "Life at Sea in the Old and New Time Lines, Part 6: Lest You Drown." *Grantville Gazette* 93.

Use of the material from the last two articles is with the kind permission of Lucille Robbins, Eric's widow and the CEO and publisher of the *Grantville Gazette*, 1632, Inc.

Index

Achille 87
Ackworth, Sir Jacob 12, 15
Adventure 165
Agricola 8, 12, 133, 139
air conditioning: air filtration 98–9; cooling systems 96–8; dehumidification 98–9; heating systems 102–3; humidity 95–7; paint, effect 95; temperature 94–6
Akbar 124
Alabama (CSS) 104–5
Alabama (1910) 208
Alert (steam sloop) 21
Alexandra 194
aluminum 151, 182, 193, 197, 199, 208, 225
American navy and merchant marine 4, 19–20, 25, 28–30, 33, 37, 41–3, 45, 47, 51–3, 59, 61–2, 69, 76, 80–2, 85, 88, 90, 94, 99–101, 103, 108, 111, 117–8, 142–4, 146, 149–50, 159, 161, 168, 170–1, 175, 185–8, 190, 192, 194, 196, 199, 203, 205, 207, 209, 211–2, 214, 217, 220–1, 224–6
ancient or prehistoric world 5, 12, 30, 34, 97, 106, 115, 118, 130, 134, 139, 155–6, 158, 163, 171, 176, 179
Andaman 125
L'Andromede (frigate) 66
Angel Gabriel (merchantman-privateer) 125
Anson, George 32, 34
Apollo (GB troopship) 212
Appert, Nicholas 57
Appier-Hanzelet, Jean 193
Archimedes 114–5, 117, 120
Arcona (Prussian) 66
Argonaute (East Indiaman) 216
Aristotle 35–6
Arklow Meadow 222
Arrow (GB sloop of war) 32
arsenic and its salts 193, 216, 218
asbestos 180, 207
Asahi (Japanese battleship) 72
ASHRAE *see under* laws and regulations
Ashurnasirapal II 156
Atocha, Nuestra Senhora de (galleon) 125, 178
Atwood, George 127

Babcock, James Francis 210
Babcock fire extinguisher 210
balsa 156, 158–9, 162
bamboo 34, 115, 158–9, 186
Basil, Saint 36
Batavia (East Indiaman) 148, 178
Beagle 205
bedding *see* sleep
Bedford 83
Bell, John (18c) 166
Bell, John (19c) 59, 98
Bell-Coleman cycle 98
Bellerophon (GB third-rate) 118
Bentham, Samuel 32, 124, 130
Bentinck, John 135
beriberi (vitamin B1 deficiency) 67–8
Betty Johnson (steam trawler) 183
beverages, alcoholic 25–9
beverages, non-alcoholic 26, 29
Bilderbeck, C.L. 223
Blane, Gilbert 65
Blue Ridge 223
Boscawen, Adm. Edward 13
Bosquet, Abraham 157, 164
Boteler, Capt. 135, 147
Bouguer, Pierre 122
Boxer, Col. Edward 167, 193
Boyle, Robert 56
Boyton, Paul 163
Braun, Johann D. 196
Bridge, Commander Cyprian 160
British navy and merchant marine 7–8, 10–2, 19, 21–2, 26–9, 31 2, 34, 37, 49–51, 53–9, 61, 63–6, 71, 74–5, 77–8, 81–5, 88, 90, 94, 99, 101, 103–7, 116–7, 124, 130, 133, 135, 137, 141, 144, 155, 160–1, 173–6, 183, 189–92, 194, 201, 203, 205–6, 212, 219, 221
Brodie stove 58
Buckingham, Duke of 79–80
Budd, William 109
Bunker Hill 208
Burberry, Thomas 100
Burgh, John 79
Burney, James 34

Caledonia (GB) 160
Campen (East Indiaman) 178
Camperdown (GB, three-decker) 82

Capello, Capt. Lenny 77
Captain (GB) 128
carbide, calcium 182–3
carbon (filament) 185
carbon dioxide (component of air) 4, 20–23, 98–9
carbon dioxide (fumigant or asphyxiant) 209, 211, 219
carbon monoxide 4, 103, 219
cargo 17–18
cargo sweat 18
Cartier, Jacques 62–3
Castalia 123
Celebrity Flora (cruise ship) 113
Centaur 139
Chaffee, Edwin 207
Charles II, King 117
Charles W. Morgan 178
Charleston 111–12
Charmonman (pump inventor) 137
Charon 135
Chevreul, Michel Eugène 179
Childers (sloop) 171
clothing 99–103, 162–3; washing and drying 100–2
coal, charcoal, and coal oil 4, 14–5, 17, 19–20, 36–9, 49, 59, 89, 96, 98, 102, 106–8, 126, 132, 137, 139, 179–80, 182, 188, 205–6, 209, 217, 219, 224
Cochrane, Thomas 193
Cole, Edwin Louis 165
Cole, William 135
Coleridge, Samuel 33
Coles, Capt. Cowper 128
Columbia 184
Columbus, Christopher 78, 171
La Concepcion (galleon) 178
Congreve, William 193
Conoley, B.J. 222
Constitution 18, 29, 51–2, 79 82, 88, 99
Cook, Capt. James 37, 63, 65, 70
Coolidge, William David 186
copper and its salts 8, 32, 34, 36–8, 41, 58, 65–6, 103, 109, 136, 139–40, 175, 177, 181, 184, 189, 197, 199, 204–6, 225
Cordle, Capt. John 77
cork 05, 150, 156–60, 162–5, 167
La Cornelie 66

253

Cornwall 7
La Couronne (68 gun warship) 191
Cowie, Adm. Thomas Jefferson 61
Crookes, Sir William 91
Cumming, Alexander 104
Cushing 77
cyanide, hydrogen (hydrocyanic acid) 217, 220–1

Dampier, William 135
Dana, Richard Henry 78–9, 100, 141, 176, 203
Danish navy and merchant marine 7, 60, 123
Dart (GB sloop of war) 32
Davie, John 71, 74
davits *see* survival at sea, boats, launching and loading
Davy, Sir Humphry 195
DDT 217
Deane, Anthony 117
De Ayanz y Beaumont, Jeronimo 138
De Bougainville, Louis-Antoine 37
De Caus, Salomon 138
Defender (racing yacht) 126
Defiance 125
Defoe, Daniel 216
della Porta, Giovanni Battista 138
Desaguliers, John 12, 16
desalination 34–5; flash (reduced pressure) 41–2; reverse osmosis 47–8; scaling control 47; solar 43–7; thermal 35–41, 107; vapor-compression 42–3
Devonian (GB steamship) 221
Digby, Everard 171
Discovery (steam barque) 21
diseases, contagious: cholera 109, 212–4; dysentery 109,.213; influenza 212–3; Legionnaires' disease 106, 109, 213; measles 213; norovirus 109, 213; plague 213–6, 219, 221–2, 224; prevention 105–8, 111, 219; quarantine and 213–14; scarlet fever 213; smallpox 213; typhoid fever 108–9, 212; varicella 213; yellow fever 214; *see also* pest control
diseases, nutritional deficiency *see* food, vitamins and minerals
Dodgeson, George 134
Dolphin 36
Dorade box 9
Doris (frigate) 59
Doris (steam frigate) 21
Dorsetshire 37
Drake, Francis 64, 109, 192–3, 202
Draper, Henry 198
Drayton, Thomas 198
Drebbel, Cornelius 36

Dumas, Alexander 11
Duncan 128
dutch navy and merchant marine 11, 36, 38,50, 57, 59, 63, 68, 78, 82, 133, 138, 174, 176, 182, 189–91

Eclipse 8
Eijkman, Christiaan 61, 68
Ekirch, A. Roger 74
electricity 14–5, 42, 48, 108, 138, 140, 144, 173, 182–90, 192, 195–6, 198, 200, 202–6, 215
Enterprise (CV-6) 208
España 87
Essex 219
Euler, Leonhard 122, 137
Excellent 86, 209

Faraday, Michael 179
Faraday cage 205
fire, as hazard at sea 27, 100, 178, 189, 201–2, 215, 219
fire prevention and control: automatic fire alarms 205–6; automatic sprinkler 208; bulkheads 207; extinguishers 209–11; fire discipline 202–3; fire triangle 206–11; firefighter protection and training 211–2; fireproofing 206–7; halons 211; lightning protection 203–5; smothering of fire 209–11; water vs. fire 207–10
Fisgard 204
Fitzgerald (US) 71
Fitzmaurice, Maj. 197
Fleming, Adm. Clas Larsson 94
Fletcher (DD992) 47
Flying Fish (schooner) 125
food: caloric needs 48–9; cooking 58–9; fresh meat 54–6; macronutrients 60–1; preservation 56–7; provisioning en route 57–8; provisioning on station 58; quality 53–4; rations 49–53; storage 59–60; vitamins and minerals 61–8
Forbes, R.B. 161, 205
Forester, C.S. 201
Forrestal 212
Foucault, Léon 198
Fram 7
Franklin, Benjamin 130, 203
French navy and merchant marine 14, 26, 32, 50, 57, 62, 66, 71, 107, 174, 176, 189, 209
Fresnel, Augustin-Jean 192, 200

Gates, Manley F 111–2
George and Susan 138
German (or Prussian) navy and merchant marine 66, 77, 93, 164–5
Gesner, Abraham 180
Gihon, Albert Leary 28, 107
Gilbertus Anglicus 36, 62
Glasgow 27

glass 40, 43–4, 57, 107, 151, 159–60, 162, 174–5, 178, 180–1, 186–7, 189–90, 192–3, 197–200, 207
Gonzalez, Antonio 57
Goodyear, Charles 100
Greathead, Henry 165
Griffiths, Capt. Anselm John 28, 31
Griffon (British brigantine) 66
Grubbs, S.B. 215, 218, 220, 224–5
Guardian 56
gunports 5–6, 10–11, 114, 117, 119, 123, 173–6
Gunther fan 22
Guy-Lussac, Joseph Louis 179

Hadley, John 197
Hales, Stephen 6, 13–7, 36, 107
Harnold, John 109
Harrington, Sir John 104
Harriot, Thomas 118
Harris, William Snow 204–5
Hauksbee, Francis 187
Hawkins, Sir Richard 36, 64
Henderson, E.H. 223
Henry Grace à Dieu 6
Heron of Alexandria 138
Herschel, Sir John 91–2
Homer, Winslow 168–9
Hoste, Paul 122
Howe, Adm. Richard 165, 192
Howland, John 141
Hunley (submarine) 178

Independence (aircraft carrier) 42
Independence (US 74) 117–8, 136
Indiana (US warship) 205
Inflexible (GB) 137, 185
Innanimoba (steamship) 224
Iowa (battleship) 52
Irving, Doctor Charles 37
Italian navy and merchant marine 144, 190, 213
Iwami 212

Jackson (armored corvette) 56
Japanese navy and marchant marine 37–8, 61, 64, 67–8, 72, 129, 212
Jefferson, Thomas 37
Jesse, Richard 158
John Bows 126
John C. Stennis 75
John Mccain (US) 71
Johnson, Samuel 48, 53, 141
Jordan (possible pump inventor) 137
Judichaer, Olaus 117
Jukichi, Capt. 37–8

Kane, Elisha Kent 219
kapok 156, 159, 162, 209
Katherine Pleasaunce 175
Kearsarge (battleship) 14
Kearsarge (US sloop-of-war) 180
kerosene 180
Kindleberger, C.P. 89, 92

Kisbee, Thomas 163–4, 167
Konigin der Nederlanden 217

La Salle 148
Lake Erie 77
Lamb, John 37
Lamb and Nicholson stove 59
Lancaster 107
Lancaster, James 64, 109
Lang, Oliver 175
laws and regulations: American Marine Standard Rat Proofing of Ships 225; American Society of Heating, Refrigerating and Air Conditioning Engineers 22–3; Carriage of Passengers Act (1855) 19, 23, 30; Coast Guard regulations (US) 142, 145–6, 151–2, 162–4, 166, 169; International Convention for the Safety of Life at Sea (SOLAS) 129–30, 142, 149–52, 161, 164, 207; International Convention on Standards of Training, Certification and Watchkeeping for Seafarers 72; International Life-Saving Appliance Code 142; International Mechanical Code 22; Merchant Shipping Act (1876) 117; Passenger Act 1803 219; Saline Water Conversion Act 43; Steamboat Act (1852) 161; Steerage Act (1819) 19, 29; Venice Convention (1897) 214
lead and its salts 31, 105, 109, 125, 127, 131–2, 138–40, 153, 175–6, 181, 188, 190, 206, 210
Leander (cruiser) 82, 104
Legazpi, Miguel Lopez de 192
Leonard, Willie 88
Leonora 32
Leslie, David 138
Leslie, Robert Charles 93
lifeboats and liferafts *see* survival at sea, craft, survival
lighting: adequacy 173; color of instrument, bridge and deck lighting 190; compass 189; concentration 196–200; interior paint color, effect 188–9; powder room 189–90
lighting, artificial: acetylene 182–3; Argand lamp 180, 197; candles and candle holders 178–9; carbon arc 184–6, 194–6, 199–200; electric 184–8; fluorescent lights 187; gas discharge lamps 187; gas lamps 182; incandescent 185–6; lanterns 181–2; laws and regulations 192; light-emitting diodes 187–8; limelights 183–4; mantles 182; mercury-xenon arc 196; mirrors 182; oil lamps 179–81; running lights 190–3; searchlights 194–5; star shells 193; xenon arc 196

lighting, natural 173–8; gratings 176; lenses 176–8'; prisms 178; skylights 176; stern lights 175; *see also* gunports and portholes
lightning protection *see under* fire prevention and control
Lind, Dr. James 36–7, 64–6,106–8
Londonderry 3
Loran, Aleksandr 211
Louis XIV, King 193
Lucian 163
Lukin, Lionel 165
Lurting, Thomas 6
Lyle, David 166–7
Lyons, Adm. Sir Edmund 87

Machault 133–4
Macintosh, Charles 100
MacPherson, R. 159
Magnificent 107
Mahmoud 89
Mainwaring, Henry 124, 133, 135, 163
Mallison, William 157
Manby, Capt. George 166–7
Mangin, Alphonse 198. 200
Manwayring *see* Mainwaring, Henry
Marian Buzcek 222
Markus, Peter 159
Martens, Friderich 34
Mary 149
Mary (in 1662) 6
Mary Rose 6, 58, 100, 117, 119, 145–6, 217
Mather, James 148
Mayflower 11, 62, 118, 141, 178
Mead, Richard 15
Medina-Sidonia, Duke of 26
Medusa 148
Mege-Mouries, Hippolyte 57
Melville, Herman 28, 33
Merhonour 192
Merriman, C.S. 163
methane 4, 15, 17
mica 175, 181, 192
Mikasa 212
Minotaur 14, 194
Monarch (GB) 128
Moniku 222
Monitor 104, 137

Namsenfjord 116
Nancy 166
Narragansett (passenger ship) 206
NATO 49, 61
Nelseco-Clarkson exhaust evaporator 42
Nelson, Horatio 32, 206
night blindness (vitamin A deficiency) 66–7
Nipsic (US) 89
Normandie (FR liner) 209
North Carolina (US, 74 gun) 80, 82
Norwegian navy and merchant marine 72, 138, 165

Norwich 15
Novara 66

Odysseus 220
Osbridge, Lieutenant 107
Oscar Austin 77
Ottoman navy and merchant marine 89
Ouzo (yacht) 157
Ovid 170
oxygen 4, 20

paint 95, 188–9
Papin, Denis 137
Paraguay (civilian ship) 57,98
Parrott, Robert 158, 167
Pasley, Capt. 34
Peacock, James 107
Pearl (frigate) 29
Pellatt, Apsley 176
Penelon (fictional character) 114
Perry, Commodore Matthew Calbraith 89
pest control 214–5; fleas 216–7; fumigation 218–22; rat extermination 217–8; rat-guards 222–4; rat-proofing 224–5; rats 215–6; rats, fleas and plague 216
Pett, Phineas 117, 125
Phillips Fire Annihilator 209
phosphine 164, 221–2
Pilgrim 203
Platt, Hugh 64
Plimsoll, Samuel 117
Plutarch 180
Porter, Commodore Daviid 81, 219
portholes (for lighting or ventilation) 6, 10, 12, 15, 22, 29, 81, 89, 154, 175–7, 209
Portuguese navy and merchant marine 38, 126, 178
Possochob (Russian steamer) 163
Preble, Commodore Edward 25, 99
Prince 125
Protector 140
Prout, William 60
pumps 32, 132–40
Purisma Concepcion 56
pyrethrum 217

Queen Charlotte (GB first-rate) 117

Radiant (clipper) 203
Raleigh, Walter 78, 134
Red Dragon 64
Reed, Edward 128–9
Reeves 77
Regulus 106
Resolution 70
rhodium 197, 199
Rio Jachal (civilian ship) 209
Roald Amundsen 221
Robinson, Hercules 107
Rogers, Thomas 198

Ross, Daniel 83
Rouen 126
Rouppe, Louise 78
Royal Anne (100 gun) 124
Royal George 13
Royal Katherine (76 gun) 124
Royal Sovereign (GB warship) 201
Royal William 124
rubber, natural and synthetic 100, 159, 161, 163, 207
Russian navy and merchant marine 70, 161, 163
Ryder, Rear Adm. 149, 160

St. Jean d'Acre 83
St. Michael (96 gun) 125
St. Vincent, Lord (John Jervis) 71
Samuel Lacing 126
San Jacinto 77
sanitation *see* waste disposal
Schat, Ane 150
Schat Skates 153–4
Schuckert, Johann Sigmund 198
Schulz, P. 196
scurvy 62–6
scuttle-butt 28, 33, 107, 109
Sea Venture 139, 148, 178
Sedgewick, William 110–1
Seeandbee (passenger ship) 205–6
Seine (steamship) 100
Sharma, Robin 24
ship design: bulkheads 129–30, 149; bulwarks 118–9, 145–6; camber 118–9; caulking 130; closed decks 118–9; draft 116–8; floatability (buoyancy) 114–6; freeboard 116–8; health, effect on 103–5, 113; scuppers 33, 119; wood vs. steel construction 113
ship stability 31, 59–60, 119–29, 208–9; bailing and scooping 132; ballast 125–7; bilge alarm 140; damaged 129; free surface effect 31, 127, 129; girdling and furring 124–5; gz curve 127–9; inclining experiment 122; initial 122–3, 127; laws and regulations 129–30; leaks 130–2, 215; roll period 124; sallying 122; *see also* gunports and pumps
ships: *Achille* 87; *Adventure* 165; *Akbar* 124; *Alabama* (CSS) 104–5; *Alabama* (1910) 208; *Alert* (steam sloop) 21; *Alexandra* 194; *Andaman* 125; *L'Andromede* (frigate) 66; *Angel Gabriel* (merchantman-privateer) 125; *Apollo* (GB troopship) 212; *Arcona* (Prussian) 66; *Argonaute* (East Indiaman) 216; *Arklow Meadow* 222; *Arrow* (GB sloop of war) 32; *Asahi* (Japanese battleship) 72; *Atocha, Nuestra Senhora de* (galleon) 125, 178; *Batavia* (East Indiaman) 148, 178; *Beagle* 205; *Bedford* 83; *Bellerophon* (GB third-rate) 118; *Betty Johnson* (steam trawler) 183; *Blue Ridge* 223; *Bunker Hill* 208; *Caledonia* (GB) 160; *Campen* (East Indiaman) 178; *Camperdown* (GB, three-decker) 82; *Captain* (GB) 128; *Castalia* 123; *Celebrity Flora* (cruise ship) 113; *Centaur* 139; *Charles W. Morgan* 178; *Charleston* 111–12; *Charon* 135; *Childers* (sloop) 171; *Columbia* 184; *La Concepcion* (galleon) 178; *Constitution* 18, 29, 51–2, 79–82, 88, 99; *La Cornelie* 66; *Cornwall* 7; *La Couronne* (68 gun warship) 191; *Cushing* 77; *Dart* (GB sloop of war) 32; *Defender* (racing yacht) 126; *Defiance* 125; *Devonian* (GB steamship) 221; *Discovery* (steam barque) 21; *Dolphin* 36; *Doris* (frigate) 59; *Doris* (steam frigate) 21; *Dorsetshire* 37; *Duncan* 128; *Eclipse* 8; *Enterprise* (CV-6) 208; *España* 87; *Essex* 219; *Excellent* 86, 209; *Fisgard* 204; *Fitzgerald* (US) 71; *Fletcher* (DD992) 47; *Flying Fish* (schooner) 125; *Forrestal* 212; *Fram* 7; *George and Susan* 138; *Glasgow* 27; *Griffon* (British brigantine) 66; *Guardian* 56; *Henry Grace à Dieu* 6; *Hunley* (submarine) 178; *Independence* (aircraft carrier) 42; *Independence* (US 74) 117–8, 136; *Indiana* (US warship) 205; *Inflexible* (GB) 137, 185; *Innanimoba* (steamship) 224; *Iowa* (battleship) 52; *Iwami* 212; *Jackson* (armored corvette) 56; *John Bows* 126; *John C. Stennis* 75; *John Mccain* (US) 71; *Katherine Pleasaunce* 175; *Kearsarge* (battleship) 14; *Kearsarge* (US sloop-of-war) 180; *Konigin der Nederlanden* 217; *La Salle* 148; *Lake Erie* 77; *Lancaster* 107; *Leander* (cruiser) 82, 104; *Leonora* 32; *Londonderry* 3; *Machault* 133–4; *Magnificent* 107; *Mahmoud* 89; *Marian Buzcek* 222; *Mary* 149; *Mary* (in 1662) 6; *Mary Rose* 6, 58, 100, 117, 119, 145–6, 217; *Mayflower* 11, 62, 118, 141, 178; *Medusa* 148; *Merhonour* 192; *Mikasa* 212; *Minotaur* 14, 194; *Monarch* (GB) 128; *Monika* 222; *Monitor* 104, 137; *Namsenfjord* 116; *Nancy* 166; *Narragansett* (passenger ship) 206; *Nipsic* (US) 89; *Normandie* (FR liner) 209; *North Carolina* (US, 74 gun) 80, 82; *Norwich* 15; *Novara* 66; *Oscar Austin* 77; *Ouzo* (yacht) 157; *Paraguay* (civilian ship) 57, 98; *Pearl* (frigate) 29; *Pilgrim* 203; *Possochob* (Russian steamer) 163; *Prince* 125; *Protector* 140; *Purisma Concepcion* 56; *Queen Charlotte* (GB first-rate) 117; *Radiant* (clipper) 203; *Red Dragon* 64; *Reeves* 77; *Regulus* 106; *Resolution* 70; *Rio Jachal* (civilian ship) 209; *Roald Amundsen* 221; *Rouen* 126; *Royal Anne* (100 gun) 124; *Royal George* 13; *Royal Katherine* (76 gun) 124; *Royal Sovereign* (GB warship) 201; *Royal William* 124; *St. Jean d'Acre* 83; *St. Michael* (96 gun) 125; *Samuel Lacing* 126; *San Jacinto* 77; *Sea Venture* 139, 148, 178; *Seeandbee* (passenger ship) 205–6; *Seine* (steamship) 100; *Shonga* 212; *Sibyl* (28-guns) 50; *Sovereign of the Sea* (1637) 174, 191; *see also Royal Sovereign*; *Speedy* (brig-sloop) 193; *Stark* 77; *Strahleven* (civilian ship) 57, 98; *Stump* (DD978) 47; *Success* (frigate) 14; *Suo* 212; *Temeraire* 194; *Terrible* (steam frigate) 217; *Theodore Roosevelt* (US aircraft carrier) 108; *Thunderer* 175; *Tiger* (steam frigate, GB) 148; *Tijger* (merchantman) 201; *Titanic* 142, 149, 158, 165, 192, 194; *Tokiwa Mura* 221; *Trenton* (US steam sloop) 185; *Trial* 116; *Uluburun* (shipwreck) 179; *Unite* 56; *Valdez* (Exxon tanker) 71; *Valiant* (Capt. Wilmshurst) 160; *Valiant* (70 gun warship) 215; *Vandalia* (sloop-of-war) 215; *Vasa* 100, 103, 122–3, 134, 174, 220; *Vengeance* 107; *Vergulde Draeck* (jacht) 178, 182; *Victory* 10–11, 15, 56, 103, 153, 175, 206; *Victory* (cargo) 10; *Warrior* 102; *Warspite* 117; *Wasa see Vasa*; *Washington* (battleship) 53; *Washington Irving* (passenger ship) 206; *William Brown* 149; *Witte Leeuw* (East Indiaman) 178; *Zeehaen* (17c fluyt) 118
Shonga 212
Short, James 198
Sibyl (28-guns) 50
silver and its salts 36, 114, 197–9
Simond, Paul-Louis 216
Simpson, Thomas 117
sleep: fixed beds 93–4; hammocks 78–90, 101, 160–1, 216; physiological requirements 71–5; quarters 5, 14; swinging cots 90–3; *see also* watchkeeping
Smith, John 64, 103

Smith, Thomas 199
Snoeck, Aegidius 36
Snow, John 109
SOLAS *see under* laws and regulations
Sovereign of the Sea (1637) 174, 191; *see also Royal Sovereign*
Spanish navy or merchant marine 5, 26, 36, 49, 54, 62, 78, 87, 100, 130, 174, 202, 206
Speedy (brig-sloop) 193
Stark 77
steam and steamships 14–6, 20–1, 36–43, 50, 57, 66, 94–6, 100, 102–3, 109, 111, 113, 116, 119, 123, 126, 130, 137–8, 148, 161, 163, 183–5, 192, 195, 201, 206, 208–9, 217–9, 221, 224
Steinheil, Carl von 198
Stevenson, Robert 192, 197, 200
Stewart, Alexander 90
Strahleven (civilian ship) 57, 98
Stump (DD978) 47
Success (frigate) 14
sulfur, sulfur dioxide and sulfuric acd 26, 64, 100, 106, 108, 193, 207, 209, 217, 220–1, 224
Suo 212
survival at sea: abandoning ship 141; boats, launching and loading 147–50, 152–5; boats, rescue 165–6; boats, ship's workboats 146–8; craft, survival (lifeboats and liferafts) 146, 148–52; distress signals 164–5; laws and regulations 142, 145–6, 151–2, 162–4, 166, 169; life buoys 161–5; line-throwing devices 166–9; man overboard 145–6; personal flotation devices 146, 155–65; preparation 146–7; prepositioned floating shelters 155; risk of death from immersion 142–3; safety aloft 143–5; safety on deck 145; shipwreck 146–7; survival suits 147, 162–3; training 170–2
Sutton, Samuel 15–7
Swedish navy and merchant marine 14, 61, 64, 100, 122–3, 138
Szent-Gyorgi, Albert 47

Takaki Kanehiro 67
tallow 132, 178–9
Taylor (oump inventor from Southampton) 134
Taylor, Halsey 109

Taylor, Maj. J. 223
Telkes, Maria 35, 44–5
Temeraire 194
Terrible (steam frigate) 217
Theodore Roosevelt (US aircraft carrier) 108
Thomson, Prof. 137
Thunderer 175
Tiger (steam frigate, GB) 148
Tijger (merchantman) 201
Tipping's sand plates 166
Titanic 142, 149, 158, 165, 192, 194
Tokiwa Mura 221
Trenton (US steam sloop) 185
Trevithick, Richard 32
Trial 116
Triewald, Martin 14
Truscott, Capt. 32
Tudor, Frederick 57, 97
Tuke, Henry Scott 140
tungsten 186

Uluburun (shipwreck) 179
Unite 56
Upton, Francis Robbins 205

Valdez (Exxon tanker) 71
Valiant (Capt. Wilmshurst) 160
Valiant (70 gun warship) 215
Vandalia (sloop-of-war) 215
Van de Velde the Younger, William 144
Van Leent, Frederik Johannes 68
Van Linschoten, Jan Huygen 36
Vasa 100, 103, 122–3, 134, 174, 220
Vegetius 171
Vengeance 107
ventilation: adequacy 3–4, 18–24; air streak 17; cargo 17–18; circulation issues 5–7, 10–12, 17–18; cross 10; laws and regulations 19, 22–23; one-way 11; problem ships 2, 14; stack effect 6
ventilation, artificial (forced) 13–17; bellows 13–14, 17; fans 10, 12–13, 17, 22–3; heat-induced 15–17; motive power 14–15; systems and sizing 16–17, 23
ventilation, natural 4–12; *see also* portholes and gunports
ventilator: blackout 10, 15; cowl vent 6–7, 10; Dorade box 9; mushroom 10; scuttles (in gunport lids) 174; torpedo 19
Vergulde Draeck (jacht) 178, 182
Victory 10–11, 15, 56, 103, 153, 175, 206
Victory (cargo) 10

vinegar 31, 50, 52, 54, 64, 219
vitamins *see* food: vitamins and minerals
von Liebig, Justus 198

Walker, Ezekiel 199
Walker, Commander William 32
Ward, John 157
warfarin 218
Warrior 102
Warspite 117
Wasa see Vasa
washing machines 102
Washington (battleship) 53
Washington Irving (passenger ship) 206
waste disposal 103–5, 112–3
watchkeeping 69–71; anti-fatigue measures 77; laws and regulations 73; modern schedules 75–7; traditional schedule 69–71
water: food and beverage sources 26–7; handling 28, 109–12; physiological requirements 25, 28; rations 26–9; ship allotment 29–30; storage 28, 30–3; supplemental sources 33; treatment 105–8; watering of ships 30; *see also* desalination
water, distilled 36
water fountain 109
water, waste 112–3
Waterhouse, Addison Goodyear 42
watering of ships 30
Webb, Silas 88–9
Wehr, Thomas 74
Westall, William 222
whale oil 179–80, 184
Wilde, H. 194
Wilkinson, John 116, 141, 156–7, 159
William, Duke of Normandy 192
William Brown 149
Wilmshurst, Capt. 160
Wilson, Charles 43–4
Wilson, George 171
wind-sails 7
Witsen, Nicolaes 118
Witte Leeuw (East Indiaman) 178
Woodall, James 64
Wouldhave, William 165
Wynmann, Nikolaus 156, 163, 171

Zeehaen (17c fluyt) 118
zinc and its salts 17, 181, 188